Lecture Notes in Computer Science 12719

More information about this subseries at http://www.springer.com/series/7408

Kirstin Peters · Tim A. C. Willemse (Eds.)

Formal Techniques for Distributed Objects, Components, and Systems

41st IFIP WG 6.1 International Conference, FORTE 2021
Held as Part of the 16th International Federated Conference
on Distributed Computing Techniques, DiSCoTec 2021
Valletta, Malta, June 14–18, 2021
Proceedings

 Springer

Editors
Kirstin Peters 🆔
TU Darmstadt
Darmstadt, Germany

Tim A. C. Willemse 🆔
Eindhoven University of Technology
Eindhoven, The Netherlands

ISSN 0302-9743   ISSN 1611-3349   (electronic)
Lecture Notes in Computer Science
ISBN 978-3-030-78088-3   ISBN 978-3-030-78089-0   (eBook)
https://doi.org/10.1007/978-3-030-78089-0

LNCS Sublibrary: SL2 – Programming and Software Engineering

This Springer imprint is published by the registered company Springer Nature Switzerland AG
The registered company address is: Gewerbestrasse 11, 6330 Cham, Switzerland

Foreword

The 16th International Federated Conference on Distributed Computing Techniques (DisCoTec 2021) took place during June 14–18, 2021. It was organized by the Department of Computer Science at the University of Malta, but was held online due to the abnormal circumstances worldwide affecting physical travel. The DisCoTec series is one of the major events sponsored by the International Federation for Information Processing (IFIP), the European Association for Programming Languages and Systems (EAPLS) and the Microservices Community. It comprises three conferences:

- *COORDINATION*, the IFIP WG 6.1 23rd International Conference on Coordination Models and Languages;
- *DAIS*, the IFIP WG 6.1 21st International Conference on Distributed Applications and Interoperable Systems;
- *FORTE*, the IFIP WG 6.1 41st International Conference on Formal Techniques for Distributed Objects, Components, and Systems.

Together, these conferences cover a broad spectrum of distributed computing subjects, ranging from theoretical foundations and formal description techniques to systems research issues. As is customary, the event also included several plenary sessions in addition to the individual sessions of each conference, which gathered attendants from the three conferences. These included joint invited speaker sessions and a joint session for the best papers from the three conferences. Associated with the federated event, four satellite events took place:

- *DisCoTec Tool*, a tutorial session promoting mature tools in the field of distributed computing;
- *ICE*, the 14th International Workshop on Interaction and Concurrency Experience;
- *FOCODILE*, the 2nd International Workshop on Foundations of Consensus and Distributed Ledgers;
- *REMV*, the 1st Robotics, Electronics, and Machine Vision Workshop.

I would like to thank the Program Committee chairs of the different events for their help and cooperation during the preparation of the conference, and the Steering Committee and Advisory Boards of DisCoTec and its conferences for their guidance and support. The organization of DisCoTec 2021 was only possible thanks to the dedicated work of the Organizing Committee, including Caroline Caruana and Jasmine Xuereb (publicity chairs), Duncan Paul Attard and Christian Bartolo Burlo (workshop chairs), Lucienne Bugeja (logistics and finances), and all the students and colleagues who volunteered their time to help. I would also like to thank the invited speakers for their excellent talks. Finally, I would like to thank IFIP WG 6.1, EAPLS and the Microservices Community for sponsoring this event, Springer's Lecture Notes in Computer Science team for their support and sponsorship, EasyChair for providing the reviewing framework, and the University of Malta for providing the support and infrastructure to host the event.

June 2021   Adrian Francalanza

Preface

This volume contains the papers presented at the 41st IFIP WG 6.1 International Conference on Formal Techniques for Distributed Objects, Components, and Systems (FORTE 2021), held as one of three main conferences of the 16th International Federated Conference on Distributed Computing Techniques (DisCoTec 2021), during June 14–18, 2021. The conference was hosted by the University of Malta but took place online due to the ongoing COVID-19 pandemic.

FORTE is a well-established forum for fundamental research on theory, models, tools, and applications for distributed systems, with special interest in

- Software quality, reliability, availability, and safety
- Security, privacy, and trust in distributed and/or communicating systems
- Service-oriented, ubiquitous, and cloud computing systems
- Component- and model-based design
- Object technology, modularity, and software adaptation
- Self-stabilization and self-healing/organizing
- Verification, validation, formal analysis, and testing of the above

The Program Committee received a total of 26 submissions, written by authors from 18 different countries. Of these, 13 papers were selected for inclusion in the scientific program. Each submission was reviewed by at least three Program Committee members with the help of 20 external reviewers in selected cases. The selection of accepted submissions was based on electronic discussions via the EasyChair conference management system.

As Program Committee, we actively contributed to the selection of the keynote speakers for DisCoTec 2021:

- Gilles Fedak, iExec, France
- Mira Mezini, Technical University of Darmstadt, Germany
- Alexandra Silva, University College London, UK

This year DisCoTec also included a tutorial session of four invited tutorials. This volume includes the following tutorial papers:

- Tutorial: Designing Distributed Software in mCRL2
- Better Late than Never or: Verifying Asynchronous Components at Runtime

We wish to thank all the authors of submitted papers, all the members of the Program Committee for their thorough evaluations of the submissions, and the external reviewers who assisted the evaluation process. We are also indebted to the Steering Committee of FORTE for their advice and suggestions. Last but not least, we thank the DisCoTec general chair, Adrian Francalanza, and his organization team for their hard, effective work in providing an excellent environment for FORTE 2021 and all other conferences and workshops, in spite of the pandemic troubles.

June 2021   Kirstin Peters
Tim A. C. Willemse

Organization

Program Committee

Luís Soares Barbosa	University of Minho, Portugal
Jiří Barnat	Masaryk University, Czech Republic
Pedro R. D'Argenio	Universidad Nacional de Córdoba, Argentina
Mila Dalla Preda	University of Verona, Italy
Wan Fokkink	Vrije Universiteit Amsterdam, Netherlands
Daniele Gorla	University of Rome "La Sapienza", Italy
Artem Khyzha	Tel Aviv University, Israel
Barbara König	University of Duisburg-Essen, Germany
Bas Luttik	Eindhoven University of Technology, Netherlands
Stephan Merz	Inria, France
Roland Meyer	TU Braunschweig, Germany
Mohammadreza Mousavi	University of Leicester, UK
Thomas Neele	Royal Holloway, University of London, UK
Ana-Maria Oprescu	University of Amsterdam, Netherlands
Catuscia Palamidessi	Inria, France
Kirstin Peters	TU Darmstadt, Germany
Anna Philippou	University of Cyprus, Cyprus
Jorge A. Pérez	University of Groningen, Netherlands
Anne Remke	WWU Münster, Germany
Kristin Yvonne Rozier	Iowa State University, USA
Cristina Seceleanu	Mälardalen University, Sweden
Maurice H. ter Beek	ISTI-CNR, Italy
Simone Tini	University of Insubria, Italy
Rob van Glabbeek	Data61 - CSIRO, Australia
Björn Victor	Uppsala University, Sweden
Georg Weissenbacher	Vienna University of Technology, Austria
Tim A. C. Willemse	Eindhoven University of Technology, Netherlands

Additional Reviewers

Backeman, Peter	Mallet, Frederic
Bunte, Olav	Mazzanti, Franco
Crafa, Silvia	Mennicke, Stephan
Genest, Blaise	Montesi, Fabrizio
Helfrich, Martin	Neves, Renato
Horne, Ross	Padovani, Luca
Jehl, Leander	Ponce-De-Leon, Hernan
Kempa, Brian	Ryan, Megan
Khakpour, Narges	van den Heuvel, Bas
Labella, Anna	Wijs, Anton

Contents

Tutorials

Full Papers

On Bidirectional Runtime Enforcement

Luca Aceto[1,2], Ian Cassar[2,3], Adrian Francalanza[3(✉)],
and Anna Ingólfsdóttir[2]

[1] Gran Sasso Science Institute, L'Aquila, Italy
[2] Department of Computer Science, ICE-TCS,
Reykjavík University, Reykjavík, Iceland
[3] Department of Computer Science, University of Malta, Msida, Malta
adrian.francalanza@um.edu.mt

Abstract. Runtime enforcement is a dynamic analysis technique that instruments a monitor with a system in order to ensure its correctness as specified by some property. This paper explores *bidirectional* enforcement strategies for properties describing the input and output behaviour of a system. We develop an operational framework for bidirectional enforcement and use it to study the enforceability of the safety fragment of Hennessy-Milner logic with recursion (sHML). We provide an automated synthesis function that generates correct monitors from sHML formulas, and show that this logic is enforceable via a specific type of bidirectional enforcement monitors called action disabling monitors.

1 Introduction

Runtime enforcement (RE) [18,32] is a dynamic verification technique that uses *monitors* to analyse the runtime behaviour of a system-under-scrutiny (SuS) and transform it in order to conform to some correctness *specification*. The seminal work in RE [11,27,32,33,37] models the behaviour of the SuS as a *trace* of *arbitrary* actions. Crucially, it assumes that the monitor can either *suppress* or *replace* any trace action and, whenever possible, *insert* additional actions into the trace. This work has been effectively used to implement *unidirectional* enforcement approaches [5,9,19,28] that monitor the trace of *outputs* produced by the SuS as illustrated by Fig. 1(a). In this setup, the monitor is instrumented with the SuS to form a *composite system* (represented by the dashed enclosure in Fig. 1) and is tasked with transforming the output behaviour of the SuS to ensure its correctness. For instance, an erroneous output β of the SuS is intercepted by

This work was partly supported by the projects "TheoFoMon: Theoretical Foundations for Monitorability" (nr.163406-051), "Developing Theoretical Foundations for Runtime Enforcement" (nr.184776-051) and "MoVeMnt: Mode(l)s of Verification and Monitorability" (nr.217987-051) of the Icelandic Research Fund, by the Italian MIUR project PRIN 2017FTXR7S IT MATTERS "Methods and Tools for Trustworthy Smart Systems", by the EU H2020 RISE programme under the Marie Skłodowska-Curie grant agreement nr. 778233, and by the Endeavour Scholarship Scheme (Malta), part-financed by the European Social Fund (ESF) - Operational Programme II – 2014–2020.

© IFIP International Federation for Information Processing 2021
Published by Springer Nature Switzerland AG 2021
K. Peters and T. A. C. Willemse (Eds.): FORTE 2021, LNCS 12719, pp. 3–21, 2021.
https://doi.org/10.1007/978-3-030-78089-0_1

the monitor and transformed into β', to stop the error from propagating to the surrounding environment.

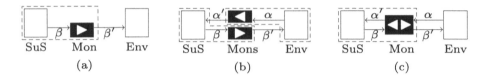

Fig. 1. Enforcement instrumentation setups.

Despite its merits, unidirectional enforcement lacks the power to enforce properties involving the *input* behaviour of the SuS. Arguably, these properties are harder to enforce: unlike outputs, inputs are instigated by the environment and not the SuS itself, meaning that the SuS possesses only partial control over them. Moreover, even when the SuS can control when certain inputs can be supplied (*e.g.*, by opening a communication port, or by reading a record from a database *etc.*), the environment still determines the provided payload.

Broadly, there are two approaches to enforce bidirectional properties at runtime. Several bodies of work employ two monitors attached at the output side of each (diadic) interacting party [12,17,26]. As shown in Fig. 1(b), the extra monitor is attached to the *environment* to analyse its outputs before they are passed on as *inputs* to the SuS. While this approach is effective, it assumes that a monitor can actually be attached to the environment (which is often inaccessible).

By contrast, Fig. 1(c) presents a less explored *bidirectional enforcement* approach where the monitor analyses the entire behaviour of the SuS without the need to instrument the environment. The main downside of this alternative setup is that it enjoys limited control over the SuS's inputs. As we already argued, the monitor may be unable to enforce a property that could be violated by an input action with an invalid payload value. In other cases, the monitor might need to adopt a different enforcement strategy to the ones that are conventionally used for enforcing output behaviour in a unidirectional one.

This paper explores how existing monitor transformations—namely, suppressions, insertions and replacements—can be repurposed to work for bidirectional enforcement, *i.e.*, the setup in Fig. 1(c). Since inputs and outputs must be enforced differently, we find it essential to distinguish between the monitor's transformations and their resulting effect on the visible behaviour of the composite system. This permits us to study the enforceability of properties defined via the safety subset sHML of the well-studied branching-time logic μHML [8,31,36] (a reformulation of the modal μ-calculus [29]). Our contributions are:

(i) A general instrumentation framework for bidirectional enforcement (Fig. 4) that is parametrisable by any system whose behaviour can be modelled as a labelled transition system. The framework subsumes the one presented in previous work [5] and differentiates between input and output actions.

Syntax

$$\varphi, \psi \in \text{sHML} ::= \text{tt} \quad (\text{truth}) \quad | \quad \text{ff} \quad (\text{falsehood}) \quad | \quad \bigwedge_{i \in I} \varphi_i \; (\text{conjunction})$$
$$| \; [p, c]\varphi \; (\text{necessity}) \; | \; \max X.\varphi \; (\text{greatest fp.}) \; | \; X \quad (\text{fp. variable})$$

Semantics

$$[\![\text{tt}, \rho]\!] \stackrel{\text{def}}{=} \text{Sys} \qquad\qquad [\![\text{ff}, \rho]\!] \stackrel{\text{def}}{=} \emptyset \qquad\qquad [\![X, \rho]\!] \stackrel{\text{def}}{=} \rho(X)$$
$$[\![\textstyle\bigwedge_{i \in I} \varphi_i, \rho]\!] \stackrel{\text{def}}{=} \bigcap_{i \in I} [\![\varphi_i, \rho]\!] \qquad [\![\max X.\varphi, \rho]\!] \stackrel{\text{def}}{=} \bigcup \{S \mid S \subseteq [\![\varphi, \rho[X \mapsto S]]\!]\}$$
$$[\![\, [p, c]\varphi, \rho]\!] \stackrel{\text{def}}{=} \{s \mid \forall \alpha, r, \sigma \cdot (s \stackrel{\alpha}{\Rightarrow} r \text{ and } \text{match}(p, \alpha) = \sigma \text{ and } c\sigma \Downarrow \text{true}) \text{ implies } r \in [\![\varphi\sigma, \rho]\!]\}$$

Fig. 2. The syntax and semantics for sHML, the safety fragment of μHML.

(ii) A novel definition formalising what it means for a monitor to *adequately enforce* a property in a bidirectional setting (Definitions 2 and 6). These definitions are parametrisable with respect to an instrumentation relation, an instance of which is given by our enforcement framework of Fig. 4.

(iii) A new result showing that the subclass of *disabling monitors* suffices to bidirectionally enforce any property expressed as an sHML formula (Theorem 1). A by-product of this result is a synthesis function (Definition 8) that generates a disabling monitor for any sHML formula.

Full proofs and additional details can to be found at [6,13].

2 Preliminaries

The Model. We assume a countable set of communication ports $a, b, c \in \text{Port}$, a set of values $v, w \in \text{Val}$, and partition the set of actions Act into *inputs* $a?v \in \text{iAct}$, and *outputs* $a!v \in \text{oAct}$ where $\text{iAct} \cup \text{oAct} = \text{Act}$. Systems are described as *labelled transition systems* (LTSs); these are triples $\langle \text{Sys}, \text{Act} \cup \{\tau\}, \rightarrow \rangle$ consisting of a set of *system states*, $s, r, q \in \text{Sys}$, a set of *visible actions*, $\alpha, \beta \in \text{Act}$, along with a distinguished silent action $\tau \notin \text{Act}$ (where $\mu \in \text{Act} \cup \{\tau\}$), and a *transition* relation, $\rightarrow \; \subseteq (\text{Sys} \times (\text{Act} \cup \{\tau\}) \times \text{Sys})$. We write $s \stackrel{\mu}{\longrightarrow} r$ in lieu of $(s, \mu, r) \in \rightarrow$, and $s \stackrel{\alpha}{\Rightarrow} r$ to denote weak transitions representing $s(\stackrel{\tau}{\longrightarrow})^* \cdot \stackrel{\alpha}{\longrightarrow} r$ where r is called the α-derivative of s. For convenience, we use the syntax of the regular fragment of value-passing CCS [23] to concisely describe LTSs. Traces $t, u \in \text{Act}^*$ range over (finite) sequences of *visible* actions. We write $s \stackrel{t}{\Rightarrow} r$ to denote a sequence of *weak* transitions $s \stackrel{\alpha_1}{\Longrightarrow} \ldots \stackrel{\alpha_n}{\Longrightarrow} r$ where $t = \alpha_1 \ldots \alpha_n$ for some $n \geq 0$; when $t = \varepsilon$, $s \stackrel{\varepsilon}{\Rightarrow} r$ means $s \stackrel{\tau}{\longrightarrow}^* r$. Additionally, we represent system runs as *explicit traces* that include τ-actions, $t_\tau, u_\tau \in (\text{Act} \cup \{\tau\})^*$ and write $s \stackrel{\mu_1 \ldots \mu_n}{\longrightarrow} r$ to denote a sequence of strong transitions $s \stackrel{\mu_1}{\longrightarrow} \ldots \stackrel{\mu_n}{\longrightarrow} r$. The function $\text{sys}(t_\tau)$ returns a canonical system that exclusively produces the sequence of actions defined by t_τ. E.g., $\text{sys}(a?3.\tau.a!5)$ produces the process $a?x.\tau.a!5.\text{nil}$. We consider states in our system LTS modulo the classic notion of *strong bisimilarity* [23,38] and write $s \sim r$ when states s and r are bisimilar.

The Logic. The behavioral properties we consider are described using sHML [7, 22], a subset of the value passing μHML [24,36] that uses *symbolic actions* of the form (p,c) consisting of an action pattern p and a condition c. Symbolic actions abstract over concrete actions using *data variables* $x, y, z \in$ DVAR that occur free in the constraint c or as binders in the pattern p. Patterns are subdivided into input $(x)?(y)$ and output $(x)!(y)$ patterns where (x) binds the information about the port on which the interaction has occurred, whereas (y) binds the payload; $\mathbf{bv}(p)$ denotes the set of binding variables in p whereas $\mathbf{fv}(c)$ represents the set of free variables in condition c. We assume a (partial) *matching function* match(p, α) that (when successful) returns a substitution σ mapping bound variables in p to the corresponding values in α; by replacing every occurrence (x) in p with $\sigma(x)$ we get the matched action α. The *filtering condition*, c, is evaluated *wrt.* the substitution returned by successful matches, written as $c\sigma \Downarrow v$ where $v \in \{\text{true}, \text{false}\}$.

Whenever a symbolic action (p, c) is *closed*, i.e., $\mathbf{fv}(c) \subseteq \mathbf{bv}(p)$, it denotes the *set* of actions $[\![(p, c)]\!] \stackrel{\text{def}}{=} \{ \alpha \mid \exists \sigma \cdot \text{match}(p, \alpha) = \sigma \text{ and } c\sigma \Downarrow \text{true} \}$. Following standard value-passing LTS semantics [23,34], our systems have *no control* over the data values supplied via inputs. Accordingly, we assume a well-formedness constraint where the condition c of an input symbolic action, $((x)?(y),c)$, *cannot* restrict the values of binder y, i.e., $y \notin \mathbf{fv}(c)$. As a shorthand, whenever a condition in a symbolic action equates a bound variable to a specific value we embed the equated value within the pattern, e.g., $((x)!(y), x = \text{a} \land y = 3)$ and $((x)?(y), x = \text{a})$ become $(\text{a}!3,\text{true})$ and $(\text{a}?(y),\text{true})$; we also elide true conditions, and just write $(\text{a}!3)$ and $(\text{a}?(y))$ in lieu of $(\text{a}!3,\text{true})$ and $(\text{a}?(y),\text{true})$.

Figure 2 presents the sHML syntax for some countable set of logical variables $X, Y \in$ LVAR. The construct $\bigwedge_{i \in I} \varphi_i$ describes a *compound* conjunction, $\varphi_1 \land \ldots \land \varphi_n$, where $I = \{1, .., n\}$ is a finite set of indices. The syntax also permits recursive properties using greatest fixpoints, $\max X.\varphi$, which bind free occurrences of X in φ. The central construct is the (symbolic) universal modal operator, $[p, c]\varphi$, where the binders $\mathbf{bv}(p)$ bind the free data variables in c and φ. We occasionally use the notation (_) to denote "don't care" binders in the pattern p, whose bound values are not referenced in c and φ. We also assume that all fixpoint variables, X, are guarded by modal operators.

Formulas in sHML are interpreted over the system powerset domain where $S \in \mathcal{P}(\text{SYS})$. The semantic definition of Fig. 2, $[\![\varphi, \rho]\!]$, is given for *both* open and closed formulas. It employs a valuation from logical variables to sets of states, $\rho \in (\text{LVAR} \rightarrow \mathcal{P}(\text{SYS}))$, which permits an inductive definition on the structure of the formulas; $\rho' = \rho[X \mapsto S]$ denotes a valuation where $\rho'(X) = S$ and $\rho'(Y) = \rho(Y)$ for all other $Y \neq X$. The only non-standard case is that for the universal modality formula, $[p, c]\varphi$, which is satisfied by any system that either *cannot* perform an action α that matches p while satisfying condition c, or for any such matching action α with substitution σ, its derivative state satisfies the continuation $\varphi\sigma$. We consider formulas modulo associativity and commutativity of $\land$, and unless stated explicitly, we assume *closed* formulas, i.e., without free logical and data variables. Since the interpretation of a closed φ is independent of the valuation ρ we write $[\![\varphi]\!]$ in lieu of $[\![\varphi, \rho]\!]$. A system s *satisfies* formula φ whenever $s \in [\![\varphi]\!]$, and a formula φ is *satisfiable*, when $[\![\varphi]\!] \neq \emptyset$.

We find it convenient to define the function *after*, describing how an sHML formula *evolves* in reaction to an action μ. Note that, for the case $\varphi = [p, c]\psi$, the formula returns $\psi\sigma$ when μ matches successfully the symbolic action (p, c) with σ, and tt otherwise, to signify a trivial satisfaction. We lift the *after* function to (explicit) traces in the obvious way, i.e., $after(\varphi, t_\tau)$ is equal to $after(after(\varphi, \mu), u_\tau)$ when $t_\tau = \mu u_\tau$ and to φ when $t_\tau = \varepsilon$. Our definition of *after* is justified vis-a-vis the semantics of Fig. 2 via Proposition 1; it will play a role in defining our notion of enforcement in Sect. 4.

Definition 1. *We define the function* $after : (\text{sHML} \times \text{ACT} \cup \{\tau\}) \to \text{sHML}$ *as:*

$$after(\varphi, \alpha) \stackrel{\text{def}}{=} \begin{cases} \varphi & \text{if } \varphi \in \{\text{tt}, \text{ff}\} \\ after(\varphi'\{^\varphi/_X\}, \alpha) & \text{if } \varphi = \max X.\varphi' \\ \bigwedge_{i \in I} after(\varphi_i, \alpha) & \text{if } \varphi = \bigwedge_{i \in I} \varphi_i \\ \psi\sigma & \text{if } \varphi = [p, c]\psi \text{ and } \exists\sigma \cdot (match(p, \alpha) = \sigma \wedge c\sigma \Downarrow true) \\ \text{tt} & \text{if } \varphi = [p, c]\psi \text{ and } \nexists\sigma \cdot (match(p, \alpha) = \sigma \wedge c\sigma \Downarrow true) \end{cases}$$

$$after(\varphi, \tau) \stackrel{\text{def}}{=} \varphi \qquad\qquad \blacksquare$$

Proposition 1. *For every system state s, formula φ and action α, if $s \in [\![\varphi]\!]$ and $s \stackrel{\alpha}{\Longrightarrow} s'$ then $s' \in [\![after(\varphi, \alpha)]\!]$.* $\qquad\qquad\square$

Example 1. The *safety* property φ_1 *repeatedly* requires that *every* input request that is made on a port that is *not* b, cannot be followed by another input on the same port in succession. However, following this input it allows a *single* output answer on the same port in response, followed by the logging of the serviced request by outputting a notification on a dedicated port b. We note how the channel name bound to x is used to constrain sub-modalities. Similarly, values bound to y_1 and y_2 are later referenced in condition $y_3 = (\log, y_1, y_2)$.

$$\varphi_1 \stackrel{\text{def}}{=} \max X.[((x)?(y_1), x \neq b)]([(x?(\_))]\text{ff} \wedge [(x!(y_2))]\varphi_1')$$
$$\varphi_1' \stackrel{\text{def}}{=} ([(x!(\_))]\text{ff} \wedge [(b!(y_3), y_3 = (\log, y_1, y_2))]X)$$

Consider the systems s_a, s_b and s_c (where $s_\text{cls} \stackrel{\text{def}}{=} (\text{b}?z.\text{if } z = \text{cls then nil else } X)$).

$$s_\text{a} \stackrel{\text{def}}{=} \text{rec } X.((\text{a}?x.y := \text{ans}(x).\text{a}!y.\text{b}!(\log, x, y).X) + s_\text{cls}) \qquad\qquad s_\text{c} \stackrel{\text{def}}{=} \text{a}?y.s_\text{a}$$
$$s_\text{b} \stackrel{\text{def}}{=} \text{rec } X.((\text{a}?x.y := \text{ans}(x).\text{a}!y.(\underline{\text{a}!y}.\text{b}!(\log, x, y).s_\text{a} + \text{b}!(\log, x, y).X)) + s_\text{cls})$$

s_a implements a request-response server that repeatedly inputs values (for some domain VAL) on port a, a$?x$, for which it internally computes an answer and assigns it to the data variable y, $y := \text{ans}(x)$. It then outputs the answer on port a in response to each request, a$!y$, and finally logs the serviced request by outputting the triple $(\log, x, y)$ on port b, b$!(\log, x, y)$. It terminates whenever it inputs a close request cls from port b, i.e., b$?z$ when $z = \text{cls}$.

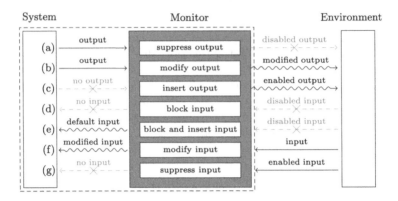

Fig. 3. Bidirectional enforcement via suppression, insertion and replacement.

Systems s_b and s_c are similar to s_a but define additional behaviour: s_c requires a startup input, $a?y$, before behaving as s_a, whereas s_b occasionally provides a redundant (underlined) answer prior to logging a serviced request. Using the semantics of Fig. 2, one can verify that $s_a \in [\![\varphi_1]\!]$, $s_c \notin [\![\varphi_1]\!]$ because of $s_c \xrightarrow{a?v_1.a?v_2}$, and $s_b \notin [\![\varphi_1]\!]$ since we have $s_b \xrightarrow{a?v_1.a!ans(v_1).a!ans(v_1)}$ (for some values v_1 and v_2). ∎

3   A Bidirectional Enforcement Model

Bidirectional enforcement seeks to transform the entire (visible) behaviour of the SuS in terms of input and output actions; this contrasts with unidirectional approaches that only modify output traces. In this richer setting, it helps to differentiate between the transformations performed by the monitor (*i.e.*, insertions, suppressions and replacements), and the way they can be used to affect the resulting behaviour of the composite system. In particular, we say that an action that can be performed by the SuS has been *disabled* when it is no longer visible in the resulting composite system (consisting of the SuS and the monitor). Dually, a visible action is *enabled* when the composite system can execute it while the SuS cannot. SuS actions are *adapted* when either their payload differs from that of the composite system, or when the action is rerouted through a different port.

We argue that implementing action enabling, disabling and adaptation differs according to whether the action is an input or an output; see Fig. 3. Enforcing actions instigated by the SuS—such as outputs—is more straightforward. Figure 3(a), (b) and (c) *resp.* state that disabling an output can be achieved by suppressing it, adapting an output amounts to replacing the payload or redirecting it to a different port, whereas output enabling can be attained via an insertion. However, enforcing actions instigated by the environment such as inputs is harder. In Fig. 3(d), we propose to *disable* an input by concealing the input port. Since this may block the SuS from progressing, the instrumented monitor may additionally *insert* a default input to unblock the system, Fig. 3(e). Input *adaptation*, Fig. 3(f), is also attained via a *replacement* (applied in the opposite direction to the output case). Inputs can also be *enabled* (when the SuS is unable to carry them out), Fig. 3(g), by having the monitor accept the input in question and then *suppress* it: from the environment's perspective, the input would be effected.

Syntax

$$m, n \in \text{TRN} ::= (p, c, p').m \quad | \quad \sum_{i \in I} m_i \ (I \text{ is a finite index set}) \quad | \quad \text{rec}\, X.m \quad | \quad X$$

Dynamics

$$\text{ESEL} \frac{m_j \xrightarrow{\gamma \blacktriangleright \gamma'} n_j}{\sum_{i \in I} m_i \xrightarrow{\gamma \blacktriangleright \gamma'} n_j} \ j \in I \qquad \text{EREC} \frac{m\{^{\text{rec}\, X.m}/_X\} \xrightarrow{\gamma \blacktriangleright \gamma'} n}{\text{rec}\, X.m \xrightarrow{\gamma \blacktriangleright \gamma'} n}$$

$$\text{ETRN} \frac{\text{match}(p, \gamma) = \sigma \quad c\sigma \Downarrow \text{true} \quad \gamma' = \pi\sigma}{(p, c, \pi).m \xrightarrow{\gamma \blacktriangleright \gamma'} m\sigma}$$

Instrumentation

$$\text{BITRNO} \frac{s \xrightarrow{b!w} s' \quad m \xrightarrow{(b!w)\blacktriangleright(a!v)} n}{m[s] \xrightarrow{a!v} n[s']} \qquad \text{BITRNI} \frac{m \xrightarrow{(a?v)\blacktriangleright(b?w)} n \quad s \xrightarrow{b?w} s'}{m[s] \xrightarrow{a?v} n[s']}$$

$$\text{BIDISO} \frac{s \xrightarrow{a!v} s' \quad m \xrightarrow{(a!v)\blacktriangleright\bullet} n}{m[s] \xrightarrow{\tau} n[s']} \qquad \text{BIDISI} \frac{m \xrightarrow{\bullet\blacktriangleright(a?v)} n \quad s \xrightarrow{a?v} s'}{m[s] \xrightarrow{\tau} n[s']}$$

$$\text{BIENO} \frac{m \xrightarrow{\bullet\blacktriangleright(a!v)} n}{m[s] \xrightarrow{a!v} n[s]} \qquad \text{BIENI} \frac{m \xrightarrow{(a?v)\blacktriangleright\bullet} n}{m[s] \xrightarrow{a?v} n[s]} \qquad \text{BIASY} \frac{s \xrightarrow{\tau} s'}{m[s] \xrightarrow{\tau} m[s']}$$

$$\text{BIDEF} \frac{s \xrightarrow{a!v} s' \quad m \xrightarrow{a!v}\!\!\!\!\not\;\; \quad \forall b \in \text{PORT}, w \in \text{VAL} \cdot m \xrightarrow{\bullet\blacktriangleright b!w}\!\!\!\!\not\;}{m[s] \xrightarrow{a!v} \text{id}[s']}$$

Fig. 4. A bidirectional instrumentation model for enforcement monitors.

Figure 4 presents an operational model for the bidirectional instrumentation proposal of Fig. 3 in terms of (symbolic) transducers[1]. Transducers, $m, n \in \text{TRN}$, are monitors that define *symbolic transformation triples*, (p,c,π), consisting of an action *pattern* p, *condition* c, and a *transformation action* π. Conceptually, the action pattern and condition determine the range of system (input or output)

---

[1] These transducers were originally introduced in [5] for unidirectional enforcement.

actions upon which the transformation should be applied, while the transformation action specifies the transformation that should be applied. The symbolic transformation pattern p is an extended version of those definable in symbolic actions, that may also include $\bullet$; when $p = \bullet$, it means that the monitor can act independently from the system to insert the action specified by the transformation action. Transformation actions are possibly open actions (*i.e.*, actions with possibly free variable such as $x?v$ or $a!x$) or the special action $\bullet$; the latter represents the suppression of the action specified by p. We assume a well-formedness constraint where, for every $(p, c, \pi).m$, p and π cannot both be $\bullet$, and when neither is, they are of the *same* type *i.e.*, an input (*resp.* output) pattern and action. Examples of well-formed symbolic transformations are $(\bullet, \mathsf{true}, a?v)$, $((x)!(y), \mathsf{true}, \bullet)$ and $((x)!(y), \mathsf{true}, a!v)$.

The monitor transition rules in Fig. 4 assume closed terms, *i.e.*, every *transformation-prefix transducer* of the form $(p, c, \pi).m$ must obey the constraint $\big(\mathbf{fv}(c) \cup \mathbf{fv}(\pi) \cup \mathbf{fv}(m)\big) \subseteq \mathbf{bv}(p)$ and similarly for recursion variables X and $\mathsf{rec}\,X.m$. Each transformation-prefix transducer yields an LTS with labels of the form $\gamma \blacktriangleright \gamma'$, where $\gamma, \gamma' \in (\mathrm{ACT} \cup \{\bullet\})$. Intuitively, transition $m \xrightarrow{\gamma \blacktriangleright \gamma'} n$ denotes the way that a transducer in state m *transforms* the action γ into γ' while transitioning to state n. The transducer action $\alpha \blacktriangleright \beta$ represents the *replacement* of α by β, $\alpha \blacktriangleright \alpha$ denotes the *identity* transformation, whereas $\alpha \blacktriangleright \bullet$ and $\bullet \blacktriangleright \alpha$ respectively denote the *suppression* and *insertion* transformations of action α. The key transition rule in Fig. 4 is ETRN. It states that the transformation-prefix transducer $(p, c, \pi).m$ transforms action γ into a (potentially) different action γ' and reduces to state $m\sigma$, whenever γ matches pattern p, *i.e.*, $\mathsf{match}(p, \gamma) = \sigma$, and satisfies condition c, *i.e.*, $c\sigma \Downarrow \mathsf{true}$. Action γ' results from instantiating the free variables in π as specified by σ, *i.e.*, $\gamma' = \pi\sigma$. The remaining rules for selection (ESEL) and recursion (EREC) are standard. We employ the shorthand notation $m \xrightarrow{\gamma}\!\!\!\!\!/\;$ to mean $\nexists\gamma', n$ such that $m \xrightarrow{\gamma \blacktriangleright \gamma'} n$. Moreover, for the semantics of Fig. 4, we can encode the identity monitor, id, as $\mathsf{rec}\,Y.((x)!(y), \mathsf{true}, x!y).Y + ((x)?(y), \mathsf{true}, x?y).Y$. As a shorthand notation, we write $(p, c).m$ instead of $(p, c, \pi).m$ when all the binding occurrences (x) in p correspond to free occurrences x in π, thus denoting an identity transformation. Similarly, we elide c whenever $c = \mathsf{true}$.

The first contribution of this work lies in the new *instrumentation relation* of Fig. 4, linking the behaviour of the SuS s with that of a monitor m: the term $m[s]$ denotes their composition as a *monitored system*. Crucially, the instrumentation rules in Fig. 4 give us a semantics in terms of an LTS over the actions $\mathrm{ACT} \cup \{\tau\}$, in line with the LTS semantics of the SuS. Following Fig. 3(b), rule BITRNO states that if the SuS transitions with an output $b!w$ to s' and the transducer can *replace* it with $a!v$ and transition to n, the *adapted* output can be externalised so that the composite system $m[s]$ transitions over $a!v$ to $n[s']$. Rule BIDISO states that if s performs an output $a!v$ that the monitor *can suppress*, the instrumentation withholds this output and the composite system silently transitions; this amounts to action *disabling* as outlined in Fig. 3(a). Rule BIENO is dual, and it *enables* the output $a!v$ on the SuS as outlined in Fig. 3(c): it augments the composite system $m[s]$ with an output $a!v$ whenever m can *insert* $a!v$, independently of the behaviour of s.

Rule BIDEF is analogous to standard rules for premature monitor termination [1, 20–22], and accounts for underspecification of transformations. We, however, restrict defaulting (termination) to output actions performed by the SuS exclusively, *i.e.*, a monitor only defaults to id when it cannot react to or enable a system output. By forbidding the monitor from defaulting upon unspecified inputs, the monitor is able to *block* them from becoming part of the composite system's behaviour. Hence, any input that the monitor is unable to react to, *i.e.*, $m \xrightarrow{a?v \blacktriangleright \gamma}$, is considered as being *invalid and blocked* by default. This technique is thus used to implement Fig. 3(d). To avoid disabling valid inputs unnecessarily, the monitor must therefore explicitly define symbolic transformations that cover *all* the valid inputs of the SuS. Note, that rule BIASY still allows the SuS to silently transition independently of m. Following Fig. 3(f), rule BITRNI adapts inputs, provided the SuS can accept the adapted input. Similarly, rule BIENI *enables* an input on a port a as described in Fig. 3(g): the composite system accepts the input while suppressing it from the SuS. Rule BIDISI allows the monitor to generate a default input value v and forward it to the SuS on a port a, thereby unblocking it; externally, the composite system silently transitions to some state, following Fig. 3(e).

Example 2. Consider the following action disabling transducer m_d, that repeatedly disables every output performed by the system via the branch $((\_)!(\_), \bullet).Y$. In addition, it limits inputs to those on port b via the input branch $(b?(\_)).Y$; inputs on other ports are disabled since none of the relevant instrumentation rules in Fig. 4 can be applied.

$$m_d \stackrel{\text{def}}{=} \text{rec } Y.(b?(\_)).Y + ((\_)!(\_), \bullet).Y$$

When instrumented with s_c from Example 1, m_d blocks its initial input, *i.e.*, we have $m_d[s_c] \xrightarrow{\alpha}$ for any α. In the case of s_b, the composite system $m_d[s_b]$ can only input requests on port b, such as the termination request $m_d[s_b] \xrightarrow{b?cls} m_d[\text{nil}]$.

$$m_{dt} \stackrel{\text{def}}{=} \text{rec } X.(((x)?(y_1), x{\neq}b).(((x_1)?(\_), x_1 \neq x).\text{id} + (x!(y_2)).m'_{dt}) + (b?(\_)).\text{id}$$

$$m'_{dt} \stackrel{\text{def}}{=} (x!(\_),\bullet).m_d + ((\_)?(\_)).\text{id} + (b!(y_3), y_3{=}(\log, y_1, y_2)).X$$

By defining branch $(b?(\_)).\text{id}$, the more elaborate monitor m_{dt} (above) allows the SuS to immediately input on port b (possibly carrying a termination request). At the same time, the branch prefixed by $((x)?(y_1), x{\neq}b)$ permits the SuS to input the first request via any port $x \neq b$, subsequently blocking inputs on the same port x (without deterring inputs on other ports) via the input branch $((x_1)?(\_), x_1 \neq x).\text{id}$. In conjunction to this branch, m_{dt} defines another branch $(x!(y_2)).m'_{dt}$ to allow outputs on the port bound to variable x. The continuation monitor m'_{dt} then defines the suppression branch $(x!(\_),\bullet).m_d$ by which it disables any *redundant* response that is output following the first one. Since it also defines branches $(b!(y_3), y_3 = (\log, y_1, y_2)).X$ and $((\_)?(\_)).\mathrm{id}$, it does not affect log events or further inputs that occur immediately after the first response.

When instrumented with system s_c from Example 1, m_{dt} allows the composite system to perform the first input but then blocks the second one, permitting only input requests on channel b, e.g., $m_{dt}[s_c] \xrightarrow{a?v} \cdot \xrightarrow{b?cls} \mathrm{id}[\mathrm{nil}]$. It also disables the first redundant response of system s_b while transitioning to m_d, which proceeds to suppress every subsequent output (including log actions) while blocking every other port except b, i.e., $m_{dt}[s_b] \xrightarrow{a?v} \cdot \xRightarrow{a!w} \cdot \xrightarrow{\tau} m_d[b!(\log, v, w).s_a] \xrightarrow{\tau}$ $m_d[s_a] \xrightarrow{a?v}$ (for every port a where $a \neq b$ and any value v). Rule IDEF allows it to default when handling unspecified outputs, e.g., for system $b!(\log, v, w).s_a$ the composite system can still perform $m_{dt}[b!(\log, v, w).s_a] \xrightarrow{b!(\log,v,w)} \mathrm{id}[s_a]$.

$$m_{det} \stackrel{\text{def}}{=} \mathrm{rec}\, X.(((x)?(y_1), x \neq b).m'_{det} + (b?(\_)).\mathrm{id})$$

$$m'_{det} \stackrel{\text{def}}{=} \mathrm{rec}\, Y_1.(\underline{(\bullet, x?v_{def}).Y_1} + (x!(y_2)).m''_{det} + ((x_1)?(\_)), x_1 \neq x).\mathrm{id}$$

$$m''_{det} \stackrel{\text{def}}{=} \mathrm{rec}\, Y_2.(\underline{(x!(\_)), x \neq b, \bullet).Y_2} + (b!(y_3), y_3 = (\log, y_1, y_2)).X + ((\_)?(\_)).\mathrm{id})$$

Monitor m_{det} (above) is similar to m_{dt} but instead employs a loop of suppressions (underlined in m''_{det}) to disable further responses until a log or termination input is made. When composed with s_b, it permits the log action to go through:

$$m_{det}[s_b] \xrightarrow{a?v} \cdot \xRightarrow{a!w} \cdot \xrightarrow{\tau} m''_{det}[b!(\log, v, w).s_b] \xrightarrow{b!(\log,v,w)} m_{det}[s_b].$$

m_{det} also defines a branch prefixed by the insertion transformation $(\bullet, x?v_{def})$ (underlined in m'_{det}) where v_{def} is a default input domain value. This permits the instrumentation to silently unblock the SuS when this is waiting for a request following an unanswered one. In fact, when instrumented with s_c, m_{det} not only forbids invalid input requests, but it also (internally) unblocks s_c by supplying the required input via the added insertion branch. This allows the composite system to proceed, as shown below (where $s'_a \stackrel{\text{def}}{=} y := \mathrm{ans}(v_{def}).a!y.b!(\log, v_{def}, y).s_a$):

$$m_{det}[s_c] \xrightarrow{a?v} \mathrm{rec}\, Y.((\bullet, a?v_{def}).Y + (a!(y_2)).m''_{det} + (b?(\_)).\mathrm{id})[s_a]$$
$$\xrightarrow{\tau} \mathrm{rec}\, Y.((\bullet, a?v_{def}).Y + (a!(y_2)).m''_{det} + (b?(\_)).\mathrm{id})[s'_a]$$
$$\xRightarrow{a!\mathrm{ans}(v_{def}).b!(\log,v_{def},y)} m_{det}[s_a] \qquad \blacksquare$$

Although in this paper we mainly focus on action disabling monitors, using our model one can also define action enabling and adaptation monitors.

Example 3. Consider now transducers m_e and m_a below:

$$m_e \stackrel{\text{def}}{=} ((x)?(y), x \neq b, \bullet).(\bullet, x!\mathrm{ans}(y)).(\bullet, b!(\log, y, \mathrm{ans}(y))).\mathrm{id}$$

$$m_a \stackrel{\text{def}}{=} \mathrm{rec}\, X.(b?(y), a?y).X + (a!(y), b!y).X.$$

Once instrumented, m_e first uses a suppression to enable an input on any port $x \neq b$ (but then gets discarded). It then automates a response by inserting an answer followed by a log action. Concretely, when composed with $r \in \{s_\mathbf{b}, s_\mathbf{c}\}$ from Example 1, the execution of the composite system can only start as follows, for some channel name $\mathbf{c} \neq \mathbf{b}$, values v and $w = \mathsf{ans}(v)$:

$$m_\mathbf{e}[r] \xrightarrow{\mathbf{c}?v} (\bullet, \mathbf{c}!w).(\bullet, \mathbf{b}!(\log, v, w)).\mathsf{id}[r] \overset{\mathbf{c}!w}{\Longrightarrow} (\bullet, \mathbf{b}!(\log, v, w)).\mathsf{id}[r] \xrightarrow{\mathbf{b}!(\log,v,w)} \mathsf{id}[r].$$

By contrast, $m_\mathbf{a}$ uses action adaptation to redirect the inputs and outputs from the SuS through port $\mathbf{b}$: it allows the composite system to exclusively input values on port $\mathbf{b}$ forwarding them to the SuS on port $\mathbf{a}$, and dually allowing outputs from the SuS on port $\mathbf{a}$ to rerout them to port $\mathbf{b}$. As a result, a composite system can *only* communicate on port $\mathbf{b}$. E.g., $m_\mathbf{a}[s_\mathbf{c}] \xrightarrow{\mathbf{b}?v_1} m_\mathbf{a}[s_\mathbf{a}] \xrightarrow{\mathbf{b}?v_2.\mathbf{b}!w_2.\mathbf{b}!(\log,v_2,w_2)} m_\mathbf{a}[s_\mathbf{a}]$ and $m_\mathbf{a}[s_\mathbf{b}] \xrightarrow{\mathbf{b}?v_1.\mathbf{b}!w_1.\mathbf{b}!(\log,v_1,w_1)} m_\mathbf{a}[s_\mathbf{b}]$. ∎

4 Enforcement

We are concerned with extending the enforceability result obtained in prior work [5] to the extended setting of bidirectional enforcement. The *enforceability* of a logic rests on the relationship between the semantic behaviour specified by the logic on the one hand, and the ability of the operational mechanism (that of Sect. 3 in this case) to enforce the specified behaviour on the other.

Definition 2 (Enforceability [5]). *A formula φ is enforceable iff there exists a transducer m such that m adequately enforces φ. A logic $\mathcal{L}$ is enforceable iff every formula $\varphi \in \mathcal{L}$ is enforceable.* ∎

Since we have limited control over the SuS that a monitor is composed with, "*m adequately enforces φ*" should hold for *any* (instrumentable) system. In [5] we stipulate that any notion of adequate enforcement should at least entail soundness.

Definition 3 (Sound Enforcement [5]). *Monitor m soundly enforces a satisfiable formula φ, denoted as $\mathsf{senf}(m, \varphi)$, iff for every state $s \in \mathrm{SYS}$, it is the case that $m[s] \in [\![\varphi]\!]$.* ∎

Example 4. Although showing that a monitor soundly enforces a formula should consider *all* systems, we give an intuition based on $s_\mathbf{a}, s_\mathbf{b}, s_\mathbf{c}$ for formula φ_1 from Example 1 (restated below) where $s_\mathbf{a} \in [\![\varphi_1]\!]$ (hence $[\![\varphi_1]\!] \neq \emptyset$) and $s_\mathbf{b}, s_\mathbf{c} \notin [\![\varphi_1]\!]$.

$$\varphi_1 \overset{\text{def}}{=} \max X.[((x)?(y_1), x \neq \mathbf{b})]([(x?(\_))]\mathsf{ff} \wedge [(x!(y_2))]\varphi_1')$$
$$\varphi_1' \overset{\text{def}}{=} ([(x!(\_))]\mathsf{ff} \wedge [(\mathbf{b}!(y_3), y_3 = (\log, y_1, y_2))]X)$$

Recall the transducers $m_\mathbf{e}, m_\mathbf{a}, m_\mathbf{d}, m_\mathbf{dt}$ and $m_\mathbf{det}$ from Example 2:

- $m_\mathbf{e}$ is *unsound* for φ_1. When composed with $s_\mathbf{b}$, it produces two consecutive output replies (underlined), meaning that $m_\mathbf{e}[s_\mathbf{b}] \notin [\![\varphi_1]\!]$: $m_\mathbf{e}[s_\mathbf{b}] \overset{t_\mathbf{e}^1}{\Longrightarrow}$ $\mathsf{id}[s_\mathbf{b}]$ where $t_\mathbf{e}^1 \overset{\text{def}}{=} \mathbf{c}?v_1.\mathbf{c}!\mathsf{ans}(v_1).\mathbf{b}!(\log, v_1, \mathsf{ans}(v_1)).\mathbf{a}?v_2.\underline{\mathbf{a}!w_2}.\underline{\mathbf{a}!w_2}$. Similarly,

$m_e[s_c] \notin [\![\varphi_1]\!]$ since the $m_e[s_c]$ executes the erroneous trace with two consecutive inputs on port a (underlined): $c?v_1.c!ans(v_1).b!(log, v_1, ans(v_1)).\underline{a?v_2.a?v_3}$. This demonstrates that $m_e[s_c]$ can still input two consecutive requests on port a (underlined). Either one of these counter examples *disproves* senf(m_e, φ_1).

– m_a turns out to be *sound* for φ_1 because once instrumented, the resulting composite system is adapted to only interact on port b. In fact we have $m_a[s_a], m_a[s_b], m_a[s_c] \in [\![\varphi_1]\!]$. Monitors m_d, m_{dt} and m_{det} are also *sound* for φ_1. Whereas, m_d prevents the violation of φ_1 by also blocking all input ports except b, m_{dt} and m_{det} achieve the same goal by disabling the invalid consecutive requests and answers that occur on a specific port (except b). ∎

By itself, sound enforcement is a weak criterion because it does not regulate the *extent* to which enforcement is applied. More specifically, although m_d from Example 2 is sound, it needlessly modifies the behaviour of s_a even though s_a satisfies φ_1: by blocking the initial input of s_a, m_d causes it to block indefinitely. The requirement that a monitor should not modify the behaviour of a system that satisfies the property being enforced can be formalised using a transparency criterion.

Definition 4 (Transparent Enforcement [5]). *A monitor m transparently enforces a formula* φ, tenf(m, φ), *iff for all* $s \in$ SYS, $s \in [\![\varphi]\!]$ *implies* $m[s] \sim s$. □

Example 5. As argued earlier, s_a suffices to disprove tenf(m_d, φ_1). Monitor m_a from Example 3 also breaches Definition 4: although $s_a \in [\![\varphi_1]\!]$, we have $m_a[s_a] \not\sim s_a$ since for any value v and w, $s_a \xrightarrow{b?v} \cdot \xrightarrow{b!w}$ but $m_a[s_a] \xrightarrow{b?v} \cdot \xcancel{\xrightarrow{b!w}}$. By contrast, monitors m_{dt} and m_{det} turn out to satisfy Definition 4 as they only intervene when it becomes apparent that a violation will occur. For instance, they only disable inputs on a specific port, as a precaution, following an unanswered request on the same port, and they only disable the redundant responses that are produced after the first response to a request. ∎

By some measures, Definition 4 is still a relatively weak requirement since it only limits transparency requirements to well-behaved systems, and disregards enforcement behaviour for systems that violate a property. For instance, consider monitor m_{dt} from Example 2 and system s_b from Example 1. At runtime s_b can exhibit the following invalid behaviour: $s_b \xrightarrow{t_1} b!(log, v, w).s_a$ where $t_1 \stackrel{\text{def}}{=} a?v.a!w.a!w$. In order to rectify this violating behaviour wrt. formula φ_1, it suffices to use a monitor that disables *one* of the responses in t_1, i.e., a!w. Following this disabling, no further modifications are required since the SuS reaches a state that does not violate the remainder of the formula φ_1, i.e., $b!(log, v, w).s_a \in [\![after(\varphi_1, t'_1)]\!]$. However, when instrumented with m_{dt}, this monitor does not only disable the invalid response, namely $m_{dt}[s_b] \xrightarrow{a?v.a!w.} m_d[b!(log, v, w).s_a]$, but subsequently disables every other action by reaching m_d, $m_d[b!(log, v, w).s_a] \xrightarrow{\tau} m_d[s_a]$. To this end, we introduce the novel requirement of *eventual transparency*.

Definition 5 (Eventually Transparent Enforcement). *Monitor* m *enforces property* φ *in an eventually transparent way,* $\textbf{evtenf}(m, \varphi)$, *iff for all systems* s, s', *traces* t *and monitors* m', $m[s] \stackrel{t}{\Longrightarrow} m'[s']$ *and* $s' \in [\![after(\varphi, t)]\!]$ *imply* $m'[s'] \sim s'$. ∎

Example 6. We have already argued why $m_{\mathbf{dt}}$ does not adhere to eventual transparency via the counterexample $s_{\mathbf{b}}$. This is not the case for $m_{\mathbf{det}}$. Although the universal quantification over all systems and traces make it hard to prove this property, we get an intuition of why this is the case from $s_{\mathbf{b}}$: when $m_{\mathbf{det}}[s_{\mathbf{b}}] \xrightarrow{a?v_1.a!w_1} \cdot \xrightarrow{\tau} m''_{\mathbf{det}}[b!(\log, v_1, w_1).s_{\mathbf{a}}]$ we have $b!(\log, v_1, w_1).s_{\mathbf{a}} \in [\![after(\varphi_1, a?v_1.a!w_1)]\!]$ and that $m''_{\mathbf{det}}[b!(\log, v_1, w_1).s_{\mathbf{a}}] \sim b!(\log, v_1, w_1).s_{\mathbf{a}}$. ∎

Corollary 1. *For all monitors* $m \in \mathrm{TRN}$ *and properties* $\varphi \in \mathrm{sHML}$, $\textbf{evtenf}(m, \varphi)$ *implies* $\textbf{tenf}(m, \varphi)$. □

Along with Definition 3 (soundness), Definition 5 (eventual transparency) makes up our definition for "m *(adequately) enforces* φ". From Corollary 1, it follows that is definition is stricter than the one given in [5].

Definition 6 (Adequate Enforcement). *A monitor* m *(adequately) enforces property* φ *iff it adheres to* (i) *soundness, Definition 3, and* (ii) *eventual transparency, Definition 5.* ∎

5   Synthesising Action Disabling Monitors

Although Definition 2 enables us to rule out erroneous monitors that purport to enforce a property, the universal quantifications over all systems in Definitions 3 and 5 make it difficult to prove that a monitor does indeed enforce a property correctly in a bidirectional setting. Establishing that a formula is enforceable, Definition 6, involves a further existential quantification over a monitor that enforces it correctly. Moreover, establishing the enforceability of a logic entails yet another universal quantification, on all the formulas in the logic.

We address these problems through an *automated synthesis procedure* that produces an enforcement monitor for *every* sHML formula. We also show that the synthesised monitors are correct, according to Definition 6. For a unidirectional setting, it has been shown that monitors that only administer *omissions* are expressive enough to enforce *safety properties* [5, 19, 25, 32]. Analogously, for our bidirectional case, we restrict ourselves to action disabling monitors and show that they can enforce *any* property expressed in terms of sHML.

Our synthesis procedure is compositional, meaning that the monitor synthesis of a composite formula is defined in terms of the enforcement monitors generated from its constituent sub-formulas. Compositionality simplifies substantially our correctness analysis of the generated monitors (*e.g.*, we can use standard inductive proof techniques). In order to ease a compositional definition, our synthesis procedure is defined in terms of a variant of sHML called sHML$_{\mathbf{nf}}$: it is a normalised syntactic subset of sHML that is still as expressive as sHML [2]. An automated procedure to translate an sHML formula into a corresponding sHML$_{nf}$ one (with the same semantic meaning) is given in [2,5].

Definition 7 (sHML Normal Form). *The set of normalised* sHML *formulas is generated by the following grammar:*

$$\varphi, \psi \in \text{sHML}_{nf} ::= \text{tt} \quad | \quad \text{ff} \quad | \quad \bigwedge_{i \in I} [p_i, c_i]\varphi_i \quad | \quad X \quad | \quad \max X.\varphi .$$

In addition, sHML$_{nf}$ *formulas are required to satisfy the following conditions:*

1. *Every branch in $\bigwedge_{i \in I} [p_i, c_i]\varphi_i$, must be* disjoint, *i.e., for every $i, j \in I$, $i \neq j$ implies $[\![(p_i, c_i)]\!] \cap [\![(p_j, c_j)]\!] = \emptyset$.*
2. *For every $\max X.\varphi$ we have $X \in \boldsymbol{fv}(\varphi)$.* ∎

In a (closed) sHML$_{nf}$ formula, the basic terms tt and ff can never appear unguarded unless they are at the top level (e.g., we can never have $\varphi \wedge$ff or $\max X_0. \ldots \max X_n.$ff). Modal operators are combined with conjunctions into one construct $\bigwedge_{i \in I} [p_i, c_i]\varphi_i$ that is written as $[p_0, c_0]\varphi_0 \wedge \ldots \wedge [p_n, c_n]\varphi_n$ when $I = \{0, \ldots, n\}$ and simply as $[p_0, c_0]\varphi$ when $|I| = 1$. The conjunct modal guards must also be *disjoint* so that *at most one* necessity guard can satisfy any particular visible action. Along with these restrictions, we still assume that sHML$_{nf}$ fixpoint variables are guarded, and that for every $((x)?(y), c)$, $y \notin \boldsymbol{fv}(c)$.

Example 7. The formula φ_3 defines a recursive property stating that, following an input on port a (carrying any value), prohibits that the system outputs a value of 4 (on any port), unless the output is made on port a with a value that is not equal to 3 (in which cases, it recurses).

$$\varphi_3 \stackrel{\text{def}}{=} \max X.[((x_1)?(y_1), x_1{=}\text{a})] \left(\begin{array}{c} [((x_2)!(y_2), x_2{=}\text{a} \wedge y_2{\neq}3)]X \\ \wedge\ [((x_3)!(y_3), y_3{=}4)]\text{ff} \end{array} \right)$$

φ_3 is not an sHML$_{nf}$ formula since its conjunction is not disjoint (e.g., the action a!4 satisfies both branches). Still, we can reformulate φ_3 as $\varphi_3' \in$ sHML$_{nf}$:

$$\varphi_3' \stackrel{\text{def}}{=} \max X.[((x_1)?(y_1), x_1{=}\text{a})] \left(\begin{array}{c} [((x_4)!(y_4), x_4{=}\text{a} \wedge y_4{\neq}4)]X \\ \wedge\ [((x_4)!(y_4), x_4{=}\text{a} \wedge y_4{=}4)]\text{ff} \end{array} \right)$$

where x_4 and y_4 are fresh variables. ∎

Our monitor synthesis function in Definition 8 converts an sHML$_{nf}$ formula φ into a transducer m. This conversion also requires information regarding the input ports of the SuS, as this is used to add the necessary insertion branches that silently unblock the SuS at runtime. The synthesis function must therefore be supplied with this information in the form of a *finite* set of input ports $\Pi \subset \text{PORT}$, which then relays this information to the resulting monitor.

Definition 8. *The synthesis function* $(\!|-|\!) : \mathrm{sHML}_{nf} \times \mathcal{P}_{fin}(\mathrm{PORT}) \to \mathrm{TRN}$ *is defined inductively as:*

$$(\!|X, \Pi|\!) \stackrel{\mathrm{def}}{=} X \qquad (\!|\mathsf{tt}, \Pi|\!) \stackrel{\mathrm{def}}{=} (\!|\mathsf{ff}, \Pi|\!) \stackrel{\mathrm{def}}{=} id \qquad (\!|\max X.\varphi, \Pi|\!) \stackrel{\mathrm{def}}{=} rec\,X.(\!|\varphi, \Pi|\!)$$

$$(\!|\varphi = \bigwedge_{i \in I}[(p_i, c_i)]\varphi_i, \Pi|\!) \stackrel{\mathrm{def}}{=} rec\,Y.\left(\sum_{i \in I}\begin{cases} dis(p_i, c_i, Y, \Pi) & if\ \varphi_i = \mathsf{ff} \\ (p_i, c_i).(\!|\varphi_i, \Pi|\!) & otherwise \end{cases}\right) + def(\varphi)$$

$$where\quad dis(p, c, m, \Pi) \quad\stackrel{\mathrm{def}}{=}\quad \begin{cases} (p, c, \bullet).m & if\ p = (x)!(y) \\ \sum_{b \in \Pi}(\bullet, c\{b/x\}, b?v_{def}).m & if\ p = (x)?(y) \end{cases} \quad and$$

$$def(\bigwedge_{i \in I}[((x_i)?(y_i), c_i)]\varphi_i \wedge \psi) \quad\stackrel{\mathrm{def}}{=}\quad \begin{cases} ((\_)?(\_)).id & when\ I = \emptyset \\ ((x)?(y), \bigwedge_{i \in I}(\neg c_i\{x/x_i, y/y_i\})).id & otherwise \end{cases}$$

where ψ has no conjuncts starting with an input modality, variables x and y are fresh, and v_{def} is a default value. ∎

The definition above assumes a bijective mapping between formula variables and monitor recursion variables. Normalised conjunctions, $\bigwedge_{i \in I}[p_i, c_i]\varphi_i$, are synthesised as a *recursive summation* of monitors, *i.e.*, $rec\,Y.\sum_{i \in I} m_i$, where Y is fresh, and every branch m_i can be one of the following:

(i) when m_i is derived from a branch of the form $[p_i, c_i]\varphi_i$ where $\varphi_i \neq \mathsf{ff}$, the synthesis produces a monitor with the *identity transformation* prefix, (p_i, c_i), followed by the monitor synthesised from the continuation φ_i, *i.e.*, $(\!|\varphi_i, \Pi|\!)$;

(ii) when m_i is derived from a violating branch of the form $[p_i, c_i]\mathsf{ff}$, the synthesis produces an *action disabling transformation* via $dis(p_i, c_i, Y, \Pi)$.

Specifically, in clause (*ii*) the dis function produces either a *suppression transformation*, $(p_i, c_i, \bullet)$, when p_i is an *output* pattern, $(x_i)!(y_i)$, or a *summation of insertions*, $\sum_{b \in \Pi}(\bullet, c_i\{b/x_i\}, b?v_{def}).m_i$, when p_i is an *input* pattern, $(x_i)?(y_i)$. The former signifies that the monitor must react to and suppress every matching (invalid) system output thus stopping it from reaching the environment. By not synthesising monitor branches that react to the erroneous input, the latter allows the monitor to hide the input synchronisations from the environment. At the same time, the synthesised insertion branches insert a default domain value v_{def} on every port $\mathsf{a} \in \Pi$ whenever the branch condition $c_i\{b/x_i\}$ evaluates to true at runtime. This stops the monitor from blocking the resulting composite system unnecessarily.

This blocking mechanism can, however, block *unspecified* inputs, *i.e.*, those that do not satisfy any modal necessity in the normalised conjunction. This is undesirable since the unspecified actions do not contribute towards a safety violation and, instead, lead to its trivial satisfaction. To prevent this, the *default monitor* $def(\varphi)$ is also added to the resulting summation. Concretely, the def function produces a *catch-all* identity monitor that forwards an input to the SuS whenever it satisfies the negation of *all* the conditions associated with modal necessities for input patterns in the normalised conjunction. This condition is constructed for a normalised conjunction of the form $\bigwedge_{i \in I}[((x_i)?(y_i), c_i)]\varphi_i \wedge \psi$ (assuming that ψ does not include further input modalities). Otherwise, if none of the conjunct modalities define an input pattern, every input is allowed, *i.e.*, the default monitor becomes $((\_)?(\_)).\mathsf{id}$, which transitions to id after forwarding the input to the SuS.

Example 8. Recall (the full version of) formula φ_1 from Example 1.

$$\varphi_1 \overset{\text{def}}{=} \max X.[((x)?(y_1), x{\neq}\mathsf{b})]([((x_1)?(\_), x_1{=}x)]\mathsf{ff} \wedge [((x_2)!(y_2), x_2{=}x)]\varphi_1')$$

$$\varphi_1' \overset{\text{def}}{=} ([((x_3)!(\_), x_3{=}x)]\mathsf{ff} \wedge [((x_4)!(y_3), x_4{=}\mathsf{b} \wedge y_3{=}(\log, y_1, y_2))]X)$$

For any arbitrary set of ports Π, the synthesis of Definition 8 produces the following monitor.

$$m_{\varphi_1} \overset{\text{def}}{=} \mathsf{rec}\, X.\mathsf{rec}\, Z.(((x)?(y_1), x{\neq}\mathsf{b}).\mathsf{rec}\, Y_1.m_{\varphi_1}') + ((x_{\mathsf{def}})?(\_), x_{\mathsf{def}} = \mathsf{b}).\mathsf{id}$$

$$m_{\varphi_1}' \overset{\text{def}}{=} \sum_{\mathsf{a} \in \Pi}(\bullet, \mathsf{a}{=}x, \mathsf{a}?v_{\mathsf{def}}).Y_1 + ((x_2)!(y_2), x_2{=}x).\mathsf{rec}\, Y_2.m_{\varphi_1}'' + ((x_{\mathsf{def}})?(\_), x_{\mathsf{def}}{\neq}x).\mathsf{id}$$

$$m_{\varphi_1}'' \overset{\text{def}}{=} ((x_3)!(\_), x_3{=}x, \bullet).Y_2 + ((x_4)!(y_3), x_4{=}\mathsf{b} \wedge y_3{=}(\log, y_1, y_2)).X + ((\_)?(\_)).\mathsf{id}$$

Monitor m_{φ_1} can be optimised by removing redundant recursive constructs such as $\mathsf{rec}\, Z.\_$ that are introduced mechanically by our synthesis.   ∎

Monitor m_{φ_1} from Example 8 (with $(\!|\varphi_1, \Pi|\!) = m_{\varphi_1}$) is very similar to $m_{\mathbf{det}}$ of Example 2, differing only in how it defines its insertion branches for unblocking the SuS. For instance, if we consider $\Pi = \{\mathsf{b}, \mathsf{c}\}$, $(\!|\varphi_1, \Pi|\!)$ would synthesise two insertion branches, namely $(\bullet, \mathsf{b} = x, \mathsf{b}?v_{\mathsf{def}})$ and $(\bullet, \mathsf{c} = x, \mathsf{c}?v_{\mathsf{def}})$. By contrast, $m_{\mathbf{det}}$ attains the same result more succinctly via the single insertion branch $(\bullet, x?v_{\mathsf{def}})$. Importantly, our synthesis provides the witness monitors needed to show enforceability.

Theorem 1 (Enforceability). sHML *is bidirectionally enforceable using the monitors and instrumentation of Fig. 4.*   □

6   Conclusions and Related Work

This work extends the framework presented in the precursor to this work [5] to the setting of bidirectional enforcement where observable actions such as inputs and outputs require different treatment. We achieve this by:

1. augmenting substantially our instrumentation relation (Fig. 4);
2. refining our definition of enforcement to incorporate transparency over violating systems (Definition 6); and
3. providing a more extensive synthesis function (Definition 8) that is proven correct (Theorem 1).

Future work. There are a number of possible avenues for extending our work. One immediate step would be the implementation of the monitor operational model presented in Sect. 3 together with the synthesis function described in Sect. 5. This effort should be integrated it within the detectEr tool suite [10,14–16]. This would allow us to assess the overhead induced by our proposed bidirectional monitoring [4]. Another possible direction would be the development of behavioural theories for the transducer operational model presented in Sect. 3, along the lines of the refinement preorders studied in earlier work on sequence recognisers [3,20,21]. Finally, applications of the theory, along the lines of [30] are also worth exploring.

Related work. As we discussed already in the Introduction, most work on RE assumes a trace-based view of the SuS [32,33,39], where few distinguish between actions with different control profiles (*e.g.*, inputs versus outputs). Although shields [28] can analyse both input and output actions, they still perform unidirectional enforcement and only modify the data associated with the output actions. The closest to our work is that by Pinisetty *et al.* [35], who consider bidirectional RE, modelling the system as a trace of input-output pairs. However, their enforcement is limited to replacements of payloads and their setting is too restrictive to model enforcements such as action rerouting and the closing of ports. Finally, Lanotte *et al.* [30] employ similar synthesis techniques and correctness criteria to ours (Definitions 3 and 4) to generate enforcement monitors for a timed setting.

References

1. Aceto, L., Achilleos, A., Francalanza, A., Ingólfsdóttir, A.: A framework for parameterized monitorability. In: Baier, C., Dal Lago, U. (eds.) FoSSaCS 2018. LNCS, vol. 10803, pp. 203–220. Springer, Cham (2018). https://doi.org/10.1007/978-3-319-89366-2_11
2. Aceto, L., Achilleos, A., Francalanza, A., Ingólfsdóttir, A., Kjartansson, S.Ö.: Determinizing monitors for HML with recursion. J. Log. Algebraic Methods Program. **111**, (2020). https://doi.org/10.1016/j.jlamp.2019.100515
3. Aceto, L., Achilleos, A., Francalanza, A., Ingólfsdóttir, A., Lehtinen, K.: The Best a Monitor Can Do. In: CSL. LIPIcs, vol. 183, pp. 7:1–7:23. Schloss Dagstuhl (2021). https://doi.org/10.4230/LIPIcs.CSL.2021.7
4. Aceto, L., Attard, D.P., Francalanza, A., Ingólfsdóttir, A.: On benchmarking for concurrent runtime verification. FASE 2021. LNCS, vol. 12649, pp. 3–23. Springer, Cham (2021). https://doi.org/10.1007/978-3-030-71500-7_1
5. Aceto, L., Cassar, I., Francalanza, A., Ingólfsdóttir, A.: On Runtime Enforcement via Suppressions. In: CONCUR. vol. 118, pp. 34:1–34:17. Schloss Dagstuhl (2018). https://doi.org/10.4230/LIPIcs.CONCUR.2018.34
6. Aceto, L., Cassar, I., Francalanza, A., Ingólfsdóttir, A.: On bidirectional enforcement. Technical report Reykjavik University (2020). http://icetcs.ru.is/theofomon/bidirectionalRE.pdf
7. Aceto, L., Ingólfsdóttir, A.: Testing Hennessy-Milner logic with recursion. In: Thomas, W. (ed.) FoSSaCS 1999. LNCS, vol. 1578, pp. 41–55. Springer, Heidelberg (1999). https://doi.org/10.1007/3-540-49019-1_4

8. Aceto, L., Ingólfsdóttir, A., Larsen, K.G., Srba, J.: Reactive Systems: Modelling, Specification and Verification. Cambridge University Press, NY, USA (2007)
9. Alur, R., Černý, P.: Streaming Transducers for Algorithmic Verification of Single-pass List-processing Programs. In: POPL, pp. 599–610. ACM (2011). https://doi.org/10.1145/1926385.1926454
10. Attard, D.P., Francalanza, A.: A monitoring tool for a branching-time logic. In: Falcone, Y., Sánchez, C. (eds.) RV 2016. LNCS, vol. 10012, pp. 473–481. Springer, Cham (2016). https://doi.org/10.1007/978-3-319-46982-9_31
11. Bielova, N., Massacci, F.: Do you really mean what you actually enforced?-edited automata revisited. J. Inf. Secur. 10(4), 239–254 (2011). https://doi.org/10.1007/s10207-011-0137-2
12. Bocchi, L., Chen, T.C., Demangeon, R., Honda, K., Yoshida, N.: Monitoring networks through multiparty session types. TCS 669, 33–58 (2017)
13. Cassar, I.: Developing Theoretical Foundations for Runtime Enforcement. Ph.D. thesis, University of Malta and Reykjavik University (2021)
14. Cassar, I., Francalanza, A., Aceto, L., Ingólfsdóttir, A.: eAOP: an aspect oriented programming framework for Erlang. In: Erlang. ACM SIGPLAN (2017)
15. Cassar, I., Francalanza, A., Attard, D.P., Aceto, L., Ingólfsdóttir, A.: A Suite of Monitoring Tools for Erlang. In: RV-CuBES. Kalpa Publications in Computing, vol. 3, pp. 41–47. EasyChair (2017)
16. Cassar, I., Francalanza, A., Said, S.: Improving Runtime Overheads for detectEr. In: FESCA. EPTCS, vol. 178, pp. 1–8 (2015)
17. Chen, T.-C., Bocchi, L., Deniélou, P.-M., Honda, K., Yoshida, N.: Asynchronous distributed monitoring for multiparty session enforcement. In: Bruni, R., Sassone, V. (eds.) TGC 2011. LNCS, vol. 7173, pp. 25–45. Springer, Heidelberg (2012). https://doi.org/10.1007/978-3-642-30065-3_2
18. Falcone, Y., Fernandez, J.-C., Mounier, L.: Synthesizing Enforcement Monitors w.r.t. the safety-progress classification of properties. In: Sekar, R., Pujari, A.K. (eds.) ICISS 2008. LNCS, vol. 5352, pp. 41–55. Springer, Heidelberg (2008). https://doi.org/10.1007/978-3-540-89862-7_3
19. Falcone, Y., Fernandez, J.C., Mounier, L.: What can you verify and enforce at runtime? J. Softw. Tools Technol. Transf. 14(3), 349 (2012)
20. Francalanza, A.: Consistently-Detecting Monitors. In: CONCUR. LIPIcs, vol. 85, pp. 8:1–8:19. Dagstuhl, Germany (2017). https://doi.org/10.4230/LIPIcs.CONCUR.2017.8
21. Francalanza, A.: A theory of monitors. Inf. Comput 104704 (2021). https://doi.org/10.1016/j.ic.2021.104704
22. Francalanza, A., Aceto, L., Ingólfsdóttir, A.: Monitorability for the Hennessy-Milner logic with recursion. Formal Methods Syst. Des. 51(1), 87–116 (2017)
23. Hennessy, M., Lin, H.: Proof systems for message-passing process algebras. Formal Aspects Comput. 8(4), 379–407 (1996). https://doi.org/10.1007/BF01213531
24. Hennessy, M., Liu, X.: A modal logic for message passing processes. Acta Inf. 32(4), 375–393 (1995). https://doi.org/10.1007/BF01178384
25. van Hulst, A.C., Reniers, M.A., Fokkink, W.J.: Maximally permissive controlled system synthesis for non-determinism and modal logic. Discr. Event Dyn. Syst. 27(1), 109–142 (2017)
26. Jia, L., Gommerstadt, H., Pfenning, F.: Monitors and blame assignment for higher-order session types. In: POPL, pp. 582–594. ACM, NY, USA (2016)
27. Khoury, R., Tawbi, N.: Which security policies are enforceable by runtime monitors? A survey. Comput. Sci. Rev. 6(1), 27–45 (2012). https://doi.org/10.1016/j.cosrev.2012.01.001

28. Könighofer, B., et al.: Shield synthesis. Formal Methods Syst. Des. **51**(2), 332–361 (2017). https://doi.org/10.1007/s10703-017-0276-9
29. Kozen, D.C.: Results on the propositional μ-calculus. Theor. Comput. Sci. **27**, 333–354 (1983)
30. Lanotte, R., Merro, M., Munteanu, A.: Runtime enforcement for control system security. In: CSF, pp. 246–261. IEEE (2020). https://doi.org/10.1109/CSF49147.2020.00025
31. Larsen, K.G.: Proof systems for satisfiability in Hennessy-Milner logic with recursion. Theor. Comput. Sci. **72**(2), 265–288 (1990). https://doi.org/10.1016/0304-3975(90)90038-J
32. Ligatti, J., Bauer, L., Walker, D.: Edit automata: enforcement mechanisms for runtime security policies. J. Inf. Secur. **4**(1), 2–16 (2005). https://doi.org/10.1007/s10207-004-0046-8
33. Ligatti, J., Bauer, L., Walker, D.: Run-time enforcement of nonsafety policies. ACM Trans. Inf. Syst. Secur. **12**(3), 19:1–19:41 (2009)
34. Milner, R., Parrow, J., Walker, D.: A calculus of mobile processes. I. Inf. Comput. **100**(1), 1–40 (1992). https://doi.org/10.1016/0890-5401(92)90008-4
35. Pinisetty, S., Roop, P.S., Smyth, S., Allen, N., Tripakis, S., Hanxleden, R.V.: Runtime enforcement of cyber-physical systems. ACM Trans. Embed. Comput. Syst. **16**(5s), 1–25 (2017)
36. Rathke, J., Hennessy, M.: Local model checking for value-passing processes (extended abstract). In: Abadi, M., Ito, T. (eds.) TACS 1997. LNCS, vol. 1281, pp. 250–266. Springer, Heidelberg (1997). https://doi.org/10.1007/BFb0014555
37. Sakarovitch, J.: Elements of Automata Theory. Cambridge University Press, New York, NY, USA (2009)
38. Sangiorgi, D.: Introduction to Bisimulation and Coinduction. Cambridge University Press, New York, NY, USA (2011)
39. Schneider, F.B.: Enforceable security policies. ACM Trans. Inf. Syst. Secur. (TISSEC) **3**(1), 30–50 (2000)

A Multi-agent Model for Polarization Under Confirmation Bias in Social Networks

Mário S. Alvim[1]([⊠]), Bernardo Amorim[1], Sophia Knight[2], Santiago Quintero[3], and Frank Valencia[4,5]

[1] Department of Computer Science, UFMG, Belo Horizonte, Brazil
msalvim@dcc.ufmg.br
[2] Department of Computer Science, University of Minnesota Duluth, Duluth, USA
[3] LIX, École Polytechnique de Paris, Paris, France
[4] CNRS-LIX, École Polytechnique de Paris, Paris, France
[5] Pontificia Universidad Javeriana Cali, Cali, Colombia

Abstract. We describe a model for polarization in multi-agent systems based on Esteban and Ray's standard measure of polarization from economics. Agents evolve by updating their beliefs (opinions) based on an underlying influence graph, as in the standard DeGroot model for social learning, but under a *confirmation bias*; i.e., a discounting of opinions of agents with dissimilar views. We show that even under this bias polarization eventually vanishes (converges to zero) if the influence graph is strongly-connected. If the influence graph is a regular symmetric circulation, we determine the unique belief value to which all agents converge. Our more insightful result establishes that, under some natural assumptions, if polarization does not eventually vanish then either there is a disconnected subgroup of agents, or some agent influences others more than she is influenced. We also show that polarization does not necessarily vanish in weakly-connected graphs under confirmation bias. We illustrate our model with a series of case studies and simulations, and show how it relates to the classic DeGroot model for social learning.

Keywords: Polarization · Confirmation bias · Multi-agent systems · Social networks

1 Introduction

Distributed systems have changed substantially in the recent past with the advent of social networks. In the previous incarnation of distributed computing [22] the

Mário S. Alvim and Bernardo Amorim were partially supported by CNPq, CAPES and FAPEMIG. Santiago Quintero and Frank Valencia were partially supported by the ECOS-NORD project FACTS (C19M03).

© IFIP International Federation for Information Processing 2021
Published by Springer Nature Switzerland AG 2021
K. Peters and T. A. C. Willemse (Eds.): FORTE 2021, LNCS 12719, pp. 22–41, 2021.
https://doi.org/10.1007/978-3-030-78089-0_2

emphasis was on consistency, fault tolerance, resource management and related topics; these were all characterized by *interaction between processes*. What marks the new era of distributed systems is an emphasis on the flow of epistemic information (facts, beliefs, lies) and its impact on democracy and on society at large.

Indeed in social networks a group may shape their beliefs by attributing more value to the opinions of outside influential figures. This cognitive bias is known as *authority bias* [32]. Furthermore, in a group with uniform views, users may become extreme by reinforcing one another's opinions, giving more value to opinions that confirm their own preexisting beliefs. This is another common cognitive bias known as *confirmation bias* [4]. As a result, social networks can cause their users to become radical and isolated in their own ideological circle causing dangerous splits in society [5] in a phenomenon known as *polarization* [4].

There is a growing interest in the development of models for the analysis of polarization and social influence in networks [6, 8, 9, 12, 14, 15, 19, 20, 28, 31, 34, 35, 37]. Since polarization involves non-terminating systems with *multiple agents* simultaneously exchanging information (opinions), concurrency models are a natural choice to capture the dynamics of polarization.

The Model. In fact, we developed a multi-agent model for polarization in [3], inspired by linear-time models of concurrency where the state of the system evolves in discrete time units (in particular [27, 33]). In each time unit, the agents *update* their beliefs about the proposition of interest taking into account the beliefs of their neighbors in an underlying weighted *influence graph*. The belief update gives more value to the opinion of agents with higher influence (*authority bias*) and to the opinion of agents with similar views (*confirmation bias*). Furthermore, the model is equipped with a *polarization measure* based on the seminal work in economics by Esteban and Ray [13]. The polarization is measured at each time unit and it is 0 if all agents' beliefs fall within an interval of agreement about the proposition. The contributions in [3] were of an experimental nature and aimed at exploring how the combination of influence graphs and cognitive biases in our model can lead to polarization.

In the current paper we prove claims made from experimental observations in [3] using techniques from calculus, graph theory, and flow networks. The main goal of this paper is identifying how networks and beliefs are structured, for agents subject to confirmation bias, when polarization *does not* disappear. Our results provide insight into the phenomenon of polarization, and are a step toward the design of robust computational models and simulation software for human cognitive and social processes.

The closest related work is that on DeGroot models [9]. These are the standard linear models for social learning whose analysis can be carried out by linear techniques from Markov chains. A novelty in our model is that its update function extends the classical update from DeGroot models with confirmation bias. As we shall elaborate in Sect. 5 the extension makes the model no longer linear and thus mathematical tools like Markov chains do not seem applicable. Our model incorporates a polarization measure in a model for social learning and extends classical convergence results of DeGroot models to the confirmation bias case.

Main Contributions. The following are the main theoretical results established in this paper. Assuming confirmation bias and some natural conditions about belief values: (1) If polarization does not disappear then either there is disconnected subgroup of agents, or some agent influences others more than she is influenced, or all the agents are initially radicalized (i.e., each individual holds the most extreme value either in favor or against of a given proposition). (2) Polarization eventually disappears (converges to zero) if the influence graph is strongly-connected. (3) If the influence graph is a regular symmetric circulation we determine the unique belief value all agents converge to.

Organization. In Sect. 2 we introduce the model and illustrate a series of examples and simulations, uncovering interesting new insights and complex characteristics of the believe evolution. The theoretical contributions (1–3) above are given in Sects. 3 and 4. We discuss DeGroot and other related work in Sects. 5 and 6. Full proofs can be found in the corresponding technical report [2]. An implementation of the model in Python and the simulations are available on Github [1].

2   The Model

Here we refine the polarization model introduced in [3], composed of static and dynamic elements. We presuppose basic knowledge of calculus and graph theory [11,38].

Static Elements of the Model. *Static elements* of the model represent a snapshot of a social network at a given point in time. They include the following components:

- A (finite) set $\mathcal{A} = \{0, 1, \ldots, n-1\}$ of $n \geq 1$ *agents*.
- A *proposition* p of interest, about which agents can hold beliefs.
- A *belief configuration* $B{:}\mathcal{A}{\to}[0,1]$ s.t. each value B_i is the instantaneous confidence of agent $i \in \mathcal{A}$ in the veracity of proposition p. Extreme values 0 and 1 represent a firm belief in, respectively, the falsehood or truth of p.
- A *polarization measure* $\rho{:}[0,1]^{\mathcal{A}}{\to}\mathbb{R}$ mapping belief configurations to real numbers. The value $\rho(B)$ indicates how polarized belief configuration B is.

There are several polarization measures described in the literature. In this work we adopt the influential measure proposed by Esteban and Ray [13].

Definition 1 (Esteban-Ray Polarization). *Consider a set* $\mathcal{Y} = \{y_0, y_1, \ldots, y_{k-1}\}$ *of size* k, *s.t. each* $y_i \in \mathbb{R}$. *Let* $(\pi, y) = (\pi_0, \pi_1, \ldots, \pi_{k-1}, y_0, y_1, \ldots, y_{k-1})$ *be a* distribution *on* $\mathcal{Y}$ *s.t.* π_i *is the frequency of value* $y_i \in \mathcal{Y}$ *in the distribution.*[1] *The* Esteban-Ray *(ER) polarization measure is defined as* $\rho_{ER}(\pi, y) = K \sum_{i=0}^{k-1} \sum_{j=0}^{k-1} \pi_i^{1+\alpha} \pi_j |y_i - y_j|$, *where* $K{>}0$ *is a constant, and typically* $\alpha{\approx}1.6$.

---

[1] W.l.o.g. we can assume the values of π_i are all non-zero and add up to 1.

The higher the value of $\rho_{ER}(\pi, y)$, the more polarized distribution (π, y) is. The measure captures the intuition that polarization is accentuated by both intra-group homogeneity and inter-group heterogeneity. Moreover, it assumes that the total polarization is the sum of the effects of individual agents on one another. The measure can be derived from a set of intuitively reasonable axioms [13].

Note that ρ_{ER} is defined on a discrete distribution, whereas in our model a general polarization metric is defined on a belief configuration $B{:}\mathcal{A}{\rightarrow}[0,1]$. To apply ρ_{ER} to our setup we convert the belief configuration B into an appropriate distribution (π, y).

Definition 2 (*k*-bin polarization). *Let D_k be a discretization of the interval $[0,1]$ into $k{>}0$ consecutive non-overlapping, non-empty intervals (bins) $I_0, I_1, \ldots, I_{k-1}$. We use the term* borderline points *of D_k to refer to the endpoints of $I_0, I_1, \ldots, I_{k-1}$ different from 0 and 1. We assume an underlying discretization D_k throughout the paper.*

Given D_k and a belief configuration B, define the distribution (π, y) as follows. Let $\mathcal{Y} = \{y_0, y_1, \ldots, y_{k-1}\}$ where each y_i is the mid-point of I_i, and let π_i be the fraction of agents having their belief in I_i. The polarization measure ρ of B is $\rho(B) = \rho_{ER}(\pi, y)$.

Notice that when there is consensus about the proposition p of interest, i.e., when all agents in belief configuration B hold the same belief value, we have $\rho(B) = 0$. This happens exactly when all agents' beliefs fall within the same bin of the underlying discretization D_k. The following property is an easy consequence from Definition 1 and Definition 2.

Proposition 1 (Zero Polarization). *Let $D_k = I_0, I_1, \ldots, I_{k-1}$ be the discretization of $[0,1]$ in Definition 2. Then $\rho(B) = 0$ iff there exists $m \in \{0, \ldots, k-1\}$ s.t. for all $i \in \mathcal{A}$, $B_i \in I_m$.*

Dynamic Elements of the Model. *Dynamic elements* formalize the evolution of agents' beliefs as they interact over time and are exposed to different opinions. They include:

- A *time frame* $\mathcal{T}{=}\{0,1,2,\ldots\}$ representing the discrete passage of time.
- A *family of belief configurations* $\{B^t{:}\mathcal{A}{\rightarrow}[0,1]\}_{t \in \mathcal{T}}$ s.t. each B^t is the belief configuration of agents in $\mathcal{A}$ w.r.t. proposition p at time step $t \in \mathcal{T}$.
- A *weighted directed graph* $\mathcal{I}{:}\mathcal{A} \times \mathcal{A}{\rightarrow}[0,1]$. The value $\mathcal{I}(i,j)$, written $\mathcal{I}_{i,j}$, represents the *direct influence* that agent i has on agent j, or the *weight* i carries with j. A higher value means stronger weight. Conversely, $\mathcal{I}_{i,j}$ can also be viewed as the *trust* or *confidence* that j has on i. We assume that $\mathcal{I}_{i,i} = 1$, meaning that agents are self-confident. We shall often refer to $\mathcal{I}$ simply as the *influence* (graph) $\mathcal{I}$.

We distinguish, however, the direct influence $\mathcal{I}_{i,j}$ that i has on j from the *overall effect* of i in j's belief. This effect is a combination of various factors, including direct influence, their current opinions, the topology of the influence graph, and how agents reason. This overall effect is captured by the update function below.

– An *update function* $\mu{:}(B^t,\mathcal{I}){\mapsto}B^{t+1}$ mapping belief configuration B^t at time t and influence graph $\mathcal{I}$ to new belief configuration B^{t+1} at time $t{+}1$. This function models the evolution of agents' beliefs over time. We adopt the following premises.

(i) **Agents present some Bayesian reasoning:** Agents' beliefs are updated in every time step by combining their current belief with a *correction term* that incorporates the new evidence they are exposed to in that step –i.e., other agents' opinions. More precisely, when agent j interacts with agent i, the former affects the latter moving i's belief towards j's, proportionally to the difference $B_j^t{-}B_i^t$ in their beliefs. The intensity of the move is proportional to the influence $\mathcal{I}_{j,i}$ that j carries with i. The update function produces an overall correction term for each agent as the average of all other agents' effects on that agent, and then incorporates this term into the agent's current belief.[2] The factor $\mathcal{I}_{j,i}$ allows the model to capture *authority bias* [32], by which agents' influences on each other may have different intensities (by, e.g., giving higher weight to an authority's opinion).

(ii) **Agents may be prone to confirmation bias:** Agents may give more weight to evidence supporting their current beliefs while discounting evidence contradicting them, independently from its source. This behavior in known in the psychology literature as *confirmation bias* [4], and is captured in our model as follows. When agent j interacts with agent i, the update function moves agent i's belief toward that of agent j, proportionally to the influence $\mathcal{I}_{j,i}$ of j on i, but with a caveat: the move is stronger when j's belief is similar to i's than when it is dissimilar.

The premises above are formally captured in the following update-function.

Definition 3 (Confirmation-bias). *Let B^t be a belief configuration at time $t \in \mathcal{T}$, and $\mathcal{I}$ be an influence graph. The confirmation-bias update-function is the map $\mu^{CB}{:}(B^t,\mathcal{I}) \mapsto B^{t+1}$ with B^{t+1} given by $B_i^{t+1} = B_i^t + 1/|\mathcal{A}_i|\sum_{j\in\mathcal{A}_i} \beta_{i,j}^t \mathcal{I}_{j,i}(B_j^t - B_i^t)$, for every agent $i \in \mathcal{A}$, where $\mathcal{A}_i = \{j \in \mathcal{A} \mid \mathcal{I}_{j,i}{>}0\}$ is the set of neighbors of i and $\beta_{i,j}^t = 1{-}|B_j^t{-}B_i^t|$ is the confirmation-bias factor of i w.r.t. j given their beliefs at time t.*

The expression $1/|\mathcal{A}_i|\sum_{j\in\mathcal{A}_i} \beta_{i,j}^t \mathcal{I}_{j,i}(B_j^t - B_i^t)$ in Definition 3 is a *correction term* incorporated into agent i's original belief B_i^t at time t. The correction is the average of the effect of each neighbor $j \in \mathcal{A}_i$ on agent i's belief at that time step. The value B_i^{t+1} is the resulting updated belief of agent i at time $t{+}1$.

The confirmation-bias factor $\beta_{i,j}^t$ lies in the interval $[0,1]$, and the lower its value, the more agent i discounts the opinion provided by agent j when

---

[2] Note that this assumption implies that an agent has an influence on himself, and hence cannot be used as a "puppet" who immediately assumes another's agent's belief.

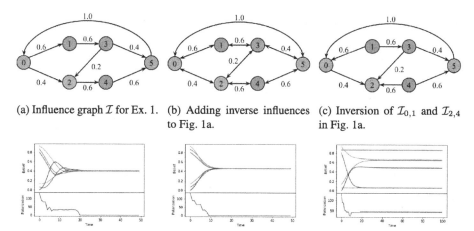

(a) Influence graph $\mathcal{I}$ for Ex. 1.   (b) Adding inverse influences to Fig. 1a.   (c) Inversion of $\mathcal{I}_{0,1}$ and $\mathcal{I}_{2,4}$ in Fig. 1a.

(d) Beliefs and pol. for Fig. 1a.   (e) Beliefs and pol. for Fig. 1b.   (f) Belief and pol. for Fig. 1c.

Fig. 1. Influence graphs and evolution of beliefs and polarization for Example 1.

incorporating it. It is maximum when agents' beliefs are identical, and minimum they are extreme opposites.

Remark 1 (Classical Update: Authority Non-Confirmatory Bias). In this paper we focus on confirmation-bias update and, unless otherwise stated, assume the underlying function is given by Definition 3. Nevertheless, in Sects. 4 and 5 we will consider a *classical update* $\mu^C : (B^t, \mathcal{I}) \mapsto B^{t+1}$ that captures non-confirmatory authority-bias and is obtained by replacing the confirmation-bias factor $\beta_{i,j}^t$ in Definition 3 with 1. That is, $B_i^{t+1} = B_i^t + 1/|\mathcal{A}_i| \sum_{j \in \mathcal{A}_i} \mathcal{I}_{j,i}(B_j^t - B_i^t)$. (We refer to this function as *classical* because it is closely related to the standard update function of the DeGroot models for social learning from Economics [9]. This correspondence will be formalized in Sect. 5.)

2.1   Running Example and Simulations

We now present a running example and several simulations that motivate our theoretical results. Recall that we assume $\mathcal{I}_{i,i} = 1$ for every $i \in \mathcal{A}$. For simplicity, in all figures of influence graphs we omit self-loops.

In all cases we compute the polarization measure (Definition 2) using a discretization D_k of $[0, 1]$ for $k = 5$ bins, each representing a possible general position w.r.t. the veracity of the proposition p of interest: *strongly against*, $[0, 0.20)$; *fairly against*, $[0.20, 0.40)$; *neutral/unsure*, $[0.40, 0.60)$; *fairly in favour*, $[0.60, 0.80)$; and *strongly in favour*, $[0.80, 1]$.[3] We set parameters $\alpha = 1.6$, as

---

[3] Recall from Definition 2 that our model allows arbitrary discretizations D_k –i.e., different number of bins, with not-necessarily uniform widths– depending on the scenario of interest.

suggested by Esteban and Ray [13], and $K = 1\,000$. In all definitions we let $\mathcal{A} = \{0, 1, \ldots, n-1\}$, and $i, j \in \mathcal{A}$ be generic agents.

As a running example we consider the following hypothetical situation.

Example 1 (Vaccine Polarization). Consider the sentence "vaccines are safe" as the proposition p of interest. Assume a set $\mathcal{A}$ of 6 agents that is initially *extremely polarized* about p: agents 0 and 5 are absolutely confident, respectively, in the falsehood or truth of p, whereas the others are equally split into strongly in favour and strongly against p.

Consider first the situation described by the influence graph in Fig. 1a. Nodes 0, 1 and 2 represent anti-vaxxers, whereas the rest are pro-vaxxers. In particular, note that although initially in total disagreement about p, Agent 5 carries a lot of weight with Agent 0. In contrast, Agent 0's opinion is very close to that of Agents 1 and 2, even if they do not have any direct influence over him. Hence the evolution of Agent 0's beliefs will be mostly shaped by that of Agent 5. As can be observed in the evolution of agents' opinions in Fig. 1d, Agent 0 moves from being initially strongly against to being fairly in favour of p around time step 8. Moreover, polarization eventually vanishes (i.e., becomes zero) around time 20, as agents reach the consensus of being fairly against p.

Now consider the influence graph in Fig. 1b, which is similar to Fig. 1a, but with reciprocal influences (i.e., the influence of i over j is the same as the influence of j over i). Now Agents 1 and 2 do have direct influences over Agent 0, so the evolution of Agent 0's belief will be partly shaped by initially opposed agents: Agent 5 and the anti-vaxxers. But since Agent 0's opinion is very close to that of Agents 1 and 2, the confirmation-bias factor will help keeping Agent 0's opinion close to their opinion against p. In particular, in contrast to the situation in Fig. 1d, Agent 0 never becomes in favour of p. The evolution of the agents' opinions and their polarization is shown in Fig. 1e. Notice that polarization vanishes around time 8 as the agents reach consensus but this time they are more positive about (less against) p than in the first situation.

Finally, consider the situation in Fig. 1c obtained from Fig. 1a by inverting the influences of Agent 0 over Agent 1 and Agent 2 over Agent 4. Notice that Agents 1 and 4 are no longer influenced by anyone though they influence others. Thus, as shown in Fig. 1f, their beliefs do not change over time, which means that the group does not reach consensus and polarization never disappears though it is considerably reduced.   □

The above example illustrates complex non-monotonic, overlapping, convergent, and non-convergent evolution of agent beliefs and polarization even in a small case with $n = 6$ agents. Next we present simulations for several influence graph topologies with $n = 1\,000$ agents, which illustrate more of this complex behavior emerging from confirmation-bias interaction among agents. Our theoretical results in the next sections bring insight into the evolution of beliefs and polarization depending on graph topologies.

In all simulations we limit execution to T time steps varying according to the experiment. A detailed mathematical specification of simulations can be found in the corresponding technical report [2].

	Strongly against [0,0.20)	Fairly against [0.20,0.40)	Neutral / unsure [0.40,0.60)	Fairly in favour [0.60,0.80)	Strongly in favour [0.80,1]
Uniform	👥👥👥👥	👥👥👥👥	👥👥👥👥	👥👥👥👥	👥👥👥👥
Mildly polarized		👥👥👥👥		👥👥👥👥	
Extremely polar.	👥👥👥👥				👥👥👥👥
Tripolar	👥👥👥👥		👥👥👥👥		👥👥👥👥

Fig. 2. Depiction of different initial belief configurations used in simulations.

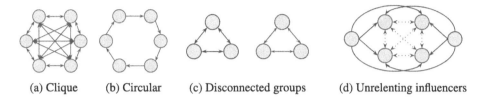

(a) Clique   (b) Circular   (c) Disconnected groups   (d) Unrelenting influencers

Fig. 3. The general shape of influence graphs used in simulations, for $n = 6$ agents.

We consider the following initial belief configurations, depicted in Fig. 2: a *uniform* belief configuration with a set of agents whose beliefs are as varied as possible, all equally spaced in the interval $[0, 1]$; a *mildly polarized* belief configuration with agents evenly split into two groups with moderately dissimilar inter-group beliefs compared to intra-group beliefs; an *extremely polarized* belief configuration representing a situation in which half of the agents strongly believe the proposition, whereas half strongly disbelieve it; and a *tripolar* configuration with agents divided into three groups.

As for influence graphs, we consider the following ones, depicted in Fig. 3:

- A *C-clique* influence graph $\mathcal{I}^{clique}$ in which each agent influences every other with constant value $C = 0.5$. This represents a social network in which all agents interact among themselves, and are all immune to authority bias.
- A *circular* influence graph $\mathcal{I}^{circ}$ representing a social network in which agents can be organized in a circle in such a way each agent is only influenced by its predecessor and only influences its successor. This is a simple instance of a balanced graph (in which each agent's influence on others is as high as the influence received, as in Definition 9 ahead), which is a pattern commonly encountered in some sub-networks.
- A *disconnected* influence graph $\mathcal{I}^{disc}$ representing a social network sharply divided into two groups in such a way that agents within the same group can considerably influence each other, but not at all the agents in the other group.
- An *unrelenting influencers* influence graph $\mathcal{I}^{unrel}$ representing a scenario in which two agents exert significantly stronger influence on every other agent than these other agents have among themselves. This could represent, e.g., a social network in which two totalitarian media companies dominate the

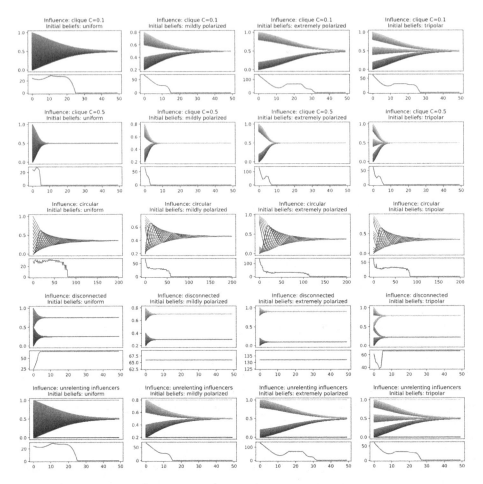

Fig. 4. Evolution of belief and polarization under confirmation bias. Horizontal axes represent time. Each row contains all graphs with the same influence graph, and each column all graphs with the same initial belief configuration. Simulations of circular influences used $n = 12$ agents, the rest used $n = 1\,000$ agents.

news market, both with similarly high levels of influence on all agents. The networks have clear agendas to push forward, and are not influenced in a meaningful way by other agents.

We simulated the evolution of agents' beliefs and the corresponding polarization of the network for all combinations of initial belief configurations and influence graphs presented above. The results, depicted in Fig. 4, will be used throughout this paper to illustrate some of our formal results. Both the Python implementation of the model and the Jupyter Notebook containing the simulations are available on Github [1].

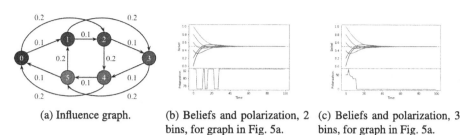

| (a) Influence graph. | (b) Beliefs and polarization, 2 bins, for graph in Fig. 5a. | (c) Beliefs and polarization, 3 bins, for graph in Fig. 5a. |

Fig. 5. Belief convergence to borderline value $1/2$. Polarization does not converge to 0 with equal-length 2 bins (Fig. 5b) and but it does with 3 equal-length bins (Fig. 5c).

3   Belief and Polarization Convergence

Polarization tends to diminish as agents approximate a *consensus*, i.e., as they (asymptotically) agree upon a common belief value for the proposition of interest. Here and in Sect. 4 we consider meaningful families of influence graphs that guarantee consensus *under confirmation bias*. We also identify fundamental properties of agents, and the value of convergence. Importantly, we relate influence with the notion of *flow* in flow networks, and use it to identify necessary conditions for polarization not converging to zero.

3.1   Polarization at the Limit

Proposition 1 states that our polarization measure on a belief configuration (Definition 2) is zero exactly when all belief values in it lie within the same bin of the underlying discretization $D_k = I_0 \ldots I_{k-1}$ of $[0, 1]$. In our model polarization converges to zero if all agents' beliefs converge to a same non-borderline value. More precisely:

Lemma 1 (Zero Limit Polarization). *Let v be a non-borderline point of D_k such that for every $i \in \mathcal{A}$, $\lim_{t \to \infty} B_i^t = v$. Then $\lim_{t \to \infty} \rho(B^t) = 0$.*

To see why we exclude the $k-1$ borderline values of D_k in the above lemma, assume $v \in I_m$ is a borderline value. Suppose that there are two agents i and j whose beliefs converge to v, but with the belief of i staying always within I_m whereas the belief of j remains outside of I_m. Under these conditions one can verify, using Definition 1 and Definition 2, that ρ will not converge to 0. This situation is illustrated in Fig. 5b assuming a discretization $D_2 = [0, 1/2), [1/2, 1]$ whose only borderline is $1/2$. Agents' beliefs converge to value $v = 1/2$, but polarization does not converge to 0. In contrast, Fig. 5c illustrates Lemma 1 for $D_3 = [0, 1/3), [1/3, 2/3), [2/3, 1].$[4]

3.2   Convergence Under Confirmation Bias in Strongly Connected Influence

We now introduce the family of *strongly-connected* influence graphs, which includes cliques, that describes scenarios where each agent has an influence over all others. Such influence is not necessarily *direct* in the sense defined next, or the same for all agents, as in the more specific cases of cliques.

Definition 4 (Influence Paths). *Let $C \in (0, 1]$. We say that i has a* direct *influence C over j, written $i \xrightarrow{C} j$, if $\mathcal{I}_{i,j} = C$.*

An influence path *is a finite sequence of distinct agents from $\mathcal{A}$ where each agent in the sequence has a direct influence over the next one. Let p be an influence path $i_0 i_1 \ldots i_n$. The size of p is $|p| = n$. We also use $i_0 \xrightarrow{C_1} i_1 \xrightarrow{C_2} \ldots \xrightarrow{C_n} i_n$ to denote p with the direct influences along this path. We write $i_0 \overset{C}{\leadsto}_p i_n$ to indicate that the product influence of i_0 over i_n along p is $C = C_1 \times \ldots \times C_n$.*

We often omit influence or path indices from the above arrow notations when they are unimportant or clear from the context. We say that i has an influence *over j if $i \leadsto j$.*

The next definition is akin to the graph-theoretical notion of strong connectivity.

Definition 5 (Strongly Connected Influence). *We say that an influence graph $\mathcal{I}$ is* strongly connected *if for all $i, j \in \mathcal{A}$ such that $i \neq j$, $i \leadsto j$.*

Remark 2. For technical reasons we assume that, *initially*, there are no two agents $i, j \in \mathcal{A}$ such that $B_i^0 = 0$ and $B_j^0 = 1$. This implies that for every $i, j \in \mathcal{A}$: $\beta_{i,j}^0 > 0$ where $\beta_{i,j}^0$ is the confirmation bias of i towards j at time 0 (See Definition 3). Nevertheless, at the end of this section we will address the cases in which this condition does not hold.

We shall use the notion of maximum and minimum belief values at a given time t.

Definition 6 (Extreme Beliefs). *Define $max^t = \max_{i \in \mathcal{A}} B_i^t$ and $min^t = \max_{i \in \mathcal{A}} B_i^t$.*

---

[4] It is worthwhile to note that this discontinuity at borderline points matches real scenarios where each bin represents a sharp action an agent takes based on his current belief value. Even when two agents' beliefs are asymptotically converging to a same borderline value from different sides, their discrete decisions will remain distinct. E.g., in the vaccine case of Example 1, even agents that are asymptotically converging to a common belief value of 0.5 will take different decisions on whether or not to vaccinate, depending on which side of 0.5 their belief falls. In this sense, although there is convergence in the underlying belief values, there remains polarization w.r.t. real-world actions taken by agents.

It is worth noticing that *extreme agents* – i.e., those holding extreme beliefs– do not necessarily remain the same across time steps. Figure 1d illustrates this point: Agent 0 goes from being the one most against the proposition of interest at time $t = 0$ to being the one most in favour of it around $t = 8$. Also, the third row of Fig. 4 shows simulations for a circular graph under several initial belief configurations. Note that under all initial belief configurations different agents alternate as maximal and minimal belief holders.

Nevertheless, in what follows will show that the beliefs of all agents, under strongly-connected influence and confirmation bias, converge to the same value since the difference between min^t and max^t goes to 0 as t approaches infinity. We begin with a lemma stating a property of the confirmation-bias update: *The belief value of any agent at any time is bounded by those from extreme agents in the previous time unit.*

Lemma 2 (Belief Extremal Bounds). *For every $i \in \mathcal{A}$, $min^t \leq B_i^{t+1} \leq max^t$.*

The next corollary follows from the assumption in Remark 2 and Lemma 2.

Corollary 1. *For every $i, j \in \mathcal{A}$, $t \geq 0$: $\beta_{i,j}^t > 0$.*

Note that monotonicity does not necessarily hold for belief evolution. This is illustrated by Agent 0's behavior in Fig. 1d. However, it follows immediately from Lemma 2 that $min^{\cdot}$ and $max^{\cdot}$ are monotonically increasing and decreasing functions of t.

Corollary 2 (Monotonicity of Extreme Beliefs). *$max^{t+1} \leq max^t$ and $min^{t+1} \geq min^t$ for all $t \in \mathbb{N}$.*

Monotonicity and the bounding of $max^{\cdot}$, $min^{\cdot}$ within $[0, 1]$ lead us, via the Monotonic Convergence Theorem [38], to the existence of *limits for beliefs of extreme agents.*

Theorem 1 (Limits of Extreme Beliefs). *There are $U, L \in [0, 1]$ s.t. $\lim_{t \to \infty} max^t = U$ and $\lim_{t \to \infty} min^t = L$.*

We still need to show that U and L are the same value. For this we prove a distinctive property of agents under strongly connected influence graphs: the belief of any agent at time t will influence every other agent by the time $t+|\mathcal{A}|-1$. This is precisely formalized below in Lemma 3. First, however, we introduce some bounds for confirmation-bias, influence as well as notation for the limits in Theorem 1.

Definition 7 (Min Factors). *Define $\beta_{min} = \min_{i,j \in \mathcal{A}} \beta_{i,j}^0$ as the minimal confirmation bias factor at $t = 0$. Also let $\mathcal{I}_{min}$ be the smallest positive influence in $\mathcal{I}$. Furthermore, let $L = \lim_{t \to \infty} min^t$ and $U = \lim_{t \to \infty} max^t$.*

Notice that since min^t and max^t do not get further apart as the time t increases (Corollary 2), $\min_{i,j \in \mathcal{A}} \beta_{i,j}^t$ is a non-decreasing function of t. Therefore β_{min} acts as a lower bound for the confirmation-bias factor in every time step.

Proposition 2. $\beta_{min} = \min_{i,j \in \mathcal{A}} \beta_{i,j}^t$ for every $t > 0$.

The factor β_{min} is used in the next result to establish that the belief of agent i at time t, the minimum confirmation-bias factor, and the maximum belief at t act as bound of the belief of j at $t+|p|$, where p is an influence path from i and j.

Lemma 3 (Path bound). If $\mathcal{I}$ is strongly connected:

1. Let p be an arbitrary path $i \overset{C}{\leadsto}_p j$. Then $B_j^{t+|p|} \leq max^t + C\beta_{min}^{|p|}/|\mathcal{A}|^{|p|}(B_i^t - max^t)$.
2. Let $\mathbf{m}^t \in \mathcal{A}$ be an agent holding the least belief value at time t and p be a path such that $\mathbf{m}^t \leadsto_p i$. Then $B_i^{t+|p|} \leq max^t - \delta$, with $\delta = (\mathcal{I}_{min}\beta_{min}/|\mathcal{A}|)^{|p|}(U-L)$.

Next we establish that all beliefs at time $t+|\mathcal{A}|-1$ are smaller than the maximal belief at t by a factor of at least ϵ depending on the minimal confirmation bias, minimal influence and the limit values L and U.

Lemma 4. Suppose that $\mathcal{I}$ is strongly-connected.

1. If $B_i^{t+n} \leq max^t - \gamma$ and $\gamma \geq 0$ then $B_i^{t+n+1} \leq max^t - \gamma/|\mathcal{A}|$.
2. $B_i^{t+|\mathcal{A}|-1} \leq max^t - \epsilon$, where ϵ is equal to $(\mathcal{I}_{min}\beta_{min}/|\mathcal{A}|)^{|\mathcal{A}|-1}(U-L)$.

Lemma 4(2) states that max$^.$ decreases by at least ϵ after $|A|-1$ steps. Therefore, after $m(|A|-1)$ steps it should decrease by at least $m\epsilon$.

Corollary 3. If $\mathcal{I}$ is strongly connected, $max^{t+m(|A|-1)} \leq max^t - m\epsilon$ for ϵ in Lemma 4.

We can now state that in strongly connected influence graphs extreme beliefs eventually converge to the same value. The proof uses Corollary 1 and Corollary 3 above.

Theorem 2. If $\mathcal{I}$ is strongly connected then $\lim_{t\to\infty} max^t = \lim_{t\to\infty} min^t$.

Combining Theorem 2, the assumption in Remark 2 and the Squeeze Theorem, we conclude that for strongly-connected graphs, all agents' beliefs converge to the same value.

Corollary 4. If $\mathcal{I}$ is strongly connected then for all $i, j \in \mathcal{A}$, $\lim_{t\to\infty} B_i^t = \lim_{t\to\infty} B_j^t$.

The Extreme Cases. We assumed in Remark 2 that there were no two agents i, j s.t. $B_i^t = 0$ and $B_j^t = 1$. Theorem 3 below addresses the situation in which this does not happen. More precisely, it establishes that under confirmation-bias update, in any strongly-connected, non-radical society, agents' beliefs eventually converge to the same value.

Definition 8 (Radical Beliefs). *An agent* $i \in \mathcal{A}$ *is called* radical *if* $B_i = 0$ *or* $B_i = 1$. *A belief configuration* B *is* radical *if every* $i \in \mathcal{A}$ *is radical.*

Theorem 3 (Confirmation-Bias Belief Convergence). *In a strongly connected influence graph and under the confirmation-bias update-function, if* B^0 *is not radical then for all* $i, j \in \mathcal{A}$, $\lim_{t \to \infty} B_i^t = \lim_{t \to \infty} B_j^t$. *Otherwise for every* $i \in \mathcal{A}$, $B_i^t = B_i^{t+1} \in \{0, 1\}$.

We conclude this section by emphasizing that belief convergence is not guaranteed in non strongly-connected graphs. Figure 1c from the vaccine example shows such a graph where neither belief convergence nor zero-polarization is obtained.

4 Conditions for Polarization

We now use concepts from flow networks to identify insightful necessary conditions for polarization never disappearing. Understanding the conditions when polarization *does not* disappear under confirmation bias is one of the main contributions of this paper.

Balanced Influence: Circulations. The following notion is inspired by the *circulation problem* for directed graphs (or flow network) [11]. Given a graph $G = (V, E)$ and a function $c{:}E{\to}\mathbb{R}$ (called *capacity*), the problem involves finding a function $f{:}E{\to}\mathbb{R}$ (called *flow*) such that: (1) $f(e) \leq c(e)$ for each $e \in E$; and (2) $\sum_{(v,w) \in E} f(v, w) = \sum_{(w,v) \in E} f(w, v)$ for all $v \in V$. If such an f exists it is called a *circulation* for G and c.

Thinking of flow as influence, the second condition, called *flow conservation*, corresponds to requiring that each agent influences others as much as is influenced by them.

Definition 9 (Balanced Influence). *We say that* $\mathcal{I}$ *is* balanced *(or a circulation) if every* $i \in \mathcal{A}$ *satisfies the constraint* $\sum_{j \in \mathcal{A}} \mathcal{I}_{i,j} = \sum_{j \in \mathcal{A}} \mathcal{I}_{j,i}$.

Cliques and circular graphs, where all (non-self) influence values are equal, are balanced (see Fig. 3b). The graph of our vaccine example (Fig. 1) is a circulation that it is neither a clique nor a circular graph. Clearly, influence graph $\mathcal{I}$ is balanced if it is a solution to a circulation problem for some $G = (\mathcal{A}, \mathcal{A} \times \mathcal{A})$ with capacity $c{:}\mathcal{A} \times \mathcal{A}{\to}[0, 1]$.

Next we use a fundamental property from flow networks describing flow conservation for graph cuts [11]. Interpreted in our case it says that any group of agents $A{\subseteq}\mathcal{A}$ influences other groups as much as they influence A.

Proposition 3 (Group Influence Conservation). *Let* $\mathcal{I}$ *be balanced and* $\{A, B\}$ *be a partition of* $\mathcal{A}$. *Then* $\sum_{i \in A} \sum_{j \in B} \mathcal{I}_{i,j} = \sum_{i \in A} \sum_{j \in B} \mathcal{I}_{j,i}$.

We now define *weakly connected influence*. Recall that an undirected graph is *connected* if there is path between each pair of nodes.

Definition 10 (Weakly Connected Influence). *Given an influence graph $\mathcal{I}$, define the undirected graph $G_{\mathcal{I}} = (\mathcal{A}, E)$ where $\{i, j\} \in E$ if and only if $\mathcal{I}_{i,j} > 0$ or $\mathcal{I}_{j,i} > 0$. An influence graph $\mathcal{I}$ is called* weakly connected *if the undirected graph $G_{\mathcal{I}}$ is connected.*

Weakly connected influence relaxes its strongly connected counterpart. However, every balanced, weakly connected influence is strongly connected as implied by the next lemma. Intuitively, circulation flows never leaves strongly connected components.

Lemma 5. *If $\mathcal{I}$ is balanced and $\mathcal{I}_{i,j} > 0$ then $j \leadsto i$.*

Conditions for Polarization. We have now all elements to identify conditions for permanent polarization. The convergence for strongly connected graphs (Theorem 3), the polarization at the limit lemma (Lemma 1), and Lemma 5 yield the following noteworthy result.

Theorem 4 (Conditions for Polarization). *Suppose that $\lim_{t \to \infty} \rho(B^t) \neq 0$. Then either: (1) $\mathcal{I}$ is not balanced; (2) $\mathcal{I}$ is not weakly connected; (3) B^0 is radical; or (4) for some borderline value v, $\lim_{t \to \infty} B_i^t = v$ for each $i \in \mathcal{A}$.*

Hence, at least one of the four conditions is necessary for the persistence of polarization. If (1) then there must be at least one agent that influences more than what he is influenced (or vice versa). This is illustrated in Fig. 1c from the vaccine example, where Agent 2 is such an agent. If (2) then there must be isolated subgroups of agents; e.g., two isolated strongly-connected components the members of the same component will achieve consensus but the consensus values of the two components may be very different. This is illustrated in the fourth row of Fig. 4. Condition (3) can be ruled out if there is an agent that is not radical, like in all of our examples and simulations. As already discussed, (4) depends on the underlying discretization D_k (e.g., assuming equal-length bins if v is borderline in D_k it is not borderline in D_{k+1}, see Fig. 5.).

Reciprocal and Regular Circulations. The notion of circulation allowed us to identify potential causes of polarization. In this section we will also use it to identify meaningful topologies whose symmetry can help us predict the exact belief value of convergence.

A *reciprocal* influence graph is a circulation where the influence of i over j is the same as that of j over i, i.e., $\mathcal{I}_{i,j} = \mathcal{I}_{j,i}$. Also a graph is (*in-degree*) *regular* if the in-degree of each nodes is the same; i.e., for all $i, j \in \mathcal{A}$, $|\mathcal{A}_i| = |\mathcal{A}_j|$.

As examples of regular and reciprocal graphs, consider a graph $\mathcal{I}$ where all (non-self) influence values are equal. If $\mathcal{I}$ is *circular* then it is a regular circulation, and if $\mathcal{I}$ is a *clique* then it is a reciprocal regular circulation. Also we can modify slightly our vaccine example to obtain a regular reciprocal circulation as shown in Fig. 6.

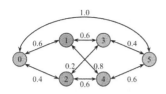

(a) Regular and reciprocal influence.

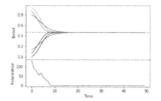

(b) Beliefs and pol. for Fig. 6a.

Fig. 6. Influence and evolution of beliefs and polar.

The importance of regularity and reciprocity of influence graphs is that their symmetry is sufficient to the determine the exact value all the agents converge to under confirmation bias: *the average of initial beliefs.* Furthermore, under classical update (see Remark 1), we can drop reciprocity and obtain the same result. The result is proven using Lemma 5, Theorem 3, Corollary 5, the squeeze theorem and by showing that $\sum_{i \in \mathcal{A}} B_i^t = \sum_{i \in \mathcal{A}} B_i^{t+1}$ using symmetries derived from reciprocity, regularity, and the fact that $\beta_{i,j}^t = \beta_{j,i}^t$.

Theorem 5 (Consensus Value). *Suppose that $\mathcal{I}$ is regular and weakly connected. If $\mathcal{I}$ is reciprocal and the belief update is confirmation-bias, or if the influence graph $\mathcal{I}$ is a circulation and the belief update is classical, then $\lim_{t \to \infty} B_i^t = {}^{1}/_{|\mathcal{A}|} \sum_{j \in \mathcal{A}} B_j^0$ for every $i \in \mathcal{A}$.*

5   Comparison to DeGroot's Model

DeGroot proposed a very influential model, closely related to our work, to reason about learning and consensus in multi-agent systems [9], in which beliefs are updated by a constant stochastic matrix at each time step. More specifically, consider a group $\{1, 2, \ldots, k\}$ of k agents, s.t. each agent i holds an initial (real-valued) opinion F_i^0 on a given proposition of interest. Let $T_{i,j}$ be a non-negative weight that agent i gives to agent j's opinion, s.t. $\sum_{j=1}^k T_{i,j} = 1$. DeGroot's model posits that an agent i's opinion F_i^t at any time $t \geq 1$ is updated as $F_i^t = \sum_{j=1}^k T_{i,j} F_i^{t-1}$. Letting F^t be a vector containing all agents' opinions at time t, the overall update can be computed as $F^{t+1} = T F^t$, where $T = \{T_{i,j}\}$ is a stochastic matrix. This means that the t-th configuration (for $t \geq 1$) is related to the initial one by $F^t = T^t F^0$, which is a property thoroughly used to derive results in the model.

When we use classical update (as in Remark 1), our model reduces to DeGroot's via the transformation $F_i^0 = B_i^0$, and $T_{i,j} = {}^{1}/_{|\mathcal{A}_i|} \mathcal{I}_{j,i}$ if $i \neq j$, or $T_{i,j} = 1 - {}^{1}/_{|\mathcal{A}_i|} \sum_{j \in \mathcal{A}_i} \mathcal{I}_{j,i}$ otherwise. Notice that $T_{i,j} \leq 1$ for all i and j, and, by construction, $\sum_{j=1}^k T_{i,j} = 1$ for all i. The following result is an immediate consequence of this reduction.

Corollary 5. *In a strongly connected influence graph $\mathcal{I}$, and under the classical update function, for all $i, j \in \mathcal{A}$, $\lim_{t \to \infty} B_i^t = \lim_{t \to \infty} B_j^t$.*

Unlike its classical counterpart, however, the confirmation-bias update (Definition 3) does not have an immediate correspondence with DeGroot's model. Indeed, this update is not linear due the confirmation-bias factor $\beta_{i,j}^t = 1-|B_j^t-B_i^t|$. This means that in our model there is no immediate analogue of the relation among arbitrary configurations and the initial one as the relation in DeGroot's model (i.e., $F^t = T^t F^0$). Therefore, proof techniques usually used in DeGroot's model (e.g., based on Markov properties) are not immediately applicable to our model. In this sense our model is an extension of DeGroot's, and we need to employ different proof techniques to obtain our results.

6   Conclusions and Other Related Work

We proposed a model for polarization and belief evolution for multi-agent systems under confirmation-bias. We showed that whenever all agents can directly or indirectly influence each other, their beliefs always converge, and so does polarization as long as the convergence value is not a borderline point. We also identified necessary conditions for polarization not to disappear, and the convergence value for some important network topologies. As future work we intend to extend our model to model evolution of beliefs and measure polarization in situations in which agents hold opinions about multiple propositions of interest.

Related Work. As mentioned in the introduction and discussed in detail in Sect. 5, the closest related work is on DeGroot models for social learning [9]. We summarize some other relevant approaches put into perspective the novelty of our approach.

Polarization. Polarization was originally studied as a psychological phenomenon in [26], and was first rigorously and quantitatively defined by economists Esteban and Ray [13]. Their measure of polarization, discussed in Sect. 2, is influential, and we adopt it in this paper. Li et al. [20], and later Proskurnikov et al. [31] modeled consensus and polarization in social networks. Like much other work, they treat polarization simply as the lack of consensus and focus on when and under what conditions a population reaches consensus. Elder's work [12] focuses on methods to avoid polarization, without using a quantitative definition of polarization. [6] measures polarization but purely as a function of network topology, rather than taking agents' quantitative beliefs and opinions into account, in agreement with some of our results.

Formal Models. Sîrbu et al. [37] use a model that updates probabilistically to investigate the effects of algorithmic bias on polarization by counting the number of opinion clusters, interpreting a single opinion cluster as consensus. Leskovec et al. [14] simulate social networks and observe group formation over time.

The Degroot models developed in [9] and used in [15] are closest to ours. Rather than examining polarization and opinions, this work is concerned with the network topology conditions under which agents with noisy data about an objective fact converge to an accurate consensus, close to the true state of the world. As already discussed the basic DeGroot models do not include confirmation bias, however [7,17,23,25,36] all generalize DeGroot-like models to include functions that can be thought of as modelling confirmation bias in different ways, but with either no measure of polarization or a simpler measure than the one we use. [24] discusses DeGroot models where the influences change over time, and [16] presents results about generalizations of these models, concerned more with consensus than with polarization.

Logic-Based Approaches. Liu et al. [21] use ideas from doxastic and dynamic epistemic logics to qualitatively model influence and belief change in social networks. Seligman et al. [34,35] introduce a basic "Facebook logic." This logic is non-quantitative, but its interesting point is that an agent's possible worlds are different social networks. This is a promising approach to formal modeling of epistemic issues in social networks. Christoff [8] extends facebook logic and develops several non-quantitative logics for social networks, concerned with problems related to polarization, such as information cascades. Young Pederson et al. [28–30] develop a logic of polarization, in terms of positive and negative links between agents, rather than in terms of their quantitative beliefs. Hunter [19] introduces a logic of belief updates over social networks where closer agents in the social network are more trusted and thus more influential. While beliefs in this logic are non-quantitative, there is a quantitative notion of influence between users.

Other Related Work. The seminal paper Huberman et al. [18] is about determining which friends or followers in a user's network have the most influence on the user. Although this paper does not quantify influence between users, it does address an important question to our project. Similarly, [10] focuses on finding most influential agents. The work on highly influential agents is relevant to our finding that such agents can maintain a network's polarization over time.

References

1. Alvim, M.S., Amorim, B., Knight, S., Quintero, S., Valencia, F.: (2020). https://github.com/Sirquini/Polarization
2. Alvim, M.S., Amorim, B., Knight, S., Quintero, S., Valencia, F.: A multi-agent model for polarization under confirmation bias in social networks, Technical report. arXiv preprint (2021)
3. Alvim, M.S., Knight, S., Valencia, F.: Toward a formal model for group polarization in social networks. In: Alvim, M.S., Chatzikokolakis, K., Olarte, C., Valencia, F. (eds.) The Art of Modelling Computational Systems: A Journey from Logic and Concurrency to Security and Privacy. LNCS, vol. 11760, pp. 419–441. Springer, Cham (2019). https://doi.org/10.1007/978-3-030-31175-9_24
4. Aronson, E., Wilson, T., Akert, R.: Social Psychology, 7th edn. Prentice Hall, Upper Saddle River (2010)
5. Bozdag, E.: Bias in algorithmic filtering and personalization. Ethics Inf. Technol. **15**, 209–227 (2013). https://doi.org/10.1007/s10676-013-9321-6

6. Calais Guerra, P., Meira Jr, W., Cardie, C., Kleinberg, R.: A measure of polarization on social media networks based on community boundaries. In: Proceedings of the 7th International Conference on Weblogs and Social Media, ICWSM 2013, pp. 215–224 (2013)

7. Cerreia-Vioglio, S., Corrao, R., Lanzani, G., et al.: Robust Opinion Aggregation and its Dynamics. IGIER, Università Bocconi (2020)

8. Christoff, Z., et al.: Dynamic logics of networks: information flow and the spread of opinion. Ph.D. thesis, Institute for Logic, Language and Computation, University of Amsterdam (2016)

9. DeGroot, M.H.: Reaching a consensus. J. Am. Stat. Assoc. 69(345), 118–121 (1974)

10. DeMarzo, P.M., Vayanos, D., Zwiebel, J.: Persuasion bias, social influence, and unidimensional opinions. Q. J. Econ. 118(3), 909–968 (2003)

11. Diestel, R.: Graph Theory, 5th edn. Springer, Heidelberg (2015). https://doi.org/10.1007/978-3-662-53622-3

12. Elder, A.: The interpersonal is political: unfriending to promote civic discourse on social media. Ethics Inf. Technol. 22, 15–24 (2019). https://doi.org/10.1007/s10676-019-09511-4

13. Esteban, J.M., Ray, D.: On the measurement of polarization. Econometrica 62(4), 819–851 (1994)

14. Gargiulo, F., Gandica, Y.: The role of homophily in the emergence of opinion controversies. arXiv preprint arXiv:1612.05483 (2016)

15. Golub, B., Jackson, M.O.: Naive learning in social networks and the wisdom of crowds. Am. Econ. Jo.: Microecon. 2(1), 112–49 (2010)

16. Golub, B., Sadler, E.: Learning in social networks. Available at SSRN 2919146 (2017)

17. Hegselmann, R., Krause, U.: Opinion dynamics and bounded confidence, models, analysis and simulation. J. Artif. Soc. Soc. Simul. 5(3), 2 (2002)

18. Huberman, B.A., Romero, D.M., Wu, F.: Social networks that matter: Twitter under the microscope. arXiv preprint arXiv:0812.1045 (2008)

19. Hunter, A.: Reasoning about trust and belief change on a social network: a formal approach. In: Liu, J.K., Samarati, P. (eds.) ISPEC 2017. LNCS, vol. 10701, pp. 783–801. Springer, Cham (2017). https://doi.org/10.1007/978-3-319-72359-4_49

20. Li, L., Scaglione, A., Swami, A., Zhao, Q.: Consensus, polarization and clustering of opinions in social networks. IEEE J. Sel. Areas Commun. 31(6), 1072–1083 (2013)

21. Liu, F., Seligman, J., Girard, P.: Logical dynamics of belief change in the community. Synthese 191(11), 2403–2431 (2014). https://doi.org/10.1007/s11229-014-0432-3

22. Lynch, N.A.: Distributed Algorithms. Morgan Kaufmann Publishers, Boston (1996)

23. Mao, Y., Bolouki, S., Akyol, E.: Spread of information with confirmation bias in cyber-social networks. IEEE Trans. Netw. Sci. Eng. 7(2), 688–700 (2020)

24. Moreau, L.: Stability of multiagent systems with time-dependent communication links. IEEE Trans. Autom. Control 50(2), 169–182 (2005)

25. Mueller-Frank, M.: Reaching consensus in social networks. IESE Research Papers D/1116, IESE Business School (2015)

26. Myers, D.G., Lamm, H.: The group polarization phenomenon. Psychol. Bull. 83, 602 (1976)

27. Nielsen, M., Palamidessi, C., Valencia, F.D.: Temporal concurrent constraint programming: denotation, logic and applications. Nord. J. Comput. 9(1), 145–188 (2002)

28. Pedersen, M.Y.: Polarization and echo chambers: a logical analysis of balance and triadic closure in social networks (2019)
29. Pedersen, M.Y., Smets, S., Ågotnes, T.: Analyzing echo chambers: a logic of strong and weak ties. In: Blackburn, P., Lorini, E., Guo, M. (eds.) LORI 2019. LNCS, vol. 11813, pp. 183–198. Springer, Heidelberg (2019). https://doi.org/10.1007/978-3-662-60292-8_14
30. Pedersen, M.Y., Smets, S., Ågotnes, T.: Further steps towards a logic of polarization in social networks. In: Dastani, M., Dong, H., van der Torre, L. (eds.) CLAR 2020. LNCS (LNAI), vol. 12061, pp. 324–345. Springer, Cham (2020). https://doi.org/10.1007/978-3-030-44638-3_20
31. Proskurnikov, A.V., Matveev, A.S., Cao, M.: Opinion dynamics in social networks with hostile camps: consensus vs. polarization. IEEE Trans. Autom. Control **61**(6), 1524–1536 (2016)
32. Ramos, V.J.: Analyzing the Role of Cognitive Biases in the Decision-Making Process. IGI Global, Hershey (2019)
33. Saraswat, V.A., Jagadeesan, R., Gupta, V.: Foundations of timed concurrent constraint programming. In: LICS, pp. 71–80. IEEE Computer Society (1994)
34. Seligman, J., Liu, F., Girard, P.: Logic in the community. In: Banerjee, M., Seth, A. (eds.) ICLA 2011. LNCS (LNAI), vol. 6521, pp. 178–188. Springer, Heidelberg (2011). https://doi.org/10.1007/978-3-642-18026-2_15
35. Seligman, J., Liu, F., Girard, P.: Facebook and the epistemic logic of friendship. CoRR abs/1310.6440 (2013)
36. Sikder, O., Smith, R., Vivo, P., Livan, G.: A minimalistic model of bias, polarization and misinformation in social networks. Sci. Rep. **10**, 1–11 (2020)
37. Sîrbu, A., Pedreschi, D., Giannotti, F., Kertész, J.: Algorithmic bias amplifies opinion polarization: a bounded confidence model. arXiv preprint arXiv:1803.02111 (2018)
38. Sohrab, H.H.: Basic Real Analysis, 2nd edn. Birkhauser, Basel (2014)

A Formalisation of SysML State Machines in mCRL2

Mark Bouwman[1]([✉]), Bas Luttik[1], and Djurre van der Wal[2]

[1] Eindhoven University of Technology, Eindhoven, The Netherlands
m.s.bouwman@tue.nl
[2] University of Twente, Enschede, The Netherlands

Abstract. This paper reports on a formalisation of the semi-formal modelling language SysML in the formal language mCRL2, in order to unlock formal verification and model-based testing using the mCRL2 toolset for SysML models. The formalisation focuses on a fragment of SysML used in the railway standardisation project EULYNX. It comprises the semantics of state machines, communication between objects via ports, and an action language called ASAL. It turns out that the generic execution model of SysML state machines can be elegantly specified using the rich data and process languages of mCRL2. This is a big step towards an automated translation as the generic model can be configured with a formal description of a specific set of state machines in a straightforward manner.

1 Introduction

The importance of correct specifications is evident for safety-critical systems such as those in the railway domain. At the same time, due to the increasing use of digital technology in those systems, specifications are getting more and more complex and harder to get completely correct. To cope with the complexity, railway engineers are gradually adopting a model-based system engineering approach for the development of their systems. EULYNX[1], an initiative of a consortium of thirteen European railway infrastructure managers, uses SysML to specify a standard for interfaces between the various components of a signalling system (signal, point, level crossing, interlocking, etc.).

The use of SysML for system requirements specification is a big step forward for the railway domain as it is significantly more precise than natural language. SysML has a fairly intuitive graphical syntax, which allows railway engineers to understand and use it without extensive training. Still, SysML is *semi-formal*: it has a well-defined syntax, but its semantics is informal and not firmly grounded in mathematics. As a consequence, system behaviour specified by a SysML model is not directly amenable to the more thorough kind of analysis that genuine *formal* methods offer.

---

[1] See https://eulynx.eu.

© IFIP International Federation for Information Processing 2021
Published by Springer Nature Switzerland AG 2021
K. Peters and T. A. C. Willemse (Eds.): FORTE 2021, LNCS 12719, pp. 42–59, 2021.
https://doi.org/10.1007/978-3-030-78089-0_3

The aim of the *FormaSig*[2] project, a collaboration of the Dutch and German railway infrastructure managers, Eindhoven University of Technology and the University of Twente, is to formalise the aforementioned EULYNX standard to the extent that delivered components conforming to the standard provably satisfy a collection of safety properties. The idea is to associate with each EULYNX SysML model a formal mCRL2 model [5,6]. Then mCRL2's model checker can be used to establish that the model satisfies the required safety properties, and automated model-based test technology can be used to reliably test compliance to the model of actual implementations (see Fig. 1). In a first case study, we have demonstrated the viability of this idea. We took the EULYNX SysML model of the Point interface, associated an mCRL2 model with it, used the mCRL2 model checker to analyse its correctness and used the model-based test tool JTorX [2] to check conformance of a SysML simulator of Point [4].

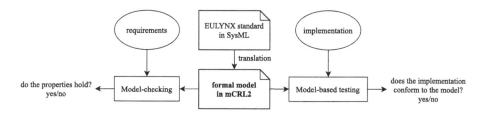

Fig. 1. FormaSig: using a formal mCRL2 model to establish that implementations conforming to the EULYNX standard satisfy properties.

The EULYNX standard is under development, and it is likely that also in the future it will be subject to changes. Hence, it is impractical to rely on manual translations from the EULYNX SysML models to mCRL2. To facilitate that model-checking and model-based test techniques will become an integral aspect of maintaining the standard, it is imperative that the translation from EULYNX SysML to mCRL2 is automated. Another benefit of having a automated translation is that, as its correctness can be rigorously examined, the likelihood of introducing mistakes in the formalised model is reduced.

How EULYNX SysML models are meant to be interpreted is specified in the EULYNX modelling standard. So far FormaSig delivers *an* interpretation of EULYNX SysML. Going forward, FormaSig also aims to increase precision of the modelling standard and become *the* official interpretation.

Implementing an automated translation from SysML to mCRL2 is, however, a nontrivial undertaking, most notably hampered by the lack of a complete and comprehensive formal semantics for SysML and the complexity of the informally described SysML execution model. Furthermore, also due to the lack of a fixed formal semantics, there are many dialects of SysML. A particular variation point is the action language, the language used to specify guards and the effects of

[2] Formal Methods in Railway Signaling Infrastructure Standardization Processes.

transitions. In EULYNX SysML all communication is performed via ports, which are referenced as variables in the action language. The action language itself is ASAL, which is tied to the PTC Windchill tool[3].

The main contribution of this paper is to present a formalisation of the informal semantics of EULYNX SysML state machines directly in mCRL2. Our formalisation consists of three parts. The first part is a generic, comprehensive formalisation of the operational semantics of UML state machines, which form the basis of EULYNX SysML state machines. This part involves formalising the notion of state hierarchy and transition selection. The second part adds an interpretation of the particular communication mechanism via ports that is used in EULYNX SysML. The third part defines an execution model for the ASAL action language. In this paper, we generalise to a class of action languages that reference ports as variables. The resulting mCRL2 specification can straightforwardly be turned into an actual formal model interpreting a particular EULYNX SysML interface by populating the relevant data types with some static details from the SysML model and generating a suitable number of instantiations of predefined processes. For the latter, we have implemented a tool that is discussed in a companion paper [22]. The resulting mCRL2 specification can be model-checked and used for model-based testing and serve as the formal model central to FormaSig idea (see Fig. 1).

The semantics of UML/SysML state machines has been formalised in preceding academic work. A number of papers describe a translation from UML state machines to PROMELA (the input language of the SPIN model checker) [11–13,18,21]. Our formalisation of transition selection draws inspiration from [13]. In [14] a structural operational semantics is presented along with a custom verification tool USM²C. The AVATAR [19] tool offers a SysML-style environment with particular focus on verifying security properties; it offers translation to UPPAAL and ProVerif. Other translations and formalisations include a translation from xUML class diagrams and state machines to mCRL2 [7,8], a translation from SysML BDDs and state machines to NuSMV [23], a formalisation of UML state machines using structured graph transformations [10] and a formalisation of UML state machines in Object-Z [9]. In [20] a translation is given from sequence diagrams to mCRL2. Our approach to formalisation differs from earlier work by specifying the generic semantics in the target formal language which can be instantiated with a specific configuration. Moreover, our formalisation includes a communication mechanism using ports.

The OMG organisation, which manages the UML and SysML standards, has also released "Precise Semantics of UML State Machines (PSSM)" [17], which gives an informal but very precise semantics. Our formalisation differs in at least one way from PSSM. We do not create completion events in order to prevent cluttering the event queue. Instead, transitions relying on a completion event have completion of the source state as an extra guard. PSSM also provides an extensive compliance test suite. In the future we would like to make a version

---

[3] https://www.ptc.com/en/products/windchill/integrity/.

of our model that adheres to PSSM and measure the effect of completion events on the state space.

This rest of the paper is organised as follows. In Sect. 2 we present the visual syntax of state machines and a summary of the run-to-completion semantics; this section can be skipped by readers familiar with state machines. In Sect. 3 we give a quick introduction to mCRL2. From Sect. 4 to Sect. 6 we go into the details of the formalisation. We conclude the paper in Sect. 7.

2   An Informal Introduction to UML State Machines

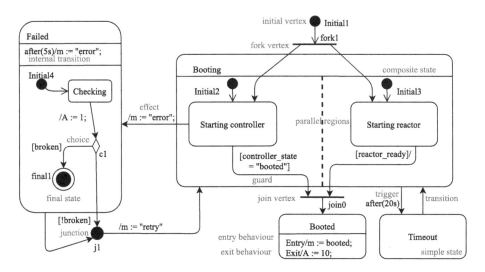

Fig. 2. Example showing all state machine constructs supported in EULYNX SysML.

Figure 2 shows an example of a state machine, with names of the various constructions added in blue. In this section, we briefly discuss the informal semantics of each construction as in the UML standard [16].

The basic constituents of state machines are *states* and *transitions*. Initial states, choice states, final states, junctions, forks and joins are also called *pseudostates*. The UML state machine formalism derives its expressiveness from these the possibility to have states and transitions nested within states, and even have transitions cross state border. Transitions may have a *trigger*, a *guard* and an *effect*. The trigger of a transition (which is optional) is an *event*; it can be a change event (notation when(x)) or a timeout event (notation after(x)).

The modeller can define behaviour that is executed upon entering or exiting a simple or composite state. *Exit behaviour* is executed before the effect of a transition, *entry behaviour* is executed after the effect of a transition. Simple and composite states can also have *internal transitions*, which do not change state (see, e.g., the state Failed in Fig. 2).

Junctions and *choice* vertices allow more concise specification of transitions that induce the same behaviour. The choice vertex c1 in Fig. 2 combines two transitions from Checking which share the common behaviour A:=1. Junctions serve a similar purpose (see junction j1 in Fig. 2) with the difference that for junctions the guards of outgoing transitions need to be checked before taking a transition to the junction, whereas for choice vertices the guards are checked when arriving at the vertex.

A state can contain other states, in which case it is called a *composite state* and the states it encloses are called *substates*. A composite state can have a *final state* (see, e.g., the state Failed in Fig. 2). Transitions from the border of a composite state can be fired regardless of the current substate, except when the transition does not have a trigger, in which case the current substate of the composite state must be a final state. A composite state may also have multiple *parallel regions*. Each region has an *initial state* and can perform local transitions independently of other regions. A transition ending at the border of a composite state with parallel regions will let each region start from its initial state. A fork indicates that a transition ends on specific states in multiple regions. Conversely, a join can begin from specific states in parallel regions.

Due to the presence of composite states, a state machine is not just in a single state but in a collection of states, a *state configuration*. A state configuration is *stable* when it does not contain pseudostates. Transitions are specified on states; a state machine may combine several transitions (as is the case with joins, forks and junctions) to perform a bigger step from one state configuration to another, which we will call a *step*. Events that occur are stored in an *event pool* until they are dispatched. A step is enabled when the specified trigger (if any) is in the event pool and all guards of transitions involved in the step evaluate to true. State machines have *run-to-completion* semantics: a state machine selects a step to execute and will completely finish executing the behaviour of the step and entry and exit behaviour before it considers performing a new step. Parallel regions may start a step simultaneously when both steps have the same trigger; in that case the state machine performs a *multi-step*.

3   Introduction to mCRL2

The mCRL2 toolset is designed to model and analyse concurrent and distributed systems. The mCRL2 language is an ACP-style process algebra and contains a rich data language based on abstract data types. The semantic interpretation of an mCRL2 model is a Labelled Transition System (LTS). By translating from SysML to mCRL2 we indirectly associate an LTS to the SysML model. The toolset contains tools for the verification of parametrised modal μ-calculus formulas, bisimulation reduction, counterexample generation, simulation and visualisation. To aid the reader in understanding the mCRL2 snippets in following sections, we will cover some basics using an example unrelated to the contents of this paper. For more information on mCRL2 we refer to https://mcrl2.org and [6].

The mCRL2 language has some primitive data types built in, such as integers, natural numbers and booleans, including common operations on them. Users can also define their own data types and operations. The code below shows an example. The *sort* Place has one *constructor*, Coordinates, with *projection functions* X and Y. Equations are treated as rewrite rules; terms that match the left hand side will be rewritten to the right hand side.

```
sort Place = struct Coordinates(X:Nat, Y: Nat);
map computeManhattanDistance: Place#Place -> Nat;
var p1, p2:Place;
eqn computeManhattanDistance(p1,p2) = abs(X(p1)-X(p2))+abs(Y(p1)-Y(p2));
```

The process definition below specifies the behaviour of the Point process; it can perform three actions: move, invite and respond. The sum operator represents a non-deterministic choice over all values of the quantified data domain. Summations over infinite data domains can be restricted by adding a condition. In the example below a condition is used to restrict a point process to move to any place on a 10 by 10 grid.

```
act move: Nat; invite, respond, meet: Place;
proc Point(p:Place) = sum new:Place. (X(new) < 10 && Y(new) < 10)
      -> move(computeManhattanDistance(p,new)).Point(new)
   + invite(p).Point(p) + sum new:Place. respond(new).Point(new);
```

The initial process expression specifies the initial state of the labelled transition associated with the specification. The example below specifies a parallel composition of two Point processes wrapped in a communication and an allow operator. Both Point processes can perform a move *action*, which is allowed by the allow operator. The invite and respond actions are not allowed and hence blocked. However, the two processes can synchronize on a *multi-action* invite|respond, which is transformed to a meet action by the communication operator, which is allowed by the allow operator. The labelled transition system (sometimes referred to as the state space) associated with this specification will have exactly 10.000 states, representing every combination of coordinates.

```
init allow({move,meet}, comm({invite|respond -> meet},
   Point(Coordinates(1,1))||Point(Coordinates(2,3))));
```

4   The Operational Semantics of State Machines

In Sect. 2 we already gave a rough sketch of the execution semantics of state machines. In this section we treat the semantics of UML state machines and its formalisation in mCRL2, including the role of the action language and some mCRL2 snippets that are illustrative of the formalisation. The model itself is available on GitHub [3]. In Sect. 5 we extend the UML semantics with EULYNX SysML specific communication over ports. In Sect. 6 we detail how to complete the model with a configuration and touch on the subject of verification.

4.1   Strategy to Formalisation

Our goal is to generically describe the semantics of state machines in mCRL2 achieving a high degree of modularity. There are several choices to be made (e.g., with respect to the granularity of interleaving, run-to-completion semantics, syntax and semantics of the action language) and we want to set up our specification in such a way that parts of it can be easily modified or replaced. A particular concern is that the specific details of a concrete state machine to be translated are separated from the generic semantics.

Due to our modular setup it is rather straightforward to configure the generic model with a specific set of communicating state machines. The user needs to do two things: 1) define the semantics of the action language and 2) encode the structure of the state machines in an mCRL2 data type and pass it as a parameter to the generic state machine process.

4.2 Abstract Action Language

The UML standard [16] is not prescriptive of the action language used to specify guards and the effect of transitions. In this paper we abstract from any particular action language.

Let `Instruction` be a sort containing all action language expressions, which we will also refer to as a *behaviour*. Let `VarName` be a sort containing all variable names. It is assumed that variables range over elements of a sort `Value`.

In order to formalise the action language semantics it may be necessary to include additional data structures, e.g., a program stack or a valuation of local variables. To encapsulate such additional data structures we introduce the notion of *execution frame*, represented by the mCRL2 sort `ExcFrame`, which will be assumed to consist of all data necessary to execute programs of the action language. We do not assume that execution of behaviour is atomic, we allow that two components interleave their execution of behaviour when they are both taking a transition. We abstract from the granularity of interleaving and simply allow an execution frame e to make a step to an execution frame e', where e' may still have behaviour waiting to be executed.

To define the semantics of the action language the user needs to add equations for the following mappings. We assume a subset of action language expressions represent predicates, which can be evaluated using `checkPredicate`.

```
sort VarValuePair = struct VarValuePair(getVariable:VarName, getValue:Value);
     Instructions = List(Instruction);
map  initializeExcFrame: Instructions#(VarName -> Value) -> ExcFrame;
     executeExcFrameCode: ExcFrame -> ExcFrame;
     checkPredicate: Instructions#(VarName -> Value) -> Bool;
     isFinished: ExcFrame -> Bool;
     getValuation: ExcFrame -> VarName -> Value;
     getVariableUpdates: ExcFrame -> List(VarValuePair);
     resetVariableUpdates: ExcFrame -> ExcFrame;
```

The mapping `getVariableUpdates` is assumed to retrieve all updates to variables that occurred during the execution of the execution frame. This field is needed for deriving change events, described in Sect. 4.5.

4.3 Representing State Machines in mCRL2

We assume that `StateName` and `CompName` have been declared as mCRL2 enumeration sorts, enumerating, respectively, all state names and all state machine identifiers occurring in the SysML model under consideration. These two sorts are part of the configuration in the model as they need to be instantiated.

We proceed by introducing the sort `StateInfo`, which is an example of mCRL2's facility to define structured sorts. By means of a structured sort, data can be concisely aggregated. An element of the sort `StateInfo` is either a triple with *constructor* `SimpleState` or with constructor `CompositeState`, both with *projection functions* `parent`, `entryAction` and `exitAction`, or it stores a single data element together with a constructor (`JoinVertex`, `JunctionVertex`, etc.).

```
StateInfo = struct SimpleState(parent: StateName, entryAction: Instructions,
    exitAction: Instructions) | CompositeState(parent: StateName,
    entryAction: Instructions, exitAction: Instructions)
  | JoinVertex(parent: StateName) | JunctionVertex(parent: StateName)
  | ForkVertex(parent: StateName) | InitialState(parent: StateName)
  | FinalState(parent: StateName) | ChoiceVertex(parent: StateName);
```

The parent of a state is stored to represent the hierarchy of states induced by composite states. A state's parent is the first enclosing composite state. We assume that the sort `StateName` has a special element `root`; states that are not enclosed in a composite state have `root` as their parent.

Our framework supports change events and timeout events, see the definition of the sort `Event` below. The event type `none` is used as placeholder for transitions without a trigger. Time is currently not modelled explicitly in our framework, even though mCRL2 does support it. Explicit timing would result in a significantly larger state space, while it is not relevant for the properties that need to be verified in the context of EULYNX. Instead, transitions with a timeout event as trigger can fire non-deterministically. The generation of change events is discussed in Sect. 4.5.

```
Event = struct none | ChangeEvent(getTriggerExpr:Instructions) | TimeoutEvent
```

The sort `Transition` (given below) is used to specify the transitions of a state machine. The Boolean `internal` is used to differentiate between selfloops and internal transitions, the latter do not induce entry and/or exit behaviour.

```
Transition = struct Transition(source:StateName, trigger:Event,
    guard:Instructions, effect:Instructions, target:StateName, internal:Bool);
```

We also define the sort `StateMachine`, which aggregates all the information we need of a state machine.

```
StateMachine = struct StateMachine(
    transitions:List(Transition),initialState:StateName,states:List(StateName),
    stateInfo: StateName -> StateInfo, initialValuation: VarName -> Value);
```

The `initialState` designates the initial state in the root of the state machine (i.e. the initial state that is not contained in a composite state). The projection functions `states` and `stateInfo` retrieve which states are present in the state machine and the associated `StateInfo`, respectively. Note that functions can be partial in mCRL2, the function `stateInfo` only needs to be defined for the state names that occur in that state machine.

Due to the hierarchy of states a state machine is 'in' a collection of states, a state configuration. A state configuration can be represented as a tree structure where the top node is not enclosed in a composite state. Parallel regions introduce nodes with multiple children. The mCRL2 excerpt below gives the definition of state configurations in the model.

```
StateConfig =
  struct StateConfig(rootState:StateName,substates:List(StateConfig));
```

An example configuration of the state machine depicted in Fig. 2 is

```
StateConfig(Booting,[StateConfig(Initial2,[]),StateConfig(Initial3,[])]).
```

4.4   Step Selection and Execution

As explained in Sect. 2, state machines make a step from one state configuration to another. Such a step could consist of multiple transitions, as is the case with junctions, joins and forks. We could in theory perform step selection by performing a reachability search across the transitions. We anticipate that this will make step selection computationally expensive. Instead, we opt to preprocess `Transitions` into `Steps`. The definition of `Step` is given below, as well as the mapping that derives `Steps` from `Transitions`. The `effect` of the step is a `ComposedBehaviour`. It allows us to create a partially ordered set of behaviours, which is needed for defining steps in the context of parallel regions (see Fig. 3).

```
sort Step = struct Step(source: StateConfig, trigger: Event,
     guard: List(Instructions),effect: ComposedBehaviour, target: StateConfig,
     internal: Bool, arrowEnd: StateName);
  ComposedBehaviour = List(InstructionOrPar);
  InstructionOrPar = struct Instruction(getInstruction:Instruction)
    | ParBehaviours(parBehaviours:List(ComposedBehaviour));
map transitionsToSteps: StateMachine -> List(Step);
```

The mCRL2 code specifying the transformation from `Transitions` to `Steps` consists of over 200 lines. Avoiding too much detail we illustrate what transformations are done. The first transformation is to create a `Step` object for every `Transition` by adding the ancestors to the source and target state.

We deal with forks by combining the outgoing transitions. The transition to the fork is changed by adding the guards of the outgoing transitions. The effects of the outgoing transitions are put in parallel (See Fig. 3). Note that in the case of a fork the step does not have a single `arrowEnd`; we assume that `StateName` has a special element `multiple` which will be used in the case of forks.

Similarly, we deal with joins by combining incoming steps and their guards. For steps ending on a composite state we add the initial state in the target. For steps from composite states there are two options: if the step has a change event as trigger then we do not add a substate to the source (step is enabled regardless of the substate); if the step does not have a trigger we require that all the parallel regions of the composite state are in a final state.

We remove junctions by introducing a step for each path over the junction.

Given a state configuration and a set of steps we can reason about which steps are enabled for firing. We will go over the restrictions for firing steps that are checked in different data equations.

The most basic requirement for selecting a step is that the source of the step must match the current state configuration. This is checked by

Fig. 3. Example steps to and from a fork.

`filterPossible`, defined below. The helper function `getAllStatesConfig` returns the set of all states that are in a state configuration. The helper function `containsPseudoState` checks whether a state configuration contains a pseudostate. Due to the run-to-completion semantics we only select a new step when we have reached a stable state configuration (i.e. a state configuration without pseudostates). For this reason we add the condition that if the current state configuration contains a pseudostate then we will only consider transitions from the pseudostate.

```
map filterPossible: List(Step)#StateConfig#StateMachine -> List(Step);
    matchState:StateConfig#StateConfig -> Bool;
var sc, sc1, sc2: StateConfig; step: Step; steps: List(Step);
    sm:StateMachine;
eqn filterPossible([], sc, sm) = [];
    filterPossible(step |> steps, sc, sm) = filterPossible(steps,sc,sm)
      ++ if(matchState(sc,source(step))
          && (containsPseudoState(sc,sm)=>containsPseudoState(source(step),sm)),
        [step], []);
    matchState(sc1,sc2) = (getAllStatesConfig(sc2)-getAllStatesConfig(sc1))=={};
```

Another requirement is that the guard evaluates to true and the trigger matches the current event that is being processed. These two checks are performed by `filterEnabled`.

```
filterEnabled: List(Step)#Event -> List(Step);
```

Another rule is that steps for which the source is lower (i.e. more deeply nested) in the state hierarchy have a higher priority than steps for which the source is higher in the state hierarchy. The mapping `filterPriority` selects the steps with the highest priority among the input. Note that there may be multiple steps on the same priority level.

```
filterPriority: List(Step) -> List(Step);
```

As mentioned earlier, a state machine can also perform a multi-step if multiple steps with the same trigger event are enabled in parallel regions. To be more precise: the state machine selects a multi-step consisting of the maximal set of non-conflicting steps, where non-conflicting means that no two steps in the set exit the same state. The mapping `multiStepPossibilities` computes all such multi-steps given a set of steps.

```
multiStepPossibilities: List(Step) -> List(List(Step));
```

Due to the way we have constructed Steps the target field of a transition is not always a complete state configuration. We leave out parallel regions in defining transitions when they do not actively contribute. If we were to include all the parallel regions in the source and target of Steps we would have to compute all variations. To construct the new state configuration computeNextState takes the target of a transition and adds the parallel regions of the current state configuration that are unaffected (i.e. not exited).

```
computeNextState: StateConfig#Step -> StateConfig;
```

computeNextState recurses through the tree structure of the target state configuration. At each level it copies over unaffected regions. It is unaffected when the region was not present in the source of the step (it was not an active participant of the step) and it is not exited by the step.

The behaviour of performing a step, i.e. an instance of Instructions, is the behaviour of the step itself combined with possible exit and entry behaviour. For internal transitions no state is entered or exited. The snippet below shows the definition of determineBehaviourStep.

```
map getEntryBehaviour: StateMachine#StateConfig#Step -> ComposedBehaviour;
    getExitBehaviour: StateMachine#StateConfig#Step -> ComposedBehaviour;
    determineBehaviourStep: StateMachine#Step#StateConfig -> ComposedBehaviour;
var sm: StateMachine; cur: StateConfig; st: Step;
eqn (!internal(st)) -> determineBehaviourStep(sm,st,cur) =
    getExitBehaviour(sm,cur,st) ++ effect(st) ++ getEntryBehaviour(sm,cur,st);
    internal(st) -> determineBehaviourStep(sm,st,cur) = effect(st);
```

Both getEntryBehaviour and getExitBehaviour compute the new state configuration after firing the transition and which states are entered/exited; subsequently they determine the order in which behaviour needs to be executed and construct a ComposedBehaviour. The order of composing entry behaviours is outside-in (top level states first) and the order of composing exit behaviours is inside-out (nested states first). To determine the order both functions recurse through the new state configuration. Behaviour of states that are entered/exited that are on the same level (parallel regions) is put in parallel.

We use a mapping computeExecutionOptions to compute all the options for what behaviour from a ComposedBehaviour can be executed next. If the head of the composed behaviour is a sequential composition of instructions it will return one execution option with all instructions up to the end of the composed behaviour or up to a parallel composition (whatever comes first). If the head of the composed behaviour is a parallel composition then we get multiple options corresponding to each parallel branch.

```
sort ExecutionOption = struct ExecutionOption(getCodeToExecute:Instructions,
        getRemainingBehavior:ComposedBehavior);
map computeExecutionOptions: ComposedBehavior -> List(ExecutionOption);
```

4.5   Change Events

A change event is generated when the content of a when(x) trigger *becomes* true. When a variable is updated we need to check which change events need to be generated. For this purpose we introduce the sort `Monitor`. A monitor stores an action language expression and the last evaluation. When we update a variable we can check which change events are generated using `deriveChangeEvents`. The mapping `updateMonitors` updates the valuation stored in the monitors.

```
sort Monitor = struct Monitor(getExpression:Instructions, getValuation:Bool);
map deriveChangeEvents: List(Monitor)#(VarName -> Value) -> List(Event);
    updateMonitors: List(Monitor)#(VarName -> Value) -> List(Monitor);
var vars: VarName -> Value; mon: Monitor; mons: List(Monitor);
eqn deriveChangeEvents(mon |> mons,vars) =
    if(checkPredicate(getExpression(mon),vars) && !getValuation(mon),
      [ChangeEvent(getExpression(mon))], []) ++ deriveChangeEvents(mons,vars);
    deriveChangeEvents([],vars) = [];
```

4.6 StateMachine Process

The state machine process uses the data operations that we described in earlier sections and uses them to specify the observable *actions* of a single state machine, which will be visible in the LTS associated to the mCRL2 model. For now we will present a slightly simplified version, which we will extend when we incorporate SysML specific communication in Sect. 5. Below we present the parameters of the process and the declaration of the observable actions (which includes the parameters of those actions).

```
act discardEvent: Event; selectMultiStep: Event#List(Step);
    executeStep: Step; executeBehaviour;
proc StateMachine(ID:CompName, SM:StateMachine, sc:StateConfig,
    eq:List(Event), steps:Set(Transition), behav:ComposedBehaviour,
    mon:List(Monitor), vars:VarName -> Value, exc:ExcFrame) = ...
```

The UML standard does not define in what order events are processed. We have opted to process events in FIFO order, hence **event_queue** is a list of events. The **StateMachine** process consists of one big alternative composition where each summand performs one action and then recurses (with updated parameters).

The observable actions are chosen to reflect decisions in the run-to-completion cycle. When both **steps** and **behav** are empty a new multi-step should be considered. If no step is enabled by the head of the event queue the process can perform a **discardEvent** action and remove it from the event queue. Alternatively, we can select a multi-step with a **selectMultiStep** action. We can now perform a **executeStep** action to start executing one of the selected steps, which updates **sc** and puts the composed behaviour of the step in **behav**. The process selects one of the execution options calculated by **computeExecutionOptions** and initializes an **ExcFrame** which is stored in **exc**. The process calls **executeExcFrameCode** and performs an **executeBehaviour** action until the execution frame is finished. Every time code is executed (and thus possibly variables are updated), it is checked whether change events can be derived. When the execution frame is finished we compute a new execution option. When the execution of **behav** is finished we select a next step from **steps**. When there are no more steps to execute the process is ready to select a new multi-step.

As an example, consider the summand that performs the **executeStep**. Note that mCRL2 allows for an abbreviated, assignment-like syntax in which only the to be updated parameters need to be mentioned in a recursive call; all other parameters of the process remain the same.

```
+ (#behavior_to_execute == 0) ->
  sum next_step:Step. (next_step in steps) -> executeStep(next_step)
    .StateMachine(steps = steps - {next_step},
      behav = determineBehaviourStep(SM,next_step,sc),
      sc = computeNextState(sc,next_step))
```

Depending on the kind of analysis that will be performed on the resulting LTS we might want different observable actions. If we would want to verify something regarding the state configuration we might want to add a self loop signalling the current state configuration. Alternatively, we might want to hide some actions by renaming them to τ, indicating that they are unobservable.

5   SysML Specific Communication

Specific to EULYNX SysML is that there are ports over which communication takes place. Internal Block Diagrams (IBDs) describe the interfaces of components by specifying the ports of components and their connections.

This paper focuses on the semantics of a set of communicating state machines. For the semantics of IBDs we refer the reader to [22]. Here we assume that we have the following communication structure. Each component has a set of ports, which are subdivided in input and output ports. An output port can be connected to multiple input ports. Both input and output ports need not be connected at all, in which case they interact with the environment. One more assumption on the action language is that ports are treated as variables: changing the variable associated to an output port leads to a communication, which updates the variable associated to the input port of the receiver.

The sort **Component** extends state machines with extra information. The sort **Channel** models the connections between ports. Both sorts are defined below. The sort **CompName** defines a finite enumeration of identifiers for components.

```
CompPortPair = struct CompPortPair(getComp: CompName, getPort: VarName);
Component = struct Component(SM: StateMachine, in_ports: List(VarName),
  out_ports:List(VarName));
Channel = struct Channel(sender: CompPortPair, receivers: List(CompPortPair));
```

To take into account communication between state machines, we modify the **StateMachine** process of Sect. 4.6 by replacing the state machine parameter with a **comp** parameter, adding a parameter **oq** and adding two extra actions:

```
act sendComp,receiveComp: CompPortPair#Value;
proc StateMachine(...,comp:Component, oq:List(VarValuePair)) = ...
```

When executing an execution frame we check whether there are updates to output ports and store those updates in output queue **oq**. When **oq** is not empty it can perform a **sendComp** action, communicating the update. At any point in time the process can receive messages via a **receiveComp** action. The summand related to receiving messages is given below.

```
sum v:Value,p:VarName. receiveComp(CompPortPair(ID,p),v)
  .StateMachine(vars = vars[p -> v],
    eq = eq ++ deriveChangeEvents(mon, vars[p->v]),
    mon = updateMonitors(mon,vars[p -> v]))
```

We want to ensure that when a value is sent on an output port, it is received by all (and only) connected input ports. This is enforced by the `Messaging` process and the allow and communication operators in the initialization process (both given below). When the number of components in a configuration is n then the allow operator and `Messaging` process should be extended with the ability to perform a `send` with up to $n-1$ `receive` actions.

```
proc Messaging(channels: List(Channel)) =
  sum ch:Channel, v:Value. (ch in channels) ->
  ((#receivers(ch) == 1) -> receiveI(sender(ch),v)|sendI(receivers(ch).0,v)
    + (#receivers(ch) == 2) -> receiveI(sender(ch),v)|sendI(receivers(ch).0,v)
      |sendI(receivers(ch).1,v)
  ).Messaging();
init allow({selectMultiStep, discardEvent, executeStep, executeBehaviour,
    send|receive, send|receive|receive},
    comm({sendComp|receiveI -> send, sendI|receiveComp -> receive},
      MessagingIntermediary||Environment
      ||StateMachine(...)||StateMachine(...) ...));
```

The central idea is that individual components need not know how ports are connected. Instead, the `Messaging` provides a 'meeting place' with which the sender and receivers synchronize. As an example, suppose some component $C1$ sends some value v on port $P1$ that should be received by two receivers $C2$ and $C3$ on ports $P2$ and $P3$, respectively. The `Messaging` process and the `StateMachine` process of the sender and the two receivers can perform the *multi-action*

```
sendComp(C1,P1,v)|receiveI(C1,P1,v)|sendI(C2,P2,v)
|sendI(C3,P3,v)|receiveComp(C2,P2,v)|receiveComp(C3,P3,v).
```

This is transformed by the communication operator to `send(C1,Port1,v)` `|receive(C2,Port2,v)|receive(C3,Port3,v)`.

Ports that are not connected to any other port are exposed to the environment, i.e. adjacent systems not included in the model. Input ports exposed to the environment can expect inputs at any moment in time. We model the environment with the `Environment` process, which can always send messages to ports in `envInputs` and receive messages from ports in `envOutputs`. Note that a connection between the environment and an exposed port must also be passed to the `Messaging` process.

```
Environment(envInputs:List(CompPortPair), envOutputs:List(CompPortPair)) =
  sum inp:CompPortPair, v:Value. (inp in envInputs)
    -> sendComp(CompPortPair(Environment,getPort(inp)),v).Environment()
  + sum out:CompPortPair, v:Value. (out in envOutputs)
    -> receiveComp(CompPortPair(Environment,getPort(out)),v).Environment();
```

6   Creating a Configuration and Model Checking

In the previous sections, we have discussed the generic parts of the mCRL2 model; in this section, we describe how to configure the model with a specific configuration and touch on the subject of model checking.

First, the enumerations `StateName`, `CompName` and `VarName` need to be instantiated. The action language needs to be defined: the sorts `Value` and `Instruction` need to be defined. Also the semantics of the action language need to be defined by extending the sort `ExcFrame` and giving defining equations for the mappings listed in Sect. 4.2. Finally, the initial process expression needs to be given, in accordance with the structure described in Sect. 5. The `Environment` and `Messaging` processes must be given appropriate parameters. For every state machine a process expression `StateMachineInit`(c, x) needs to be added, where c is a `CompName` and x is a `Component` object.

The model available on GitHub [3] contains an example configuration. This configuration contains just one component named `C1` with the state machine of Fig. 2. The initial valuation of `C1` gives `controller_state` the string "booted" as initial value and sets `reactor_ready` to true. There is one channel: component `C1` has an output port `m`, which is open to the environment.

We can use the mCRL2 model checker to verify properties expressed in the expressive parametrised first-order modal μ-calculus. For instance, we can verify that we always eventually reach the state Booted. We need to capture this property in a μ-calculus formula using the action labels of the model. Note that when `C1` enters the state Booted, it sends a message on port `m` to the environment, so the desired property is expressed by the following formula:

```
mu X. [true]X || <send(CompPortPair(C1,m),Value_String(STR_booted))
|receive(CompPortPair(Environment,m),Value_String(STR_booted))>true.
```

Using the mCRL2 toolset we can check the formula, which does not hold for the model. The toolset produces a counterexample file containing the part of state space that (dis)proves the formula. In this case we get a lasso shaped counterexample with a loop between the states Booting and Failed. The labels on the transitions in the trace are the same as in the mCRL2 model (`selectMultiStep`, `executeStep`, `executeBehaviour` and `send|receive`).

7   Discussion and Conclusion

One of the main benefits of our generic formalisation of the semantics of SysML in mCRL2 is that that it facilitates a straightforward automated translation. To have an automated translation from SysML to mCRL2 we only need to implement a tool that extracts the configuration data from a SysML model and prints the mCRL2 code as described in Sect. 6. Such a tool has recently been built and is discussed in a companion paper (see [22]).

Another benefit of directly formalizing in mCRL2 (compared to formalising in plain mathematics) is that the mCRL2 toolkit acts as an IDE. The parser and type checker of the editor root out the most obvious mistakes. Moreover, the model can be simulated when provided with a configuration of a simple set of state machines. This provides an additional way of verifying whether the semantics is as intended. There is still room for improvement of the mCRL2 toolset though: subtle mistakes in data equations can be hard to debug. Debugging techniques such as breakpoints and being able to step through term rewriting would be beneficial in this regard.

The statespace induced by a SysML model is potentially infinite as event queues can grow without bound. This happens when incoming messages trigger change events faster than the receiving component can process the events.

Since mCRL2 has an explicit-state model checker verification is no longer possible when the state space is infinite, though symbolic tools are in development [15]. The state space can be restricted by bounding the event queue, disallowing reception of messages until some events are processed. The downside of this approach is that the model 'loses' behaviour that could be analysed during model-checking. We leave it for future work to implement more sophisticated bounded event pools such as, e.g., the *controlled buffers* used in [1].

We reckon that significant optimization can be done to reduce the state space. One such optimization possibility was discovered in a case study of the EULYNX Point interface [4]. In our model all updates to variables are stored in the `vars` parameter of the `StateMachine` process. This is not always necessary; only when a variable is read in an action language expression do we really need to store the value. In particular, the variables associated to output ports are rarely referenced by the state machine. We could add a `referencedVariables` field to state machines and adjust the semantics to only remember the value of variables that are actually referenced. This would reduce the state space whilst preserving the behaviour modulo strong bisimilarity.

The UML standard does not give guidelines about the degree of interleaving in the execution of action language expressions. This ambiguity affects both the interleaving between state machines and the interleaving between parallel behaviours in a step. We would like to be able to generate mCRL2 models with varying interleaving models. The finest mode could break behaviour execution down to single instructions (such as looking up the value of a variable) and would allow the most detailed analysis. The coarsest mode could implement a run-to-completion semantics for parallel behaviour, reducing the state space. This variation can be realised by modifying the `ExecuteExcFrameCode` mapping.

Evidence provided by the mCRL2 toolkit (dis)proving properties is presented as an LTS with labels from the mCRL2 model. In the future we would like to improve usability by converting these evidence LTSs to UML sequence diagrams. This may not always be possible (or beneficial) as the evidence LTS may contain the entire state space. We reckon that some common evidence structures such as simple traces and lassos are well suited for conversion to sequence diagrams.

Concluding, we have shown how we have formalised the semantics of (SysML) state machines directly in mCRL2. The generic mCRL2 model is flexible and could be adjusted for a wide range of action languages. The step to an automated translation using our model is small and has been achieved in FormaSig.

Acknowledgement. FormaSig and, by extension, this work are fully funded by ProRail and DB Netz AG. The vision presented in this article does not necessarily reflect the strategy of DB Netz AG or ProRail, but reflects the personal views of the authors.

We also thank the anonymous reviewers for their constructive suggestions, which led to improvements of the paper.

References

1. Abdelhalim, I., Schneider, S., Treharne, H.: An integrated framework for checking the behaviour of fUML models using CSP. Int. J. Softw. Tools Technol. Transf. **15**(4), 375–396 (2013). https://doi.org/10.1007/s10009-012-0243-0
2. Belinfante, A.: JTorX: Exploring Model-Based Testing. Ph.D. thesis, University of Twente, Enschede, Netherlands (2014). http://purl.utwente.nl/publications/91781
3. Bouwman, M.: mCRL2 model capturing the generic semantics of EULYNX SysML. https://github.com/markuzzz/SysML-to-mCRL2
4. Bouwman, M., van der Wal, D., Luttik, B., Stoelinga, M., Rensink, A.: What is the point: formal analysis and test generation or a railway standard. In: Baraldi, P., Di Maio, F., Zio, E. (eds.) Proceedings of ESREL2020-PSAM15, pp. 921–928. Research Publishing, Singapore (2020). https://doi.org/10.3850/978-981-14-8593-0_4410-cd
5. Bunte, O., et al.: The mCRL2 toolset for analysing concurrent systems - improvements in expressivity and usability. In: Vojnar, T., Zhang, L. (eds.) Tools and Algorithms for the Construction and Analysis of Systems (TACAS 2019). Lecture Notes in Computer Science, vol. 11428, pp. 21–39. Springer, Cham (2019). https://doi.org/10.1007/978-3-030-17465-1_2
6. Groote, J.F., Mousavi, M.R.: Modeling and Analysis of Communicating Systems. MIT Press (2014)
7. Hansen, H.H., Ketema, J., Luttik, B., Mousavi, M.R., van de Pol, J.: Towards model checking executable UML specifications in mCRL2. ISSE **6**(1–2), 83–90 (2010). https://doi.org/10.1007/s11334-009-0116-1
8. Hansen, H.H., Ketema, J., Luttik, B., Mousavi, M.R., van de Pol, J., dos Santos, O.M.: Automated verification of executable UML models. In: Aichernig, B.K., de Boer, F.S., Bonsangue, M.M. (eds.) FMCO 2010. Lecture Notes in Computer Science, vol. 6957, pp. 225–250. Springer, Cham (2010). https://doi.org/10.1007/978-3-642-25271-6_12
9. Kim, S., Carrington, D.A.: A formal model of the UML metamodel: the UML state machine and its integrity constraints. In: Bert, D., Bowen, J.P., Henson, M.C., Robinson, K. (eds.) ZB 2002: Formal Specification and Development in Z and B. ZB 2002. Lecture Notes in Computer Science, vol. 2272, pp. 497–516. Springer, Berlin, Heidelberg (2002). https://doi.org/10.1007/3-540-45648-1_26
10. Kuske, S.: A Formal Semantics of UML State Machines Based on Structured Graph Transformation. In: Gogolla, M., Kobryn, C. (eds.) UML 2001. LNCS, vol. 2185, pp. 241–256. Springer, Heidelberg (2001). https://doi.org/10.1007/3-540-45441-1_19
11. Latella, D., Majzik, I., Massink, M.: Automatic verification of a behavioural subset of UML statechart diagrams using the SPIN model-checker. Formal Asp. Comput. **11**(6), 637–664 (1999). https://doi.org/10.1007/s001659970003
12. Lilius, J., Paltor, I.: vUML: a tool for verifying UML models. In: The 14th IEEE International Conference on Automated Software Engineering, ASE 1999, Cocoa Beach, Florida, USA, 12–15 October 1999, pp. 255–258. IEEE Computer Society (1999). https://doi.org/10.1109/ASE.1999.802301
13. Lilius, J., Paltor, I.P.: The semantics of UML state machines (1999)
14. Liu, S., Liu, Y., André, É., Choppy, C., Sun, J., Wadhwa, B., Dong, J.S.: A formal semantics for complete UML state machines with communications. In: Johnsen, E.B., Petre, L. (eds.) IFM 2013. LNCS, vol. 7940, pp. 331–346. Springer, Heidelberg (2013). https://doi.org/10.1007/978-3-642-38613-8_23

15. Neele, T., Willemse, T.A.C., Groote, J.F.: Solving parameterised boolean equation systems with infinite data through quotienting. In: Bae, K., Ölveczky, P.C. (eds.) FACS 2018. LNCS, vol. 11222, pp. 216–236. Springer, Cham (2018). https://doi.org/10.1007/978-3-030-02146-7_11

16. Object Managament Group: OMG Unified Modeling Language, version 2.5.1 (2017). https://www.omg.org/spec/UML/

17. Object Managament Group: Precise Semantics of UML State Machines (PSSM), version 1.0 (2019). https://www.omg.org/spec/PSSM/

18. Lilius, J., Paltor, I.P.: Formalising UML state machines for model checking. In: France, R., Rumpe, B. (eds.) UML 1999. LNCS, vol. 1723, pp. 430–444. Springer, Heidelberg (1999). https://doi.org/10.1007/3-540-46852-8_31

19. Pedroza, G., Apvrille, L., Knorreck, D.: AVATAR: A SysML environment for the formal verification of safety and security properties. In: 11th Annual International Conference on New Technologies of Distributed Systems, NOTERE 2011, Paris, France, 9–13 May 2011, pp. 1–10. IEEE (2011). https://doi.org/10.1109/NOTERE.2011.5957992

20. Remenska, D., et al.: From UML to process algebra and back: an automated approach to model-checking software design artifacts of concurrent systems. In: Brat, G., Rungta, N., Venet, A. (eds.) NASA Formal Methods. NFM 2013. LNCS, vol. 7871, pp. 244–260. Springer, Heidelberg (2013). https://doi.org/10.1007/978-3-642-38088-4_17

21. Schäfer, T., Knapp, A., Merz, S.: Model checking UML state machines and collaborations. Electron. Notes Theor. Comput. Sci. 55(3), 357–369 (2001). https://doi.org/10.1016/S1571-0661(04)00262-2

22. van der Wal, D., Bouwman, M., Stoelinga, M., Rensink, A.: On capturing the EULYNX railway standard with an internal DSL in Java. In: preparation for submission (2021)

23. Wang, H., Zhong, D., Zhao, T., Ren, F.: Integrating model checking with SysML in complex system safety analysis. IEEE Access 7, 16561–16571 (2019). https://doi.org/10.1109/ACCESS.2019.2892745

How Adaptive and Reliable is Your Program?

Valentina Castiglioni[1]($\boxtimes$) ⓘ, Michele Loreti[2] ⓘ, and Simone Tini[3] ⓘ

[1] Reykjavik University, Reykjavik, Iceland
valentinac@ru.is
[2] University of Camerino, Camerino, Italy
michele.loreti@unicam.it
[3] University of Insubria, Como, Italy
simone.tini@uninsubria.it

Abstract. We consider the problem of *modelling* and *verifying* the behaviour of systems characterised by a close interaction of a *program* with the *environment*. We propose to model the program-environment interplay in terms of the probabilistic modifications they induce on a set of application-relevant data, called *data space*. The behaviour of a system is thus identified with the probabilistic evolution of the initial data space. Then, we introduce a metric, called *evolution metric*, measuring the differences in the *evolution sequences* of systems and that can be used for system verification as it allows for expressing how well the program is fulfilling its tasks. We use the metric to express the properties of *adaptability* and *reliability* of a program, which allow us to identify potential critical issues of it w.r.t. changes in the initial environmental conditions. We also propose an *algorithm*, based on statistical inference, for the evaluation of the evolution metric.

1 Introduction

With the ever-increasing complexity of the digital world and diffusion of IoT systems, cyber-physical systems, and smart devices, we are witnessing the rise of software applications, henceforth *programs*, that must be able to deal with highly changing operational conditions, henceforth *environment*. Examples of such programs are the software components of unmanned vehicles, (on-line) service applications, the devices in a smart house, etc., which have to interact with other programs and heterogeneous devices, and with physical phenomena like wind, temperature, etc. Henceforth, we use the term *system* to denote the combination of the environment and the program acting on it. Hence, the *behaviour* of a system is the result of the program-environment interplay.

This work has been partially supported by the IRF project "OPEL" (grant No. 196050-051) and by the PRIN project "IT-MaTTerS" (grant No. 2017FTXR7S).

K. Peters and T. A. C. Willemse (Eds.): FORTE 2021, LNCS 12719, pp. 60–79, 2021.
https://doi.org/10.1007/978-3-030-78089-0_4

The main challenge in the analysis and verification of these systems is then the dynamical and, sometimes, unpredictable behaviour of the environment. The highly dynamical behaviour of physical processes can only be approximated in order to become computationally tractable and it can constitute a safety hazard for the devices in the system (like, e.g., an unexpected gust of wind for a drone that is autonomously setting its trajectory to avoid obstacles); some devices or programs may appear, disappear, or become temporarily unavailable; faults or conflicts may occur (like, e.g., in a smart home the program responsible for the ventilation of a room may open a window causing a conflict with the program that has to limit the noise level); sensors may introduce some measurement errors; etc. The introduction of *uncertainties* and *approximations* in these systems is therefore inevitable.

In the literature, we can find a wealth of proposals of stochastic and probabilistic models, as, e.g., *Stochastic Hybrid Systems* [8,28] and *Markov Decision Processes* [32], and *ad hoc* solutions for specific application contexts, as, e.g., establishing safety guarantees for drones flying under particular circumstances [26,41]. Yet, in these studies, either the environment is not explicitly taken into account or it is modelled only deterministically. In addition to that, due to the variety of applications and heterogeneity of systems, no general formal framework to deal with these challenges has been proposed so far. The lack of concise abstractions and of an automatic support makes the analysis and verification of the considered systems difficult, laborious, and error prone.

Our Contribution. With this paper we aim at taking a first step towards a solution of the above-mentioned challenges, with a special focus on *verification*, by developing the tools for the verification of the ability of programs to *adjust* their behaviour to the unpredictable environment. Formally, we introduce two *measures* allowing us to assess how well a given program can perform under perturbations in the environmental conditions. We call these two measures *adaptability* and *reliability*. As an example, consider a drone that is autonomously flying along a given trajectory. In this setting, a perturbation can be given by a gust of wind that moves the drone out of its trajectory. We will say that the program controlling the drone is *adaptable* if it can retrieve the initial trajectory within a suitable amount of time. In other words, we say that a program is adaptable if no matter how much its behaviour is affected by the perturbations, it is able to react to them and regain its intended behaviour within a given amount of time. On the other hand, it may be the case that the drone is able to detect the presence of a gust of wind and can oppose to it, being only slightly moved from its initial trajectory. In this case, we say that the program controlling the drone is *reliable*. Hence, *reliability* expresses the ability of a program to maintain its intended behaviour (up-to some reasonable tolerance) despite the presence of perturbations in the environment.

In order to measure the adaptability and reliability of a program, we need to be able to express *how well* it is fulfilling its tasks. The systems that we are considering are strongly characterised by a quantitative behaviour, given by both the presence of uncertainties and the data used by program and environment. It seems then natural, and reasonable, to quantify the differences in the behaviour of systems by means of a metric over data. However, in order to informally discuss our proposal of a *metric semantics* for the kind of systems that we are considering, we need first to explain how the behaviour of these systems is defined, namely to introduce our general formal model for them. Our idea is to favour the modelling of the program-environment interplay over precise specifications of the operational behaviour of a program. In the last decade, many researchers have focused their studies on formal models capturing both the *qualitative* and *quantitative* behaviour of systems: Probabilistic Automata [34], Stochastic Process Algebras [5,16,27], Labelled Markov Chains and Stochastic Hybrid Systems [8,28]. A common feature of these models is that the (quantitative, labelled) transitions expressing the computation steps directly model the behaviour of the system as a whole.

In this paper we take a ***different point of view***: we propose to model the behaviour of program and environment separately, and then explicitly represent their *interaction* in a purely *data-driven* fashion. In fact, while the environmental conditions are (partially) available to the program as a set of data, allowing it to adjust its behaviour to the current situation, the program is also able to use data to (partially) control the environment and fulfil its tasks. It is then natural to model the program-environment interplay in terms of the changes they induce on a set of application-relevant data, henceforth called *data space*. This feature will allow for a significant simplification in modelling the behaviour of the program, which can be *isolated* from that of the environment. Moreover, as common to favour *computational tractability* [1,2], we adopt a *discrete time* approach.

We can then study the behaviour of the system as a whole by analysing how data evolve in time. In our model, a *system* consists in *three distinct components*: 1. a *process* P describing the behaviour of the program, 2. a *data state* **d** describing the current state of the data space, and 3. an *environment evolution* $\mathcal{E}$ describing the effect of the environment on **d**. As we focus on the interaction with the environment, we abstract from the internal computation of the program and model only its activity on **d**. At each step, a process can *read/update* values in **d** and $\mathcal{E}$ applies on the resulting data state, providing a new data state at the next step. To deal with the uncertainties, we introduce *probability* at two levels: (i) we use the ***discrete*** *generative probabilistic model* [24] to define processes, and (ii) $\mathcal{E}$ induces a ***continuous*** *distribution* over data states. The behaviour of the system is then entirely expressed by its *evolution sequence*, i.e., the sequence of distributions over data states obtained at each step. Given the novelties of our model, as a side contribution we show that this behaviour defines a *Markov process*.

It is now reasonable to define our *metric semantics* in terms of a (time-dependent) distance on the evolution sequences of systems, which we call the *evolution metric*. The *evolution metric* will allow us to: 1. verify how well a program is fulfilling its tasks by comparing it with its specification, 2. compare the activity of different programs in the same environment, 3. compare the behaviour of one program w.r.t. different environments and changes in the initial conditions. The third feature will allow us to measure the *adaptability* and *reliability* of programs. The evolution metric will consist of two components: a *metric on data states* and the *Wasserstein metric* [39]. The former is defined in terms of a (time-dependent) *penalty function* allowing us to compare two data states only on the base of the objectives of the program. The latter lifts the metric on data states to a metric on distributions on data states. We then obtain a metric on evolution sequences by considering the maximal of the Wasserstein distances over time. We provide an *algorithm* for the estimation of the evolution sequences of systems and thus for the evaluation of the evolution metric. Following [37], the Wasserstein metric is evaluated in time $O(N \log N)$, where N is the (maximum) number of samples. We already adopted this approach in [11] in the context of finite-states self-organising collective systems, without any notion of environment or data space.

As an example of application, we use our framework to model a simple smart-room scenario. We consider two programs: the thermostat of a heating system, and an air quality controller. The former has to keep the room temperature within a desired comfort interval. The latter has to keep the quality of the air above a given threshold. We use our algorithm to evaluate the differences between two systems having the same programs but starting from different initial conditions. Finally, we apply it to measure the adaptability and reliability of the considered programs.

2   Background

Measurable Spaces. A σ-*algebra* over a set Ω is a family Σ of subsets of Ω s.t. $\Omega \in \Sigma$, and Σ is closed under complementation and under countable union. The pair (Ω, Σ) is called a *measurable space* and the sets in Σ are called *measurable sets*, ranged over by $\mathbb{A}, \mathbb{B}, \ldots$. For an arbitrary family Φ of subsets of Ω, the σ-algebra *generated* by Φ is the smallest σ-algebra over Ω containing Φ. In particular, we recall that given a topology T over Ω, the *Borel* σ-*algebra* over Ω, denoted $\mathcal{B}(\Omega)$, is the σ-algebra generated be the open sets in T. Given two measurable spaces (Ω_i, Σ_i), $i = 1, 2$, the *product* σ-*algebra* $\Sigma_1 \times \Sigma_2$ is the σ-algebra on $\Omega_1 \times \Omega_2$ generated by the sets $\{\mathbb{A}_1 \times \mathbb{A}_2 \mid \mathbb{A}_i \in \Sigma_i\}$.

Given measurable spaces $(\Omega_1, \Sigma_1), (\Omega_2, \Sigma_2)$, a function $f \colon \Omega_1 \to \Omega_2$ is said to be Σ_1-*measurable* if $f^{-1}(\mathbb{A}_2) \in \Sigma_1$ for all $\mathbb{A}_2 \in \Sigma_2$, with $f^{-1}(\mathbb{A}_2) = \{\omega \in \Omega_1 \mid f(\omega) \in \mathbb{A}_2\}$.

Probability Spaces. A *probability measure* on a measurable space (Ω, Σ) is a function $\mu \colon \Sigma \to [0, 1]$ such that: i) $\mu(\Omega) = 1$, ii) $\mu(\mathbb{A}) \geq 0$ for all $\mathbb{A} \in \Sigma$, and iii) $\mu(\bigcup_{i \in I} \mathbb{A}_i) = \sum_{i \in I} \mu(\mathbb{A}_i)$ for every countable family of pairwise disjoint measurable sets $\{\mathbb{A}_i\}_{i \in I} \subseteq \Sigma$. Then (Ω, Σ, μ) is called a *probability space*.

Notation. *With a slight abuse of terminology, we shall henceforth use the term* distribution *in place of the term probability measure.*

We let $\Delta(\Omega, \Sigma)$ denote the set of all distributions over (Ω, Σ). For $\omega \in \Omega$, the *Dirac distribution* δ_ω is defined by $\delta_\omega(\mathbb{A}) = 1$, if $\omega \in \mathbb{A}$, and $\delta_\omega(\mathbb{A}) = 0$, otherwise, for all $\mathbb{A} \in \Sigma$. For a countable set of reals $(p_i)_{i \in I}$ with $p_i \geq 0$ and $\sum_{i \in I} p_i = 1$, the *convex combination* of the distributions $\{\mu_i\}_{i \in I} \subseteq \Delta(\Omega, \Sigma)$ is the distribution $\sum_{i \in I} p_i \cdot \mu_i$ in $\Delta(\Omega, \Sigma)$ defined by $(\sum_{i \in I} p_i \cdot \mu_i)(\mathbb{A}) = \sum_{i \in I} p_i \mu_i(\mathbb{A})$, for all $\mathbb{A} \in \Sigma$. A distribution $\mu \in \Delta(\Omega, \Sigma)$ is called *discrete* if $\mu = \sum_{i \in I} p_i \cdot \delta_{\omega_i}$, with $\omega_i \in \Omega$, for some countable set of indexes I. In this case, the *support* of μ is $\mathsf{supp}(\mu) = \{\omega_i \mid i \in I\}$.

The Wasserstein Hemimetric. A *metric* on a set Ω is a function $m \colon \Omega \times \Omega \to \mathbb{R}^{\geq 0}$ s.t. $m(\omega_1, \omega_2) = 0$ iff $\omega_1 = \omega_2$, $m(\omega_1, \omega_2) = m(\omega_2, \omega_1)$, and $m(\omega_1, \omega_2) \leq m(\omega_1, \omega_3) + m(\omega_3, \omega_2)$, for all $\omega_1, \omega_2, \omega_3 \in \Omega$. We obtain a *hemimetric* by relaxing the first property to $m(\omega_1, \omega_2) = 0$ if $\omega_1 = \omega_2$, and by dropping the requirement on symmetry. A (hemi)metric m is *l-bounded* if $m(\omega_1, \omega_2) \leq l$ for all $\omega_1, \omega_2 \in \Omega$. For a (hemi)metric on Ω, the pair (Ω, m) is a *(hemi)metric space*.

In order to define a *hemimetric on distributions* we use the Wasserstein lifting [39]. We recall that a *Polish space* is a separable completely metrisable topological space.

Definition 1 (Wasserstein hemimetric). *Consider a Polish space Ω and let m be a hemimetric on Ω. For any two distributions μ and ν on $(\Omega, \mathcal{B}(\Omega))$, the* Wasserstein lifting *of m to a distance between μ and ν is defined by*

$$\mathbf{W}(m)(\mu, \nu) = \inf_{\mathfrak{w} \in \mathfrak{W}(\mu,\nu)} \int_{\Omega \times \Omega} m(\omega, \omega') \mathrm{d}\mathfrak{w}(\omega, \omega')$$

where $\mathfrak{W}(\mu, \nu)$ is the set of the couplings *of μ and ν, namely the set of joint distributions $\mathfrak{w}$ over the product space $(\Omega \times \Omega, \mathcal{B}(\Omega \times \Omega))$ having μ and ν as left and right marginal, respectively, namely $\mathfrak{w}(\mathbb{A} \times \Omega) = \mu(\mathbb{A})$ and $\mathfrak{w}(\Omega \times \mathbb{A}) = \nu(\mathbb{A})$, for all $\mathbb{A} \in \mathcal{B}(\Omega)$.*

Despite the Wasserstein distance was originally given on metrics, the Wasserstein hemimetric given above is well-defined. A formal proof of this can be found in [19] and the references therein.

Notation. *As elsewhere in the literature, we use the term* metric *in place of* hemimetric.

3   The Model

In this section, we introduce the three components of our systems, namely the *data space*, the *process* describing the behaviour of the program, and the *environment evolution* describing the effects of the environment. The following example perfectly embodies the kind of program-environment interactions we are interested in.

Example 1. We consider a *smart-room scenario* in which the program should guarantee that both the *temperature* and the *air quality* in the room are in a given *comfort zone*. The room is equipped with a *heating system* and an *air filtering system*. Both are equipped with a *sensor* and an *actuator*. In the heating system the sensor is a thermometer that reads the room temperature, while the actuator is used to turn the heater on or off. Similarly, in the air filtering system the sensor perceives the air quality, giving a value in $[0, 1]$, while the actuator activates the air exchangers. The environment models the evolution of temperature and air quality in the room, as described by the following stochastic difference equations, with sample time interval $\Delta\tau = 1$:

$$T(\tau + 1) = T(\tau) + a(e(\tau)) \cdot (T_e - T(\tau)) + h(\tau) \cdot b \cdot (T_h - T(\tau)) \quad (1)$$

$$T_s(\tau) = T(\tau) + n_t(\tau) \quad (2)$$

$$A(\tau + 1) = A(\tau) + e(\tau) \cdot q^+ \cdot (1 - A(\tau)) - (1 - e(\tau)) \cdot q^- \cdot A(\tau) \quad (3)$$

$$A_s(\tau) = A(\tau) + n_a(\tau) \quad (4)$$

Above, $T(\tau)$ and $A(\tau)$ are the room temperature and air quality at time τ, while $T_s(\tau)$ and $A_s(\tau)$ are the respective values read by sensors, which are obtained from the real ones by adding *noises* $n_t(\tau)$ and $n_a(\tau)$, that we assume to be distributed as Gaussian (normal) distributions $\mathcal{N}(0, v_t^2)$ and $\mathcal{N}(0, v_a^2)$, resp., for some suitable v_t^2 and v_a^2. Then, $h(\tau)$ and $e(\tau)$ represent the state of the actuators of the heating and air filtering system, respectively. Both take value 1 when the actuator is *on*, and 0 otherwise. Following [2,23,31], the temperature dynamics depends on two (non negative) values, $a(e(\tau))$ and b, giving the average heat transfer rates normalised w.r.t. the thermal capacity of the room. In detail, $a(e(\tau))$ is the heat loss rate from the room (through walls, windows, etc.) to the external ambient for which we assume a constant temperature T_e. In our case, this value depends on $e(\tau)$, since the loss rate increases when the air exchangers are on. Then, b is the heat transfer rate from the heater, whose temperature is the constant T_h, to the room. The air quality dynamics is similar: when the air exchangers are off, the air quality decreases with a rate q^-, while it increases of a rate q^+ when they are on.

Modelling the Data Space. We define the data space by means of a *finite* set of *variables* Var representing: i) *environmental conditions* (pressure, temperature, humidity, etc.,); ii) *values perceived by sensors* (unavoidably affected by imprecision and approximations); iii) *state of actuators* (usually elements in a discrete domain). For each $x \in$ Var we assume a measurable space $(\mathcal{D}_x, \mathcal{B}_x)$, with $\mathcal{D}_x \subseteq \mathbb{R}$ the domain of x and $\mathcal{B}_x$ the Borel σ-algebra on $\mathcal{D}_x$. Without loosing generality, we can assume that $\mathcal{D}_x$ is either a *finite set* or a *compact* subset of $\mathbb{R}$. Notably, $\mathcal{D}_x$ is a Polish space. As Var is a finite set, we can always assume it to be ordered, i.e., Var $= \{x_1, \ldots, x_n\}$ for some $n \in \mathbb{N}$.

Definition 2 (Data space). *We define the* data space *over* Var*, notation* $\mathcal{D}_{\text{Var}}$*, as the Cartesian product of the variables domains, namely* $\mathcal{D}_{\text{Var}} = \times_{i=1}^n \mathcal{D}_{x_i}$. *Then, as a σ-algebra on* $\mathcal{D}_{\text{Var}}$ *we consider the the product σ-algebra* $\mathcal{B}_{\mathcal{D}_{\text{Var}}} = \times_{i=1}^n \mathcal{B}_{x_i}$.

Example 2. The data space for the system in Example 1 is defined on the variables T, T_s, h, A, A_s and e. Their domains are $\mathcal{D}_T = \mathcal{D}_{T_s} = [t_m, t_M]$, for suitable values $t_m < t_M$, $\mathcal{D}_A = \mathcal{D}_{A_s} = [0, 1]$, and $\mathcal{D}_h = \mathcal{D}_e = \{0, 1\}$.

When no confusion arises, we will use $\mathcal{D}$ and $\mathcal{B}_{\mathcal{D}}$ in place of $\mathcal{D}_{\mathrm{Var}}$ and $\mathcal{B}_{\mathcal{D}_{\mathrm{Var}}}$, respectively. The elements in $\mathcal{D}$ are the n-ples of the form $(v_1, \ldots, v_n)$, with $v_i \in \mathcal{D}_{x_i}$, which can be also identified by means of functions $\mathbf{d} \colon \mathrm{Var} \to \mathbb{R}$ from variables to values, with $\mathbf{d}(x) \in \mathcal{D}_x$ for all $x \in \mathrm{Var}$. Each function $\mathbf{d}$ identifies a particular configuration of the data in the data space, and it is thus called a *data state*.

Definition 3 (Data state). *A* data state *is a mapping* $\mathbf{d} \colon \mathrm{Var} \to \mathbb{R}$ *from state variables to values, with* $\mathbf{d}(x) \in \mathcal{D}_x$ *for all* $x \in \mathrm{Var}$.

For simplicity, we shall write $\mathbf{d} \in \mathcal{D}$ in place of $(\mathbf{d}(x_1), \ldots, \mathbf{d}(x_n)) \in \mathcal{D}$. Since program and environment interact on the basis of the *current* values of data, we have that at each step there is a data state $\mathbf{d}$ that identifies the *current state of the data space* on which the next computation step is built. Given a data state $\mathbf{d}$, we let $\mathbf{d}[x = v]$ denote the data state $\mathbf{d}'$ associating v with x, and $\mathbf{d}(y)$ with any $y \neq x$.

Modelling Processes. We introduce a simple process calculus allowing us to specify programs that interact with a data state $\mathbf{d}$ in a given environment. We assume that the action performed by a process at a given computation step is determined probabilistically, according to the *generative* probabilistic model [24].

Definition 4 (Syntax of processes). *We let* $\mathcal{P}$ *be the set of* processes P *defined by:*

$$P ::= (\bar{e} \to \bar{x}).P' \mid \text{if } [e]\ P_1 \text{ else } P_2 \mid \sum_{i \in I} p_i \cdot P_i \mid P_1 \|_p P_2 \mid A$$
$$e ::= v \in V \mid x \in \mathrm{Var} \mid op_k(e_1, \ldots, e_k)$$

where, $V \subseteq \mathbb{R}$ *countable,* $p, p_1, \ldots$ *weights in* $[0, 1] \cap \mathbb{Q}$, I *is finite,* A *ranges over process variables,* op_k *indicates a measurable operator* $\mathbb{R}^k \to \mathbb{R}$, *and* $\bar{\ }$ *denotes a finite sequence of elements. We assume to have a single definition* $A \stackrel{def}{=} P$ *for each process variable* A. *Moreover, we require that* $\sum_{i \in I} p_i = 1$ *for any process* $\sum_{i \in I} p_i \cdot P_i$.

Process $(\bar{e} \to \bar{x}).P$ evaluates the sequence of expressions $\bar{e}$ with the current data state $\mathbf{d}$ and assigns the results $[\![\bar{e}]\!]_{\mathbf{d}}$ to the sequence of variables $\bar{x}$. We may use $\sqrt{}$ to denote the prefix $(\emptyset \to \emptyset)$. Process if $[e]\ P_1$ else P_2 behaves either as P_1 when $[\![e]\!]_{\mathbf{d}} = \top$, or as P_2 when $[\![e]\!]_{\mathbf{d}} = \bot$. Then, $\sum_{i=1}^{n} p_i \cdot P_i$ is the *generative probabilistic choice*: process P_i has probability p_i to move. The *generative probabilistic interleaving* construct $P_1 \|_p P_2$ lets the two argument processes to interleave their actions, where at each step P_1 moves with probability p and P_2 with probability $1 - p$. Process variables allow us to specify recursive behaviours by means of equations of the form $A \stackrel{def}{=} P$. To avoid Zeno behaviours we assume that all occurrences of process variables appear *guarded* by prefixing constructs in P. We assume the standard notions of *free* and *bound* process variables. A program is then a *closed* process, i.e., a process without free variables.

Formally, actions performed by a process can be abstracted in terms of the *effects* they have on the data state, i.e., via *substitutions* of the form $\theta = [x_{i_1} \leftarrow v_{i_1}, \ldots, x_{i_k} \leftarrow v_{i_k}]$, also denoted $\bar{x} \leftarrow \bar{v}$ if $\bar{x} = x_{i_1}, \ldots, x_{i_k}$ and $\bar{v} = v_{i_1}, \ldots, v_{i_k}$. Since in Definition 4 operations op_k are assumed to be measurable, we can model the effects as $\mathcal{B}_\mathcal{D}$-measurable functions $\theta \colon \mathcal{D} \to \mathcal{D}$ s.t. $\theta(\mathbf{d}) := \mathbf{d}[\bar{x} = \bar{v}]$ whenever $\theta = \bar{x} \leftarrow \bar{v}$. We denote by Θ the set of effects. The behaviour of a process can then be defined by means of a function $\mathsf{pstep} \colon \mathcal{P} \times \mathcal{D} \to \Delta(\Theta \times \mathcal{P})$ that given a process P and a data state $\mathbf{d}$ yields a *discrete* distribution over $\Theta \times \mathcal{P}$. Function pstep is defined as follows:

(PR1) $\mathsf{pstep}((\bar{e} \to \bar{x}).P', \mathbf{d}) = \delta_{(\bar{x} \leftarrow \llbracket \bar{e} \rrbracket \mathbf{d}, P')}$

(PR2) $\mathsf{pstep}(\text{if } [e] \ P_1 \text{ else } P_2, \mathbf{d}) = \begin{cases} \mathsf{pstep}(P_1, \mathbf{d}) & \text{if } \llbracket e \rrbracket \mathbf{d} = 1 \\ \mathsf{pstep}(P_2, \mathbf{d}) & \text{if } \llbracket e \rrbracket \mathbf{d} = 0 \end{cases}$

(PR3) $\mathsf{pstep}(\sum_i p_i \cdot P_i, \mathbf{d}) = \sum_i p_i \cdot \mathsf{pstep}(P_i, \mathbf{d})$

(PR4) $\mathsf{pstep}(P_1 \|_p P_2, \mathbf{d}) = p \cdot (\mathsf{pstep}(P_1, \mathbf{d}) \|_p P_2) + (1 - p) \cdot (P_1 \|_p \mathsf{pstep}(P_2, \mathbf{d}))$

(PR5) $\mathsf{pstep}(A, \mathbf{d}) = \mathsf{pstep}(P, \mathbf{d}) \qquad (\text{if } A \overset{def}{=} P).$

In rule (PR4), for $\pi \in \Delta(\Theta \times \mathcal{P})$, we let $\pi \|_p P$ (resp. $P \|_p \pi$) denote the distribution $\pi' \in \Delta(\Theta \times \mathcal{P})$ s.t.: $\pi'(\theta, P') = \pi(\theta, P'')$, whenever $P' = P'' \|_p P$ (resp. $P' = P \|_p P''$), and 0, otherwise.

Proposition 1 (Properties of process semantics). *Let $P \in \mathcal{P}$ and $\mathbf{d} \in \mathcal{D}$. Then $\mathsf{pstep}(P, \mathbf{d})$ is a discrete distribution with finite support.*

Example 3. We define a program to control the smart-room scenario of Example 1. In detail, we want to guarantee that the temperature in the room is in the interval $Z = [t_{\min}, t_{\max}] \subseteq \mathcal{D}_T$, while the air quality is above a given threshold $q_a \in \mathcal{D}_A$. The following process A^T_{off} (resp. A^T_{on}) turns the heating system *on* (resp. *off*) when the temperature acquired by the sensor goes under $t_{\min} - \varepsilon_t$ (resp. over $t_{\max} + \varepsilon_t$). The use of the tolerance ε_t guarantees that the heating system is not repeatedly turned on/off.

$$\mathsf{A}^T_{off} \overset{def}{=} \text{if } [T_s < t_{\min} - \varepsilon_t] \ (1 \to h).\mathsf{A}^T_{on} \text{ else } \sqrt{.}\mathsf{A}^T_{off}$$
$$\mathsf{A}^T_{on} \overset{def}{=} \text{if } [T_s > t_{\max} + \varepsilon_t] \ (0 \to h).\mathsf{A}^T_{off} \text{ else } \sqrt{.}\mathsf{A}^T_{on}.$$

The behaviour of program components controlling the air filtering system is similar and implemented by the following processes A^A_{off} and A^A_{on}:

$$\mathsf{A}^A_{off} \overset{def}{=} \text{if } [A_s \leq q_a - \varepsilon_a] \ (1 \to e).\mathsf{A}^A_{on} \text{ else } \sqrt{.}\mathsf{A}^A_{off}$$
$$\mathsf{A}^A_{on} \overset{def}{=} \text{if } [A_s > q_a + \varepsilon_a] \ (0 \to e).\mathsf{A}^A_{off} \text{ else } \sqrt{.}\mathsf{A}^A_{on}.$$

The composition of the programs is given by the process $\mathsf{P} = \mathsf{A}^T_{off} \|_{0.5} \mathsf{A}^A_{off}$.

Modelling the Environment. We model the action of the environment by a mapping $\mathcal{E}$, called *environment evolution*, taking a data state to a distribution over data states.

Definition 5 (Environment evolution). *An* environment evolution *is a map* $\mathcal{E}\colon \mathcal{D} \to \Delta(\mathcal{D}, \mathcal{B}_\mathcal{D})$ *s.t. for each* $\mathbb{D} \in \mathcal{B}_\mathcal{D}$ *the mapping* $\mathbf{d} \mapsto \mathcal{E}(\mathbf{d})(\mathbb{D})$ *is* $\mathcal{B}_\mathcal{D}$-*measurable.*

Due to the interaction with the program, the probability induced by $\mathcal{E}$ at the next time step *depends only* on the current state of the data space. It is then natural to assume that the behaviour of the environment is modelled as a discrete time Markov process.

Example 4. For our smart-room scenario, the environment evolution $\mathcal{E}$ can be derived directly from Eqs. (1)–(4). Notice that, in this case, randomness follows from the Gaussian noises associated with the temperature and air quality sensors.

Modelling System's Behaviour. We use the notion of *configuration* to model the state of the system at each time step.

Definition 6 (Configuration). *A* configuration *is a triple* $c = \langle P, \mathbf{d} \rangle_\mathcal{E}$*, where* P *is a process,* $\mathbf{d}$ *is a data state and* $\mathcal{E}$ *is an environment evolution. We denote by* $\mathcal{C}_{\mathcal{P}, \mathcal{D}, \mathcal{E}}$ *the set of configurations defined over* $\mathcal{P}, \mathcal{D}$ *and* $\mathcal{E}$*.*

When no confusion arises, we shall write $\mathcal{C}$ in place of $\mathcal{C}_{\mathcal{P}, \mathcal{D}, \mathcal{E}}$.

Let $(\mathcal{P}, \Sigma_\mathcal{P})$ be the measurable space of processes, where $\Sigma_\mathcal{P}$ is the power set of $\mathcal{P}$, and $(\mathcal{D}, \mathcal{B}_\mathcal{D})$ be the measurable space of data states. As $\mathcal{E}$ is fixed, we can identify $\mathcal{C}$ with $\mathcal{P} \times \mathcal{D}$ and equip it with the product σ-algebra $\Sigma_\mathcal{C} = \Sigma_\mathcal{P} \times \mathcal{B}_\mathcal{D}$: $\Sigma_\mathcal{C}$ is generated by the sets $\{\langle \mathbb{P}, \mathbb{D} \rangle_\mathcal{E} \mid \mathbb{P} \in \Sigma_\mathcal{P}, \mathbb{D} \in \mathcal{B}_\mathcal{D}\}$, where $\langle \mathbb{P}, \mathbb{D} \rangle_\mathcal{E} = \{\langle P, \mathbf{d} \rangle_\mathcal{E} \mid P \in \mathbb{P}, \mathbf{d} \in \mathbb{D}\}$.

Notation. *For* $\mu_\mathcal{P} \in \Delta(\mathcal{P}, \Sigma_\mathcal{P})$ *and* $\mu_\mathcal{D} \in \Delta(\mathcal{D}, \mathcal{B}_\mathcal{D})$ *we let* $\mu = \langle \mu_\mathcal{P}, \mu_\mathcal{D} \rangle_\mathcal{E}$ *denote the product distribution on* $(\mathcal{C}, \Sigma_\mathcal{C})$*, i.e.,* $\mu(\langle \mathbb{P}, \mathbb{D} \rangle_\mathcal{E}) = \mu_\mathcal{P}(\mathbb{P}) \cdot \mu_\mathcal{D}(\mathbb{D})$ *for all* $\mathbb{P} \in \Sigma_\mathcal{P}$ *and* $\mathbb{D} \in \mathcal{B}_\mathcal{D}$*. If* $\mu_\mathcal{P} = \delta_P$ *for some* $P \in \mathcal{P}$*, we shall denote* $\langle \delta_P, \mu_\mathcal{D} \rangle_\mathcal{E}$ *simply by* $\langle P, \mu_\mathcal{D} \rangle_\mathcal{E}$*.*

We aim to express the behaviour of a system in terms of the changes on data. We start with the *one-step* behaviour of a configuration, in which we combine the effects on the data state induced by the activity of the process (given by pstep) and the subsequent action by the environment. Formally, we define a function cstep that, given a configuration, yields a distribution on $(\mathcal{C}, \Sigma_\mathcal{C})$ (Definition 7 below). Then, we use cstep to define the *multi-step* behaviour of configuration c as a sequence $\mathcal{S}_{c,0}^\mathcal{C}, \mathcal{S}_{c,1}^\mathcal{C}, \ldots$ of distributions on $(\mathcal{C}, \Sigma_\mathcal{C})$. To this end, we show that cstep is a Markov kernel (Proposition 3 below). Finally, to abstract from processes and focus only on data, from the sequence $\mathcal{S}_{c,0}^\mathcal{C}, \mathcal{S}_{c,1}^\mathcal{C}, \ldots$, we obtain a sequence of distributions $\mathcal{S}_{c,0}^\mathcal{D}, \mathcal{S}_{c,1}^\mathcal{D}, \ldots$ on $(\mathcal{D}, \mathcal{B}_\mathcal{D})$ called the *evolution sequence* of the system (Definition 9 below).

Definition 7 (One-step semantics). *Function* $\mathsf{cstep} \colon \mathcal{C} \rightarrow \Delta(\mathcal{C}, \Sigma_{\mathcal{C}})$ *is defined for all configurations* $\langle P, \mathbf{d} \rangle_{\mathcal{E}} \in \mathcal{C}$ *by*

$$\mathsf{cstep}(\langle P, \mathbf{d} \rangle_{\mathcal{E}}) = \sum_{(\theta, P') \in \mathsf{supp}(\mathsf{pstep}(P, \mathbf{d}))} \mathsf{pstep}(P, \mathbf{d})(\theta, P') \cdot \langle P', \mathcal{E}(\theta(\mathbf{d})) \rangle_{\mathcal{E}}. \quad (5)$$

The next result follows by $\mathcal{E}(\theta(\mathbf{d})) \in \Delta(\mathcal{D}, \mathcal{B}_{\mathcal{D}})$ (Definition 5), which ensures that $\langle P', \mathcal{E}(\theta(\mathbf{d})) \rangle_{\mathcal{E}} \in \Delta(\mathcal{C}, \Sigma_{\mathcal{C}})$, and $\mathsf{pstep}(P, \mathbf{d})$ is a discrete distribution in $\Delta(\Theta \times \mathcal{P})$ (Proposition 1).

Proposition 2. *For any configuration* $c \in \mathcal{C}$, $\mathsf{cstep}(c)$ *is a distribution on* $(\mathcal{C}, \Sigma_{\mathcal{C}})$.

Since $\mathsf{cstep}(c) \in \Delta(\mathcal{C}, \Sigma_{\mathcal{C}})$ for each $c \in \mathcal{C}$, we can rewrite $\mathsf{cstep} \colon \mathcal{C} \times \Sigma_{\mathcal{C}} \rightarrow [0, 1]$, so that for each configuration $c \in \mathcal{C}$ and measurable set $\mathbb{C} \in \Sigma_{\mathcal{C}}$, $\mathsf{cstep}(c)(\mathbb{C})$ denotes the probability of reaching in one step a configuration in $\mathbb{C}$ starting from c. We can prove that cstep is the Markov kernel of the Markov process modelling our system. This follows by Proposition 2 and by proving that for each $\mathbb{C} \in \Sigma_{\mathcal{C}}$, the mapping $c \mapsto \mathsf{cstep}(c)(\mathbb{C})$ is $\Sigma_{\mathcal{C}}$-measurable for all $c \in \mathcal{C}$.

Proposition 3. *The function* cstep *is a Markov kernel.*

Hence, the multi-step behaviour of configuration c can be defined as a time homogeneous Markov process having cstep as Markov kernel and δ_c as initial distribution.

Definition 8 (Multi-step semantics). *Let* $c \in \mathcal{C}$ *be a configuration. The multi-step behaviour of* c *is the sequence of distributions* $\mathcal{S}^{\mathcal{C}}_{c,0}, \mathcal{S}^{\mathcal{C}}_{c,1}, \ldots$ *on* $(\mathcal{C}, \Sigma_{\mathcal{C}})$ *defined inductively as follows:*

$$\mathcal{S}^{\mathcal{C}}_{c,0}(\mathbb{C}) = \delta_c(\mathbb{C}), \text{ for all } \mathbb{C} \in \Sigma_{\mathcal{C}}$$

$$\mathcal{S}^{\mathcal{C}}_{c,i+1}(\mathbb{C}) = \int_{\mathcal{C}} \mathsf{cstep}(b)(\mathbb{C}) \mathrm{d}(\mathcal{S}^{\mathcal{C}}_{c,i}(b)), \text{ for all } \mathbb{C} \in \Sigma_{\mathcal{C}}.$$

We can prove that $\mathcal{S}^{\mathcal{C}}_{c,0}, \mathcal{S}^{\mathcal{C}}_{c,1}, \ldots$ are well defined, namely they are distributions on $(\mathcal{C}, \Sigma_{\mathcal{C}})$. The proof follows by an easy induction based on Proposition 3.

Proposition 4. *For any* $c \in \mathcal{C}$, *all* $\mathcal{S}^{\mathcal{C}}_{c,0}, \mathcal{S}^{\mathcal{C}}_{c,1}, \ldots$ *are distributions on* $(\mathcal{C}, \Sigma_{\mathcal{C}})$.

As the program-environment interplay can be observed only in the changes they induce on the data states, we define the *evolution sequence* of a configuration as the sequence of distributions over data states that are reached by it, step-by-step.

Definition 9 (Evolution sequence). *The* evolution sequence *of* $c = \langle P, \mathbf{d} \rangle_{\mathcal{E}}$ *is a sequence* $\mathcal{S}^{\mathcal{D}}_c \in \Delta(\mathcal{D}, \mathcal{B}_{\mathcal{D}})^{\omega}$ *of distributions over* $\mathcal{D}$ *such that* $\mathcal{S}^{\mathcal{D}}_c = \mathcal{S}^{\mathcal{D}}_{c,0} \ldots \mathcal{S}^{\mathcal{D}}_{c,n} \ldots$ *if and only if for all* $i \geq 0$ *and for all* $\mathbb{D} \in \mathcal{B}_{\mathcal{D}}$, $\mathcal{S}^{\mathcal{D}}_{c,i}(\mathbb{D}) = \mathcal{S}^{\mathcal{C}}_{c,i}(\langle P, \mathbb{D} \rangle_{\mathcal{E}})$.

4 Towards a Metric for Systems

We aim at defining a *distance* over the systems described in the previous section, called the *evolution metric*, allowing us to do the following:

1. Verify how well a program is fulfilling its tasks.
2. Establish whether one program behaves better than another one in an environment.
3. Compare the interactions of a program with different environments.

These three objectives can be naturally obtained thanks to the possibility of modelling the program in isolation from the environment typical of our model, and to our purely data-driven system semantics. Intuitively, since the behaviour of a system is entirely described by its evolution sequence, the evolution metric $\mathfrak{m}$ will indeed be defined as a distance on the evolution sequences of systems. However, in order to obtain the proper technical definition of $\mathfrak{m}$, some considerations are due.

Firstly, we notice that in most applications the tasks of the program can be expressed in a purely data-driven fashion. We can identify a set of *parameters of interest* such that, at any time step, any difference between them and the data actually obtained can be interpreted as a flaw in system behaviour. We use a *penalty function* ρ to quantify these differences. From the penalty function we can obtain a *distance on data states*, namely a 1-bounded *hemimetric* $m^{\mathcal{D}}$ expressing how much a data state $\mathbf{d}_2$ is worse than a data state $\mathbf{d}_1$ according to parameters of interests. Secondly, we recall that the evolution sequence of a system consists in a sequence of *distributions* over data states. Hence, we use the *Wasserstein metric* to lift $m^{\mathcal{D}}$ to a distance $\mathbf{W}(m^{\mathcal{D}})$ over distributions over data states. Informally, with the Wasserstein metric we can express how much worse a configuration is expected to behave w.r.t. another one at a given time. Finally, we need to lift $\mathbf{W}(m^{\mathcal{D}})$ to a distance on the entire evolution sequences of systems. For our purposes, a reasonable choice is to take the maximum over time of the pointwise (w.r.t. time) Wasserstein distances (see Remark 1 below for further details on this choice).

A Metric on Data States. We start by proposing a metric on data states, seen as *static components* in isolation from processes and environment. To this end, we introduce a *penalty function* $\rho \colon \mathcal{D} \to [0,1]$, a continuous function that assigns to each data state $\mathbf{d}$ a penalty in $[0,1]$ expressing how far the values of the parameters of interest in $\mathbf{d}$ are from their desired ones (hence $\rho(\mathbf{d}) = 0$ if $\mathbf{d}$ respects all the parameters). Since some parameters can be time-dependent, so is ρ: at any time step τ, the τ-penalty function ρ_τ compares the data states w.r.t. the values of the parameters expected at time τ.

Example 5. We recall, from Example 2, that $\mathcal{D}_T = [t_m, t_M]$ and $\mathcal{D}_A = [0,1]$. The task of our program is to keep the value of T within the comfort zone $Z = [t_{\min}, t_{\max}]$, for some $t_{\min}, t_{\max}$, and that of A above a threshold $q_a \in \mathcal{D}_A$ (cf. Example 3). Hence, we define a penalty function that assigns the penalty

0 if the value of T is in Z and that of A is greater or equal to q_a, otherwise it is proportional to how much T and A are far from Z and q_a, respectively. We let $\rho_\tau(\mathbf{d}) = \max\{\rho^T(\mathbf{d}(T)), \rho^A(\mathbf{d}(A))\}$, where $\rho^T(t)$ is 0 if $t \in [t_{\min}, t_{\max}]$ and $\frac{\max\{t - t_{\max}, t_{\min} - t\}}{\max\{t_M - t_{\max}, t_{\min} - t_m\}}$ otherwise, while $\rho^A(q) = \max\{0, q_a - q\}$.

A formal definition of the penalty function is beyond the purpose of this paper, also due to its context-dependent nature. Besides, notice that we can assume that ρ already includes some tolerances w.r.t. the exact values of the parameters in its evaluation, and thus we do not consider them. The (*timed*) *metric on data states* is then defined as the asymmetric difference between the penalties assigned to them by the penalty function.

Definition 10 (Metric on data states). *For any time step τ, let $\rho_\tau \colon \mathcal{D} \to [0,1]$ be the τ-penalty function on $\mathcal{D}$. The τ-metric on data states in $\mathcal{D}$, $m^{\mathcal{D}}_{\rho,\tau} \colon \mathcal{D} \times \mathcal{D} \to [0,1]$, is defined, for all $\mathbf{d}_1, \mathbf{d}_2 \in \mathcal{D}$, by $m^{\mathcal{D}}_{\rho,\tau}(\mathbf{d}_1, \mathbf{d}_2) = \max\{\rho_\tau(\mathbf{d}_2) - \rho_\tau(\mathbf{d}_1), 0\}$.*

Notice that $m^{\mathcal{D}}_{\rho,\tau}(\mathbf{d}_1, \mathbf{d}_2) > 0$ iff $\rho_\tau(\mathbf{d}_2) > \rho_\tau(\mathbf{d}_1)$, i.e., the penalty assigned to $\mathbf{d}_2$ is higher than that assigned to $\mathbf{d}_1$. For this reason, we say that $m^{\mathcal{D}}_{\rho,\tau}(\mathbf{d}_1, \mathbf{d}_2)$ expresses *how worse* $\mathbf{d}_2$ is than $\mathbf{d}_1$ w.r.t. the objectives of the system. It is not hard to see that for all $\mathbf{d}_1, \mathbf{d}_2, \mathbf{d}_3 \in \mathcal{D}$ we have $m^{\mathcal{D}}_{\rho,\tau}(\mathbf{d}_1, \mathbf{d}_2) \leq 1$, $m^{\mathcal{D}}_{\rho,\tau}(\mathbf{d}_1, \mathbf{d}_1) = 0$, and $m^{\mathcal{D}}_{\rho,\tau}(\mathbf{d}_1, \mathbf{d}_2) \leq m^{\mathcal{D}}_{\rho,\tau}(\mathbf{d}_1, \mathbf{d}_3) + m^{\mathcal{D}}_{\rho,\tau}(\mathbf{d}_3, \mathbf{d}_2)$, thus ensuring that $m^{\mathcal{D}}_{\rho,\tau}$ is a 1-bounded hemimetric.

Proposition 5. *Function $m^{\mathcal{D}}_{\rho,\tau}$ is a 1-bounded hemimetric on $\mathcal{D}$.*

Lifting $m^{\mathcal{D}}_{\rho,\tau}$ to Distributions. The second step to obtain the evolution metric consists in lifting $m^{\mathcal{D}}_{\rho,\tau}$ to a metric on distributions on data states. Among the several notions of lifting in the literature (see [33] for a survey), we opt for that of Wasserstein, since: i) it preserves the properties of the ground metric; ii) it allows us to deal with discrete and continuous measures; iii) it is computationally tractable via statistical inference. According to Definition 1, the Wasserstein lifting of $m^{\mathcal{D}}_{\rho,\tau}$ to a distance between two distributions $\mu, \nu \in \Delta(\mathcal{D}, \mathcal{B}_{\mathcal{D}})$ is defined by

$$\mathbf{W}(m^{\mathcal{D}}_{\rho,\tau})(\mu, \nu) = \inf_{\mathfrak{w} \in \mathfrak{W}(\mu,\nu)} \int_{\mathcal{D} \times \mathcal{D}} m^{\mathcal{D}}_{\rho,\tau}(\mathbf{d}, \mathbf{d}') \mathrm{d}\mathfrak{w}(\mathbf{d}, \mathbf{d}').$$

The Evolution Metric. We now need to lift $\mathbf{W}(m^{\mathcal{D}}_{\rho,\tau})$ to a distance on evolution sequences. To this end, we observe that the evolution sequence of a configuration includes the distributions over data states induced after *each* computation step. Thus, the time step between two distributions is determined by the program. However, it could be the case that the changes on data induced by the environment can be appreciated only along wider time intervals. Our running example is a clear instance of this situation: while we can reasonably assume that the duration of the computation steps of the thermostat is of the order of a millisecond, the variations in the temperature that can be detected in the same time interval are indeed negligible w.r.t. the program's task. A significant temperature rise or drop can be observed only in longer time. To deal with this kind of situations, we introduce the notion of *observation times*, namely a *discrete* set OT of time steps at which the modifications induced by the program-environment interplay give us useful information on the evolution of the system. Hence, a comparison of the evolution sequences based on the differences in the distributions reached at the times in OT can be considered meaningful. Moreover, considering only the differences at the observation times will favour the computational tractability of the evolution metric.

We define the evolution metric as a sort of *weighted infinity norm* of the tuple of the Wasserstein distances between the distributions in the evolution sequences. As weight we consider a non-increasing function $\lambda \colon \mathrm{OT} \to (0,1]$ expressing how much the distance at time τ affects the overall distance between configurations c_1 and c_2. We refer to λ as to the *discount function*, and to $\lambda(\tau)$ as to the *discount factor at time τ*.

Definition 11 (Evolution metric). *Assume a set* OT *of observation times and a discount function* λ. *Let* ρ *be a penalty function and let* $m_{\rho,\tau}^{\mathcal{D}}$ *be the metric on data states defined on it. Then, the* λ*-evolution metric over* ρ *and* OT *is the mapping* $\mathfrak{m}_{\rho,\mathrm{OT}}^{\lambda} \colon \mathcal{C} \times \mathcal{C} \to [0,1]$ *defined, for all configurations* $c_1, c_2 \in \mathcal{C}$, *by*

$$\mathfrak{m}_{\rho,\mathrm{OT}}^{\lambda}(c_1,c_2) = \sup_{\tau \in \mathrm{OT}} \lambda(\tau) \cdot \mathbf{W}(m_{\rho,\tau}^{\mathcal{D}})\left(\mathcal{S}_{c_1,\tau}^{\mathcal{D}}, \mathcal{S}_{c_2,\tau}^{\mathcal{D}}\right).$$

Since $m_{\rho,\tau}^{\mathcal{D}}$ is a 1-bounded hemimetric (Proposition 5) and lifting $\mathbf{W}$ preserves such a property, we can easily derive the same property for $\mathfrak{m}_{\mathrm{OT}}^{\lambda}$.

Proposition 6. *Function* $\mathfrak{m}_{\rho,\mathrm{OT}}^{\lambda}$ *is a 1-bounded hemimetric on* $\mathcal{C}$.

Notice that if λ is a *strictly* non-increasing function, then it specifies how much the distance of *future events* is mitigated and, moreover, it guarantees that to obtain upper bounds on the evolution metric only a *finite* number of observations is needed.

Remark 1. Usually, due to the presence of uncertainties, the behaviour of a system can be considered acceptable even if it differs from its intended one *up-to a certain tolerance*. Similarly, the properties of adaptability and reliability that we aim to study will check whether a program is able to perform well in a perturbed environment *up-to a given tolerance*. In this setting, the choice of defining the evolution metric as the pointwise maximal distance in time between the evolution sequences of systems is natural and reasonable: if in the worst case (the maximal distance) the program keeps the parameters of interest within the given tolerance, then its entire behaviour can be considered acceptable. However, with this approach we have that a program is only as good as its worst performance, and one could argue that there are application contexts in which our evolution metric would be less meaningful. For these reasons, we remark that we could have given a *parametric* version of Definition 11 and defining the evolution metric in terms of a generic *aggregation function* f over the tuple of Wasserstein distances. Then, one could choose the best instantiation for f according to the chosen application context. The use of a parametric definition would have not affected the technical development of our paper. However, to keep the notation and presentation as simple as possible, we opted to define $\mathfrak{m}_{\rho,\text{OT}}^{\lambda}$ directly in the weighted infinity norm form. A similar reasoning applies to the definition of the penalty function that we gave in Example 5.

5   Estimating the Evolution Metric

In this section we show how the evolution metric can be estimated via statistical techniques. Firstly, we show how we can estimate the evolution sequence of a given configuration c. Then, we evaluate the distance between two configurations c_1 and c_2 on their estimated evolution sequences.

Computing Empirical Evolution Sequences. To compute the empirical evolution sequence of a configuration c the following function EST can be used.

```
1: function EST(c, k, N)
2:     ∀i : (0 ≤ i ≤ k) : Eᵢ ← ∅
3:     counter ← 0
4:     while counter < N do
5:         (c₀, ..., cₖ) ← SIM(c, k)
6:         ∀i : Eᵢ ← Eᵢ, cᵢ
7:         counter ← counter + 1
8:     end while
9:     return E₀, ..., Eₖ
10: end function
```

Function EST(c, k, N) invokes N times function SIM, i.e., any simulation algorithm sampling a sequence of configurations $c_0, \ldots, c_k$, modelling k steps of a computation from $c = c_0$. Then, the sequence $E_0, \ldots, E_k$ is computed, where E_i is the tuple $c_i^1, \ldots, c_i^N$ of configurations observed at time i in each of the N sampled computations.

Each E_i can be used to estimate the distribution $\mathcal{S}_{c,i}^{\mathcal{C}}$. For any i, with $0 \leq i \leq k$, we let $\hat{\mathcal{S}}_{c,i}^{\mathcal{C},N}$ be the distribution s.t. for any $\mathbb{C} \in \Sigma_{\mathcal{C}}$ we have $\hat{\mathcal{S}}_{c,i}^{\mathcal{C},N}(\mathbb{C}) = \frac{|E_i \cap \mathbb{C}|}{N}$. Finally, we let $\hat{\mathcal{S}}_c^{\mathcal{D},N} = \hat{\mathcal{S}}_{c,0}^{\mathcal{D},N} \ldots \hat{\mathcal{S}}_{c,k}^{\mathcal{D},N}$ be the *empirical evolution sequence* s.t. for any measurable set of data states $\mathbb{D} \in \mathcal{B}_{\mathcal{D}}$ we have $\hat{\mathcal{S}}_{c,i}^{\mathcal{D},N}(\mathbb{D}) = \hat{\mathcal{S}}_{c,i}^{\mathcal{C},N}(\langle \mathcal{P}, \mathbb{D} \rangle_{\mathcal{E}})$. Then, by applying the weak law of large numbers to the i.i.d samples, we get that when N goes to infinite both $\hat{\mathcal{S}}_{c,i}^{\mathcal{C},N}$ and $\hat{\mathcal{S}}_{c,i}^{\mathcal{D},N}$ converge weakly to $\mathcal{S}_{c,i}^{\mathcal{C}}$ and $\mathcal{S}_{c,i}^{\mathcal{D}}$ respectively:

$$\lim_{N \to \infty} \hat{\mathcal{S}}_{c,i}^{\mathcal{C},N} = \mathcal{S}_{c,i}^{\mathcal{C}} \qquad \lim_{N \to \infty} \hat{\mathcal{S}}_{c,i}^{\mathcal{D},N} = \mathcal{S}_{c,i}^{\mathcal{D}}. \qquad (6)$$

The tool and the scripts of the examples are available (in Python) at https://github.com/quasylab/spear.

Example 6. We apply our simulation to the heating system from Sect. 3, with initial configuration $c_1 = \langle \mathsf{P}, \{T = 5.0, T_s = 5.0, h = 0, A = 0.5, A_s = 0.5, e = 0\} \rangle_{\mathcal{E}}$, where P is the process in Example 3, and $[t_{\min}, t_{\max}] = [15, 20]$. In Fig. 1 the probability distribution of the temperature after 50 steps is reported.

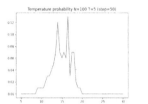
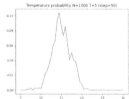
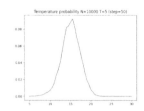

Fig. 1. Estimated distribution of the temperature after 50 steps with $N = 10^2$, $N = 10^3$ and $N = 10^4$. As comfort zone we consider the interval $[15, 20]$.

Computing Distance Between Two Configurations. Function EST allows us to collect independent samples at each time step i from 0 to a deadline k. These samples can be used to estimate the distance between two configurations c_1 and c_2. Following a similar approach to [37], to estimate the Wasserstein distance $\mathbf{W}(m_{\rho,i}^{\mathcal{D}})$ between two (unknown) distributions $\mathcal{S}_{c_1,i}^{\mathcal{D}}$ and $\mathcal{S}_{c_2,i}^{\mathcal{D}}$ we can use N independent samples $\{c_1^1, \ldots, c_1^N\}$ taken from $\mathcal{S}_{c_1,i}^{\mathcal{C}}$ and $\ell \cdot N$ independent samples $\{c_2^1, \ldots, c_2^{\ell \cdot N}\}$ taken from $\mathcal{S}_{c_2,i}^{\mathcal{C}}$. After that, we exploit the i-penalty function ρ and we consider the two sequences of values: $\{\omega_j = \rho_i(\mathbf{d}_1^j) | \langle P_1^j, \mathbf{d}_1^j \rangle_{\mathcal{E}_1} = c_1^j\}$ and $\{\nu_h = \rho_i(\mathbf{d}_2^h) | \langle P_2^h, \mathbf{d}_2^h \rangle_{\mathcal{E}_2} = c_2^h\}$. We can assume, without loss of generality, that these sequences are ordered, i.e., $\omega_j \leq \omega_{j+1}$ and $\nu_h \leq \nu_{h+1}$. The value $\mathbf{W}(m_{\rho,i}^{\mathcal{D}})(\mathcal{S}_{c_1,i}^{\mathcal{D}}, \mathcal{S}_{c_2,i}^{\mathcal{D}})$ can be approximated as $\frac{1}{\ell N} \sum_{h=1}^{\ell N} \max\{\nu_h - \omega_{\lceil \frac{h}{\ell} \rceil}, 0\}$. The next theorem, based on results in [37,40], ensures that the larger the number of samplings the closer the gap between the estimated value and the exact one.

Theorem 1. *Let $\mathcal{S}_{c_1,i}^{\mathcal{C}}, \mathcal{S}_{c_2,i}^{\mathcal{C}} \in \Delta(\mathcal{C}, \Sigma_{\mathcal{C}})$ be unknown, and ρ be a penalty function. Let $\{\omega_j = \rho_i(\mathbf{d}_1^j)\}$ and $\{\nu_h = \rho_i(\mathbf{d}_2^h)\}$ be the ordered sequences obtained from independent samples taken from $\mathcal{S}_{c_1,i}^{\mathcal{C}}$ and $\mathcal{S}_{c_2,i}^{\mathcal{C}}$, respectively. Then, it holds, a.s., that $\mathbf{W}(m_{\rho,i}^{\mathcal{D}})(\mathcal{S}_{c_1,i}^{\mathcal{D}}, \mathcal{S}_{c_2,i}^{\mathcal{D}}) = \lim_{N \to \infty} \frac{1}{\ell N} \sum_{h=1}^{\ell N} \max\{\nu_h - \omega_{\lceil \frac{h}{\ell} \rceil}, 0\}$.*

The outlined procedure is realised by functions DIST and COMPW in Fig. 2. The former takes as input the two configurations to compare, the penalty function (seen as the sequence of the i-penalty functions), the discount function λ, the bounded set OT if observation times, and the parameters N and ℓ used to obtain the samplings of computation. Function DIST collects the samples E_i of possible computations during the observation period $[0, \max_{OT}]$, where $\max_{OT}$ denotes the last observation time. Then, for each $i \in OT$, the distance at time i is computed via the function $\text{COMPW}(E_{1,i}, E_{2,i}, \rho_i)$. As the penalty function allows us to reduce the evaluation of the Wasserstein distance in $\mathbb{R}^n$ to its evaluation on $\mathbb{R}$, due to the sorting of $\{\nu_h \mid h \in [1, \ldots, \ell N]\}$ the complexity of function COMPW is $O(\ell N \log(\ell N))$ (cf. [37]).

Example 7. We change the initial value of the air quality in the configuration c_1 in Example 6, and consider $c_2 = \langle \mathsf{P}, \{T = 5.0, T_s = 5.0, h = 0, A = 0.3, A_s = 0.3, e = 0\} \rangle_{\mathcal{E}}$. Figure 3a shows the variation in time of the distance between c_1 and c_2.

```
1: function DIST(c₁, c₂, ρ, λ, OT, N, ℓ)      1: function COMPW(E₁, E₂, ρ)
2:     k ← maxOT                              2:     (⟨P₁¹, d₁¹⟩ε₁, ..., ⟨P₁ᴺ, d₁ᴺ⟩ε₁) ← E₁
3:     E₁,₁, ..., E₁,ₖ ← EST(c₁, k, N)         3:     (⟨P₂¹, d₂¹⟩ε₂, ..., ⟨P₂ℓᴺ, d₂ℓᴺ⟩ε₂) ← E₂
4:     E₂,₁, ..., E₂,ₖ ← EST(c₂, k, ℓN)        4:     ∀j : (1 ≤ j ≤ N) : ωⱼ ← ρ(d₁ʲ)
5:     m ← ∞                                   5:     ∀h : (1 ≤ h ≤ ℓN) : νₕ ← ρ(d₂ʰ)
6:     for all i ∈ OT do                      6:     re index {ωⱼ} s.t. ωⱼ ≤ ωⱼ₊₁
7:         mᵢ ← COMPW(E₁,ᵢ, E₂,ᵢ, ρᵢ)         7:     re index {νₕ} s.t. νₕ ≤ νₕ₊₁
8:         m ← min{m, λ(i) · mᵢ}              8:     return (1/ℓN) Σₕ₌₁ℓᴺ |ω⌈h/ℓ⌉ − νₕ|
9:     end for                                9: end function
10:    return m
11: end function
```

Fig. 2. Functions used to estimate the evolution metric on systems.

6   Adaptability and Reliability of Programs

In this section we exploit the evolution metric to study some dependability properties of programs, which we call *adaptability* and *reliability*, w.r.t. a data state and an environment. Both notions entail the ability of the process to induce a *similar* behaviour in systems that start from *similar* initial conditions. They differ in how time is considered.

The notion of adaptability imposes some constraints on the *long term* behaviour of systems, disregarding their possible initial dissimilarities. Given the thresholds $\eta_1, \eta_2 \in [0, 1)$ and an observable time $\tilde{\tau}$, we say that a program P is adaptable w.r.t. a data state $\mathbf{d}$ and an environment evolution $\mathcal{E}$ if whenever P starts its computation from a data state $\mathbf{d}'$ that differs from $\mathbf{d}$ for at most η_1, then we are guaranteed that the distance between the evolution sequences of the two systems after time $\tilde{\tau}$ is bounded by η_2.

Definition 12 (Adaptability). *Let $\tilde{\tau} \in \mathrm{OT}$ and $\eta_1, \eta_2 \in [0, 1)$. We say that P is $(\tilde{\tau}, \eta_1, \eta_2)$-adaptable w.r.t. the data state $\mathbf{d}$ and the environment evolution $\mathcal{E}$ if $\forall \mathbf{d}' \in \mathcal{D}$ with $m_{\rho,0}^{\mathcal{D}}(\mathbf{d}, \mathbf{d}') \leq \eta_1$ it holds $\mathfrak{m}_{\{\tau \in \mathrm{OT} | \tau \geq \tilde{\tau}\}}^{\lambda}(\langle P, \mathbf{d}\rangle_{\mathcal{E}}, \langle P, \mathbf{d}'\rangle_{\mathcal{E}}) \leq \eta_2$.*

We remark that one can always consider the data state $\mathbf{d}$ as the ideal model of the world used for the specification of P, and the data state $\mathbf{d}'$ as the real world in which P has to execute. Hence, the idea behind adaptability is that even if the initial behaviour of the two systems is quite different, P is able to reduce the gap between the real evolution and the desired one within the time threshold $\tilde{\tau}$. Notice that being $(\tilde{\tau}, \eta_1, \eta_2)$-adaptable for $\tilde{\tau} = \min\{\tau \mid \tau \in \mathrm{OT}\}$ is equivalent to being (τ, η_1, η_2)-adaptable for all $\tau \in \mathrm{OT}$.

The notion of reliability strengthens that of adaptability by bounding the distance on the evolution sequences from the beginning. A program is reliable if it guarantees that small variations in the initial conditions cause only bounded variations in its evolution.

Definition 13 (Reliability). *Let $\eta_1, \eta_2 \in [0, 1)$. We say that P is (η_1, η_2)-reliable w.r.t. the data state $\mathbf{d}$ and the environment evolution $\mathcal{E}$ if $\forall \mathbf{d}' \in \mathcal{D}$ with $m_{\rho,0}^{\mathcal{D}}(\mathbf{d}, \mathbf{d}') \leq \eta_1$ it holds $\mathfrak{m}_{\mathrm{OT}}^{\lambda}(\langle P, \mathbf{d}\rangle_{\mathcal{E}}, \langle P, \mathbf{d}'\rangle_{\mathcal{E}}) \leq \eta_2$.*

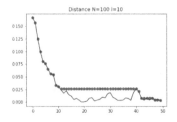 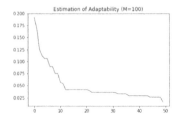

(a) In blue: pointwise distance between c_1 and c_2. In red: $m^\lambda_{\rho,\{\tau' \geq \tau\}}(c_1, c_2)$, for each τ (Ex. 7).

(b) Adaptability of P in c_1 (from Ex. 6) for $M = 100$, $\eta_1 = 0.2$ (Ex. 8).

Fig. 3. Examples of the evaluation of the evolution metric (assuming λ being the constant 1).

We can use our algorithm to verify adaptability and reliability of a given program. Given a configuration $\langle P, \mathbf{d} \rangle_\mathcal{E}$, a set OT of observation times and a given threshold $\eta_1 \geq 0$, we can sample M variations $\{\mathbf{d}_1, \ldots, \mathbf{d}_M\}$ of $\mathbf{d}$, s.t. for any i, $m^\mathcal{D}_{\rho,0}(\mathbf{d}, \mathbf{d}_i) \leq \eta_1$. Then, for each sampled data state we can estimate the distance between $c = \langle P, \mathbf{d} \rangle_\mathcal{E}$ and $c_i = \langle P, \mathbf{d}_i \rangle_\mathcal{E}$ at the different time steps in OT, namely $m^\lambda_{\{\tau \in OT | \tau \geq \tilde{\tau}\}}(c, c_i)$ for any $\tilde{\tau} \in OT$. Finally, for each $\tilde{\tau} \in OT$, we let $l_{\tilde{\tau}} = \max_i \{m^\lambda_{\{\tau \in OT | \tau \geq \tilde{\tau}\}}(c, c_i)\}$. We can observe that, for the chosen η_1, each $l_{\tilde{\tau}}$ gives us a lower bound to the $\tilde{\tau}$-adaptability of the program. Similarly, for $\tau_{\min} = \min_{OT} \tau$, $l_{\tau_{\min}}$ gives a lower bound for its reliability.

Example 8. Figure 3b shows the evaluation of l_τ for the program P in the configuration c_1 from Example 6 with parameters $M = 100$ and $\eta_1 = 0.2$. Observe that the initial perturbation is not amplified and after 12 steps it is almost absorbed. In particular, our program is $(12, 0.2, \eta_2)$-adaptable w.r.t. the data state and the environment evolution in Example 6, for any $\eta_2 \geq 0.05 + e^{12}_{\mathbf{W}}$, where $e^{12}_{\mathbf{W}}$ is the approximation error $e^{12}_{\mathbf{W}} = |\mathbf{W}(m^\mathcal{D}_{\rho,12})(\hat{\mathcal{S}}^{\mathcal{D},1000}_{c_1,12}, \hat{\mathcal{S}}^{\mathcal{D},10000}_{c_2,12}) - \mathbf{W}(m^\mathcal{D}_{\rho,12})(\mathcal{S}^\mathcal{D}_{c_1,12}, \mathcal{S}^\mathcal{D}_{c_2,12})|$. We refer the interested reader to [36, Corollary 3.5, Equation (3.10)] for an estimation of $e^{12}_{\mathbf{W}}$.

7   Concluding Remarks

As a first step for future research we will provide a simple logic, defined in the vein of *Signal Temporal Logic* (STL) [30], that can be used to specify requirements on the evolution sequences of a system. Our intuition is that we can exploit the evolution metric, and the algorithm we have proposed, to develop a quantitative model checking tool for this type of systems. Moreover, we would like to enhance the modelling of time. Firstly we could relax the timing constraints on the evolution metric by introducing a *time tolerance* and defining a *stretched evolution metric* as a Skorokhod-like metric [35], as those used for conformance testing [18]. Then, we could provide an extension of our techniques to the case in which also the program shows a continuous time behaviour.

The use of metrics for the analysis of systems stems from [17,22,29] where, in a process algebraic setting, it is argued that metrics are indeed more informative than behavioural equivalences when quantitative information on the behaviour is taken into account. The Wasserstein lifting has then found several successful applications: from the definition of *behavioural metrics* (e.g., [7,12,20]), to privacy [9,10,15] and machine learning (e.g., [4,25,38]). Usually, one can use behavioural metrics to quantify how well an implementation (I) meets its specification (S). In [14] the authors do so by setting a two players game with weighted choices, and the cost of the game is interpreted as the distance between I and S. Hence the authors propose three distance functions: *correctness*, *coverage*, and *robustness*. Correctness expresses how often I violates S, coverage is its dual, and robustness measures how often I can make an unexpected error with the resulting behaviour still meeting S. A similar game-based approach is used in [13] to define a *masking fault-tolerant* distance. Briefly, a system is masking fault-tolerant if faults do not induce any observable behaviour in the system. Hence, the proposed distance measures how many faults are tolerated by I while being masked by the states of the system. Notice that the notion of robustness from [14] and the masking fault-tolerant distance from [13] are quite different from our reliability. In fact, we are not interested in counting how many times an error occurs, but in checking whether the system is able to regain the desired behaviour after the occurrence of an error.

Systems showing an highly dynamic behaviour are usually modelled as Stochastic Hybrid Systems (SHSs) (see, e.g., [6,8]), which allow for combining in a single model the discrete, continuous and probabilistic features of systems. Our model clearly belongs to a subclass of SHSs. However, as previously outlined, our approach differs from that of SHSs since we model the program, the environment and their interaction (the data state) as three distinct components. This choice allows us to study the behaviour of the program by means of the evolution metric. It would be interesting to investigate if, and how, our method can be extended to the general class of SHSs.

We remark here that our objective in this paper was to provide some tools for the *analysis* of the interaction of a *given* program with the environment, and not for the *synthesis* of a program. However, for programs that are *controllers*, some metric-based approaches have been proposed for their synthesis [3,21]. We will then study whether our approach can be combined with some learning techniques in order to design and synthesise robust controllers.

References

1. Abate, A., D'Innocenzo, A., Benedetto, M.D.D.: Approximate abstractions of stochastic hybrid systems. IEEE Trans. Automat. Contr. **56**(11), 2688–2694 (2011)
2. Abate, A., Katoen, J., Lygeros, J., Prandini, M.: Approximate model checking of stochastic hybrid systems. Eur. J. Control. **16**(6), 624–641 (2010)
3. Abate, A., Prandini, M.: Approximate abstractions of stochastic systems: a randomized method. In: Proceedings of CDC-ECC 2011, pp. 4861–4866 (2011)
4. Arjovsky, M., Chintala, S., Bottou, L.: Wasserstein generative adversarial networks. In: Proceedings of ICML 2017, pp. 214–223 (2017)

5. Bernardo, M., Nicola, R.D., Loreti, M.: A uniform framework for modeling non-deterministic, probabilistic, stochastic, or mixed processes and their behavioral equivalences. Inf. Comput. **225**, 29–82 (2013)
6. Bloom, H.A.P., Lygeros, J. (eds.): Stochastic Hybrid Systems: Theory and Safety Critical Applications. Lecture Notes in Control and Information Sciences, vol. 337. Springer, Heidelberg (2006). https://doi.org/10.1007/11587392
7. Breugel, F.: A behavioural pseudometric for metric labelled transition systems. In: Abadi, M., de Alfaro, L. (eds.) CONCUR 2005. LNCS, vol. 3653, pp. 141–155. Springer, Heidelberg (2005). https://doi.org/10.1007/11539452_14
8. Cassandras, C.G., Lygeros, J. (eds.): Stochastic Hybrid Systems. Control Engineering, vol. 24, 1st edn. CRC Press, Boca Raton (2007)
9. Castiglioni, V., Chatzikokolakis, K., Palamidessi, C.: A logical characterization of differential privacy via behavioral metrics. In: Bae, K., Ölveczky, P.C. (eds.) FACS 2018. LNCS, vol. 11222, pp. 75–96. Springer, Cham (2018). https://doi.org/10.1007/978-3-030-02146-7_4
10. Castiglioni, V., Chatzikokolakis, K., Palamidessi, C.: A logical characterization of differential privacy. Sci. Comput. Program. **188**, 102388 (2020)
11. Castiglioni, V., Loreti, M., Tini, S.: Measuring adaptability and reliability of large scale systems. In: Margaria, T., Steffen, B. (eds.) ISoLA 2020. LNCS, vol. 12477, pp. 380–396. Springer, Cham (2020). https://doi.org/10.1007/978-3-030-61470-6_23
12. Castiglioni, V., Loreti, M., Tini, S.: The metric linear-time branching-time spectrum on nondeterministic probabilistic processes. Theor. Comput. Sci. **813**, 20–69 (2020)
13. Castro, P.F., D'Argenio, P.R., Demasi, R., Putruele, L.: Measuring masking fault-tolerance. In: Vojnar, T., Zhang, L. (eds.) TACAS 2019, Part II. LNCS, vol. 11428, pp. 375–392. Springer, Cham (2019). https://doi.org/10.1007/978-3-030-17465-1_21
14. Cerný, P., Henzinger, T.A., Radhakrishna, A.: Simulation distances. Theor. Comput. Sci. **413**(1), 21–35 (2012)
15. Chatzikokolakis, K., Gebler, D., Palamidessi, C., Xu, L.: Generalized bisimulation metrics. In: Baldan, P., Gorla, D. (eds.) CONCUR 2014. LNCS, vol. 8704, pp. 32–46. Springer, Heidelberg (2014). https://doi.org/10.1007/978-3-662-44584-6_4
16. Ciocchetta, F., Hillston, J.: Bio-PEPA: an extension of the process algebra PEPA for biochemical networks. Electron. Notes Theor. Comput. Sci. **194**(3), 103–117 (2008)
17. Desharnais, J., Gupta, V., Jagadeesan, R., Panangaden, P.: Metrics for labelled Markov processes. Theor. Comput. Sci. **318**(3), 323–354 (2004)
18. Deshmukh, J.V., Majumdar, R., Prabhu, V.S.: Quantifying conformance using the skorokhod metric. Formal Methods Syst. Design **50**(2–3), 168–206 (2017)
19. Faugeras, O.P., Rüschendorf, L.: Risk excess measures induced by hemi-metrics. Probab. Uncertain. Quant. Risk **3**(1), 1–35 (2018). https://doi.org/10.1186/s41546-018-0032-0
20. Gebler, D., Larsen, K.G., Tini, S.: Compositional bisimulation metric reasoning with probabilistic process calculi. Log. Methods Comput. Sci. **12**(4) (2016)
21. Ghosh, S., Bansal, S., Sangiovanni-Vincentelli, A.L., Seshia, S.A., Tomlin, C.: A new simulation metric to determine safe environments and controllers for systems with unknown dynamics. In: Proceedings of HSCC 2019, pp. 185–196 (2019)
22. Giacalone, A., Jou, C.C., Smolka, S.A.: Algebraic reasoning for probabilistic concurrent systems. In: Proceedings of IFIP Work, Conference on Programming, Concepts and Methods, pp. 443–458 (1990)

23. Girard, A., Gößler, G., Mouelhi, S.: Safety controller synthesis for incrementally stable switched systems using multiscale symbolic models. IEEE Trans. Automat. Contr. **61**(6), 1537–1549 (2016)
24. van Glabbeek, R.J., Smolka, S.A., Steffen, B.: Reactive, generative and stratified models of probabilistic processes. Inf. Comput. **121**(1), 59–80 (1995)
25. Gulrajani, I., Ahmed, F., Arjovsky, M., Dumoulin, V., Courville, A.C.: Improved training of Wasserstein GANs. In: Proceedings of Advances in Neural Information Processing Systems, pp. 5767–5777 (2017)
26. Heredia, G., et al.: Control of a multirotor outdoor aerial manipulator. In: Proceedings of IROS 2014, pp. 3417–3422. IEEE (2014)
27. Hillston, J., Hermanns, H., Herzog, U., Mertsiotakis, V., Rettelbach, M.: Stochastic process algebras: integrating qualitative and quantitative modelling. In: Proceedings of International Conference on Formal Description Techniques 1994. IFIP, vol. 6, pp. 449–451 (1994)
28. Hu, J., Lygeros, J., Sastry, S.: Towards a theory of stochastic hybrid systems. In: Lynch, N., Krogh, B.H. (eds.) HSCC 2000. LNCS, vol. 1790, pp. 160–173. Springer, Heidelberg (2000). https://doi.org/10.1007/3-540-46430-1_16
29. Kwiatkowska, M., Norman, G.: Probabilistic metric semantics for a simple language with recursion. In: Penczek, W., Szałas, A. (eds.) MFCS 1996. LNCS, vol. 1113, pp. 419–430. Springer, Heidelberg (1996). https://doi.org/10.1007/3-540-61550-4_167
30. Maler, O., Nickovic, D.: Monitoring temporal properties of continuous signals. In: Lakhnech, Y., Yovine, S. (eds.) FORMATS/FTRTFT -2004. LNCS, vol. 3253, pp. 152–166. Springer, Heidelberg (2004). https://doi.org/10.1007/978-3-540-30206-3_12
31. Malhame, R., Yee Chong, C.: Electric load model synthesis by diffusion approximation of a high-order hybrid-state stochastic system. IEEE Trans. Automat. Contr. **30**(9), 854–660 (1985)
32. Puterman, M.L.: Markov Decision Processes: Discrete Stochastic Dynamic Programming. Wiley Series in Probability and Statistics. Wiley, USA (2005)
33. Rachev, S.T., Klebanov, L.B., Stoyanov, S.V., Fabozzi, F.J.: The Methods of Distances in the Theory of Probability and Statistics. Springer, Heidelberg (2013). https://doi.org/10.1007/978-1-4614-4869-3
34. Segala, R.: Modeling and verification of randomized distributed real-time systems. Ph.D. thesis, Massachusetts Institute of Technology, Cambridge, MA, USA (1995)
35. Skorokhod, A.V.: Limit theorems for stochastic processes. Theory Probab. Appl. **1**, 261–290 (1956)
36. Sriperumbudur, B.K., Fukumizu, K., Gretton, A., Schölkopf, B., Lanckriet, G.R.G.: On the empirical estimation of integral probability metrics. Electron. J. Stat. **6**, 1550–1599 (2021)
37. Thorsley, D., Klavins, E.: Approximating stochastic biochemical processes with Wasserstein pseudometrics. IET Syst. Biol. **4**(3), 193–211 (2010)
38. Tolstikhin, I.O., Bousquet, O., Gelly, S., Schölkopf, B.: Wasserstein auto-encoders. In: Proceedings of ICLR 2018 (2018)
39. Vaserstein, L.N.: Markovian processes on countable space product describing large systems of automata. Probl. Peredachi Inf. **5**(3), 64–72 (1969)
40. Villani, C.: Optimal Transport: Old and New, vol. 338. Springer, Heidelberg (2008). https://doi.org/10.1007/978-3-540-71050-9
41. Virágh, C., Nagy, M., Gershenson, C., Vásárhelyi, G.: Self-organized UAV traffic in realistic environments. In: Proceedings of IROS 2016, pp. 1645–1652. IEEE (2016)

Branching Place Bisimilarity: A Decidable Behavioral Equivalence for Finite Petri Nets with Silent Moves

Roberto Gorrieri[✉]

Dipartimento di Informatica—Scienza e Ingegneria, Università di Bologna,
Mura A. Zamboni 7, 40127 Bologna, Italy
roberto.gorrieri@unibo.it

Abstract. Place bisimilarity $\sim_p$ is a behavioral equivalence for finite
Petri nets, proposed in [1] and proved decidable in [13]. In this paper
we propose an extension to finite Petri nets with silent moves of the
place bisimulation idea, yielding *branching* place bisimilarity $\approx_p$, following the intuition of branching bisimilarity [6] on labeled transition systems. We prove that $\approx_p$ is a decidbale equivalence relation. Moreover, we
argue that it is strictly finer than branching fully-concurrent bisimilarity
[12,22], essentially because $\approx_p$ does not consider as unobservable those
τ-labeled net transitions with pre-set size larger than one, i.e., those
resulting from multi-party interaction.

1 Introduction

Place bisimilarity, originating from an idea by Olderog [19] (under the name of
strong bisimilarity) and then refined by Autant, Belmesk and Schnoebelen [1],
is a behavioral equivalence over finite Place/Transition Petri nets (P/T nets,
for short), based on relations over the *finite set of net places*, rather than over
the (possibly infinite) set of net markings. This equivalence does respect the
expected causal behavior of Petri nets; in fact, van Glabbeek proved in [7] that
place bisimilarity is slightly finer than *structure preserving bisimilarity* [7], in
turn slightly finer than *fully-concurrent bisimilarity* [3]. Place bisimilarity was
proved decidable in [13] and, to date, it is the only sensible behavioral equivalence
which was proved decidable over finite Petri nets (with the exception of net
isomorphism).

This paper aims at extending the place bisimulation idea to Petri nets with
silent transitions. To this aim, we take inspiration from *branching* bisimilarity,
proposed in [6] over labeled transition systems [8,16] (LTSs, for short), a behavioral relation more appropriate than weak bisimilarity [17], as it better respects
the timing of choices.

The main problem we had to face was to properly understand if and when
a silent net transition can be really considered as potentially unobservable. In
fact, while in the theory of sequential, nondeterministic systems, modeled by
means of LTSs, all the τ-labeled transitions can, to some extent, be abstracted

© IFIP International Federation for Information Processing 2021
Published by Springer Nature Switzerland AG 2021
K. Peters and T. A. C. Willemse (Eds.): FORTE 2021, LNCS 12719, pp. 80–99, 2021.
https://doi.org/10.1007/978-3-030-78089-0_5

away, in the theory of Petri nets (and of distributed systems, in general), it is rather questionable whether this is the case. For sure a silent net transition with pre-set and post-set of size 1 may be abstracted away, as it represents some internal computation, local to a single sequential component of the distributed system. However, a τ-labeled net transition with pre-set of size 2 or more, which models a multi-party interaction, is really observable: since to establish the synchronization it is necessary to use some communication infrastructure, for sure one observer can see that such a synchronization takes place. This is, indeed, what happens over the Internet: a communication via IP is an observable event, even if the actual content of the message may be unobservable (in case it is encrypted).

For this reason, our definition of branching place bisimulation considers as potentially unobservable only the so-called τ-*sequential* transitions, i.e., those silent transitions whose pre-set and post-set have size 1. We prove that branching place bisimilarity $\approx_p$ is an equivalence relation, where the crucial step in this proof is to prove that the relational composition of two branching place bisimulations is a branching place bisimulation. Of course, $\approx_p$ is rather discriminating if compared to other behavioral semantics; in particular, we conjecture that it is strictly finer than branching fully-concurrent bisimilarity [12,22], essentially because the latter may also abstract w.r.t. silent transitions that are not τ-sequential (and also may relate markings of different size).

The main contribution of this paper is to show that $\approx_p$ is decidable for finite P/T nets. The proof idea is as follows. As a place relation $R \subseteq S \times S$ is finite if the set S of places is finite, there are finitely many place relations for a finite net. We can list all these relations, say $R_1, R_2, \ldots R_n$. It is decidable whether a place relation R_i is a branching place bisimulation by checking two *finite* conditions over a *finite* number of marking pairs: this is a non-obvious observation, as a branching place bisimulation requires that the place bisimulation game holds for the infinitely many pairs m_1 and m_2 which are *bijectively* related via R_i (denoted by $(m_1, m_2) \in R_i^\oplus$). Hence, to decide whether $m_1 \approx_p m_2$, it is enough to check, for $i = 1, \ldots n$, whether R_i is a branching place bisimulation and, in such a case, whether $(m_1, m_2) \in R_i^\oplus$.

The paper is organized as follows. Section 2 recalls the basic definitions about Petri nets. Section 3 recalls the main definitions and results about place bisimilarity. Section 4 introduces branching place bisimilarity and proves that it is an equivalence relation. Section 5 shows that $\approx_p$ is decidable. Finally, in Sect. 6 we discuss the pros and cons of branching place bisimilarity, and describe related literature and some future research.

2   Basic Definitions

Definition 1 (Multiset). *Let $\mathbb{N}$ be the set of natural numbers. Given a finite set S, a multiset over S is a function $m : S \to \mathbb{N}$. The* support set *$dom(m)$ of m is $\{s \in S \mid m(s) \neq 0\}$. The set of all multisets over S, denoted by $\mathcal{M}(S)$, is ranged over by m. We write $s \in m$ if $m(s) > 0$. The* multiplicity *of s in m is given by the number $m(s)$. The* size *of m, denoted by $|m|$, is the number $\sum_{s \in S} m(s)$, i.e., the total number of its elements. A multiset m such that $dom(m) = \emptyset$ is called* empty *and is denoted by θ. We write $m \subseteq m'$ if $m(s) \leq m'(s)$ for all $s \in S$. Multiset union $\_ \oplus \_$ is defined as follows: $(m \oplus m')(s) = m(s) + m'(s)$. Multiset difference $\_ \ominus \_$ is defined as follows: $(m_1 \ominus m_2)(s) = max\{m_1(s) - m_2(s), 0\}$. The* scalar product *of a number j with m is the multiset $j \cdot m$ defined as $(j \cdot m)(s) = j \cdot (m(s))$. By s_i we also denote the multiset with s_i as its only element. Hence, a multiset m over $S = \{s_1, \ldots, s_n\}$ can be represented as $k_1 \cdot s_1 \oplus k_2 \cdot s_2 \oplus \ldots \oplus k_n \cdot s_n$, where $k_j = m(s_j) \geq 0$ for $j = 1, \ldots, n$.* ☐

Definition 2 (Place/Transition net). *A* labeled *Place/Transition Petri net (P/T net for short) is a tuple $N = (S, A, T)$, where*

- *S is the finite set of* places, *ranged over by s (possibly indexed),*
- *A is the finite set of* labels, *ranged over by ℓ (possibly indexed), and*
- *$T \subseteq (\mathscr{M}(S) \setminus \{\theta\}) \times A \times \mathscr{M}(S)$ is the finite set of* transitions, *ranged over by t (possibly indexed).*

Given a transition $t = (m, \ell, m')$, we use the notation:

- *•t to denote its* pre-set *m (which cannot be empty) of tokens to be consumed;*
- *$l(t)$ for its* label *ℓ, and*
- *$t^\bullet$ to denote its* post-set *m' of tokens to be produced.*

Hence, transition t can be also represented as $^\bullet t \xrightarrow{l(t)} t^\bullet$. ☐

Graphically, a place is represented by a little circle and a transition by a little box. These are connected by directed arcs, which may be labeled by a positive integer, called the *weight*, to denote the number of tokens consumed (when the arc goes from a place to the transition) or produced (when the arc goes form the transition to a place) by the execution of the transition; if the number is omitted, then the weight default value is 1.

Definition 3 (Marking, P/T net system). *A multiset over S is called a* marking. *Given a marking m and a place s, we say that the place s contains $m(s)$ tokens, graphically represented by $m(s)$ bullets inside place s. A P/T net system $N(m_0)$ is a tuple (S, A, T, m_0), where (S, A, T) is a P/T net and m_0 is a marking over S, called the* initial marking. *We also say that $N(m_0)$ is a* marked net. ☐

Definition 4 (Enabling, firing sequence, transition sequence, reachable marking). *Given a P/T net $N = (S, A, T)$, a transition t is* enabled *at m, denoted by $m[t\rangle$, if $^\bullet t \subseteq m$. The execution (or firing) of t enabled at m produces the marking $m' = (m \ominus {}^\bullet t) \oplus t^\bullet$. This is written $m[t\rangle m'$. A firing sequence starting at m is defined inductively as follows:*

- *$m[\epsilon\rangle m$ is a firing sequence (where ϵ denotes an empty sequence of transitions) and*

- *if $m[\sigma\rangle m'$ is a firing sequence and $m'[t\rangle m''$, then $m[\sigma t\rangle m''$ is a firing sequence.*

If $\sigma = t_1 \ldots t_n$ (for $n \geq 0$) and $m[\sigma\rangle m'$ is a firing sequence, then there exist $m_1, \ldots, m_{n+1}$ such that $m = m_1[t_1\rangle m_2[t_2\rangle \ldots m_n[t_n\rangle m_{n+1} = m'$, and $\sigma = t_1 \ldots t_n$ is called a transition sequence *starting at m and ending at m'.*

The definition of pre-set and post-set can be extended to transition sequences as follows: $^\bullet\epsilon = \theta$, $^\bullet(t\sigma) = {}^\bullet t \oplus ({}^\bullet\sigma \ominus t^\bullet)$, $\epsilon^\bullet = \theta$, $(t\sigma)^\bullet = \sigma^\bullet \oplus (t^\bullet \ominus {}^\bullet\sigma)$.

The set of reachable markings *from m is $[m\rangle = \{m' \mid \exists\sigma.m[\sigma\rangle m'\}$.* Note that the reachable markings can be countably infinitely many. □

Definition 5 (P/T net with silent moves, τ-sequential). *A P/T net $N = (S, A, T)$ such that $\tau \in A$, where τ is the only invisible action that can be used to label transitions, is called a P/T net.*

A transition $t \in T$ is τ-sequential *if $l(t) = \tau$ and $|t^\bullet| = 1 = |{}^\bullet t|$. A P/T net N is τ-sequential if $\forall t \in T$ if $l(t) = \tau$, then t is τ-sequential.* □

Definition 6 (Idling transitions, τ-sequential (acyclic) transition sequence). *Given a P/T net $N = (S, A, T)$ with silent moves, the set of idling transitions is $I(S) = \{i(s) \mid s \in S, i(s) = (s, \tau, s)\}$. In defining silent transition sequences, we take the liberty of using also the fictitious idling transitions, so that, e.g., if $\sigma = i(s_1)i(s_2)$, then $s_1 \oplus s_2[\sigma\rangle s_1 \oplus s_2$. Given a transition sequence σ, its observable label $o(\sigma)$ is computed inductively as:*

$$o(\epsilon) = \epsilon$$

$$o(t\sigma) = \begin{cases} l(t)o(\sigma) & \text{if } l(t) \neq \tau \\ o(\sigma) & \text{otherwise.} \end{cases}$$

A transition sequence $\sigma = t_1 t_2 \ldots t_n$ (where $n \geq 1$ and some of the t_i can be idling transitions) is τ-1-sequential *if $l(t_i) = \tau$, $|t_i^\bullet| = 1 = |{}^\bullet t_i|$ for $i = 1, \ldots, n$, and $t_i^\bullet = {}^\bullet t_{i+1}$ for $i = 1, \ldots, n-1$, so that $o(\sigma) = \epsilon$ and $|\sigma^\bullet| = 1 = |{}^\bullet\sigma|$.*

A transition sequence $\sigma = \sigma_1\sigma_2 \ldots \sigma_k$ is τ-k-sequential *if σ_i is τ-1-sequential for $i = 1, \ldots, k$, $^\bullet\sigma = {}^\bullet\sigma_1 \oplus {}^\bullet\sigma_2 \oplus \ldots \oplus {}^\bullet\sigma_k$ and $\sigma^\bullet = \sigma_1^\bullet \oplus \sigma_2^\bullet \oplus \ldots \oplus \sigma_k^\bullet$, so that $o(\sigma) = \epsilon$ and $|\sigma^\bullet| = k = |{}^\bullet\sigma|$. We say that σ is τ-sequential if it is τ-k-sequential for some $k \geq 1$.*

A τ-1-sequential $\sigma = t_1 t_2 \ldots t_n$ is acyclic *if $^\bullet\sigma = m_0[t_1\rangle m_1[t_2\rangle m_2 \ldots m_{n-1}[t_n\rangle m_n = \sigma^\bullet$ and $m_i \neq m_j$ for all $i \neq j$, with $i, j \in \{1, 2, \ldots, n\}$. A τ-k-sequential $\sigma = \sigma_1\sigma_2 \ldots \sigma_k$ is acyclic if σ_i is acyclic and τ-1-sequential for $i = 1, \ldots, k$. We say that σ is an acyclic τ-sequential transition sequence if it is acyclic and τ-k-sequential for some $k \geq 1$.* □

Remark 1 (**Acyclic τ-sequential transition sequence**). The definition of acyclic τ-1-sequential transition sequence is a bit non-standard as it may allow for a cycle when the initial marking and the final one are the same. For instance, $\sigma = i(s)i(s)$ is cyclic, while the apparently cyclic subsequence $\sigma' = i(s)$ is actually acyclic, according to our definition. Note that, given a τ-1-sequential transition sequence σ, it is always possible to find an acyclic τ-1-sequential transition sequence σ' such that $^\bullet\sigma = {}^\bullet\sigma'$ and $\sigma^\bullet = \sigma'^\bullet$. For instance, if $^\bullet\sigma = m_0[t_1\rangle m_1[t_2\rangle m_2 \ldots m_{n-1}[t_n\rangle m_n = \sigma^\bullet$ and the only cycle is given by $m_i[t_{i+1}\rangle m_{i+1} \ldots m_{j-1}[t_j\rangle m_j$ with $m_i = m_j$ and $i \geq 1$, then $\sigma' = t_1 t_2 \ldots t_i t_{j+1} \ldots t_n$ is acyclic and $^\bullet\sigma = {}^\bullet\sigma'$ and $\sigma^\bullet = \sigma'^\bullet$.

Note also that, given a τ-k-sequential transition sequence $\sigma = \sigma_1 \sigma_2 \ldots \sigma_k$, it is always possible to find an acyclic τ-k-sequential transition sequence $\sigma' = \sigma'_1 \sigma'_2 \ldots \sigma'_k$, where σ'_i is the acyclic τ-1-sequential transition sequence corresponding to σ_i for $i = 1, 2, \ldots, k$, in such a way that $^\bullet\sigma = {}^\bullet\sigma'$ and $\sigma^\bullet = \sigma'^\bullet$. Finally, note that, given two markings m_1 and m_2 of equal size k, it is decidable whether there exists an acyclic τ-k-sequential transition σ such that $^\bullet\sigma = m_1$ and $\sigma^\bullet = m_2$. $\qquad\square$

Definition 7 (Interleaving Bisimulation). *Let $N = (S, A, T)$ be a P/T net. An* interleaving bisimulation *is a relation $R \subseteq \mathcal{M}(S) \times \mathcal{M}(S)$ such that if $(m_1, m_2) \in R$ then*

- *$\forall t_1$ such that $m_1[t_1\rangle m'_1$, $\exists t_2$ such that $m_2[t_2\rangle m'_2$ with $l(t_1) = l(t_2)$ and $(m'_1, m'_2) \in R$,*
- *$\forall t_2$ such that $m_2[t_2\rangle m'_2$, $\exists t_1$ such that $m_1[t_1\rangle m'_1$ with $l(t_1) = l(t_2)$ and $(m'_1, m'_2) \in R$.*

Two markings m_1 and m_2 are interleaving bisimilar, *denoted by $m_1 \sim_{int} m_2$, if there exists an interleaving bisimulation R such that $(m_1, m_2) \in R$.* $\qquad\square$

Interleaving bisimilarity was proved undecidable in [15] for P/T nets having at least two unbounded places, with a proof based on the comparison of two *sequential* P/T nets (i.e., nets not offering any concurrent behavior). Hence, interleaving bisimulation equivalence is undecidable even for the subclass of sequential finite P/T nets. Esparza observed in [5] that all the non-interleaving bisimulation-based equivalences (in the spectrum ranging from interleaving bisimilarity to fully-concurrent bisimilarity [3]) collapse to interleaving bisimilarity over sequential P/T nets. Hence, the proof in [15] applies to all these non-interleaving bisimulation equivalences as well.

Definition 8 (Branching interleaving bisimulation). *Let $N = (S, A, T)$ be a P/T net with silent moves. A* branching *interleaving bisimulation is a relation $R \subseteq \mathcal{M}(S) \times \mathcal{M}(S)$ such that if $(m_1, m_2) \in R$ then*

- *$\forall t_1$ such that $m_1[t_1\rangle m'_1$,*
    - *either $l(t_1) = \tau$ and $\exists \sigma_2$ such that $o(\sigma_2) = \epsilon$, $m_2[\sigma_2\rangle m'_2$ with $(m_1, m'_2) \in R$ and $(m'_1, m'_2) \in R$,*
    - *or $\exists \sigma, t_2$ such that $o(\sigma) = \epsilon$, $l(t_1) = l(t_2)$, $m_2[\sigma\rangle m[t_2\rangle m'_2$ with $(m_1, m) \in R$ and $(m'_1, m'_2) \in R$,*
- *and, symmetrically, $\forall t_2$ such that $m_2[t_2\rangle m'_2$.*

Two markings m_1 and m_2 are branching interleaving bisimilar, *denoted $m_1 \approx_{bri} m_2$, if there exists a branching interleaving bisimulation R that relates them.* $\qquad\square$

This definition is not a rephrasing on nets of the original definition on LTSs in [6], rather of a slight variant called *semi-branching bisimulation* [2,6], which gives rise to the same equivalence relation as the original definition but has better mathematical properties. Branching interleaving bisimilarity $\approx_{bri}$ is the largest branching interleaving bisimulation and also an equivalence relation. Of course, also branching interleaving bisimilarity is undecidable for finite P/T nets.

3   Place Bisimilarity

We now present place bisimulation, introduced in [1] as an improvement of *strong bisimulation*, a behavioral relation proposed by Olderog in [19] on safe nets which fails to induce an equivalence relation. Our definition is formulated in a slightly different way, but it is coherent with the original one. First, an auxiliary definition.

Definition 9 (Additive closure). *Given a P/T net $N = (S, A, T)$ and a place relation $R \subseteq S \times S$, we define a* marking relation $R^{\oplus} \subseteq \mathcal{M}(S) \times \mathcal{M}(S)$, *called the* additive closure *of R, as the least relation induced by the following axiom and rule.*

$$\frac{}{(\theta, \theta) \in R^{\oplus}} \qquad \frac{(s_1, s_2) \in R \quad (m_1, m_2) \in R^{\oplus}}{(s_1 \oplus m_1, s_2 \oplus m_2) \in R^{\oplus}}$$

□

Note that, by definition, two markings are related by $R^{\oplus}$ only if they have the same size; in fact, the axiom states that the empty marking is related to itself, while the rule, assuming by induction that m_1 and m_2 have the same size, ensures that $s_1 \oplus m_1$ and $s_2 \oplus m_2$ have the same size.

Proposition 1 *For each relation $R \subseteq S \times S$, if $(m_1, m_2) \in R^{\oplus}$, then $|m_1| = |m_2|$.*

□

Note also that there may be several proofs of $(m_1, m_2) \in R^{\oplus}$, depending on the chosen order of the elements of the two markings and on the definition of R. For instance, if $R = \{(s_1, s_3), (s_1, s_4), (s_2, s_3), (s_2, s_4)\}$, then $(s_1 \oplus s_2, s_3 \oplus s_4) \in R^{\oplus}$ can be proved by means of the pairs (s_1, s_3) and (s_2, s_4), as well as by means of $(s_1, s_4), (s_2, s_3)$. An alternative way to define that two markings m_1 and m_2 are related by $R^{\oplus}$ is to state that m_1 can be represented as $s_1 \oplus s_2 \oplus \ldots \oplus s_k$, m_2 can be represented as $s_1' \oplus s_2' \oplus \ldots \oplus s_k'$ and $(s_i, s_i') \in R$ for $i = 1, \ldots, k$. In fact, a naive algorithm for checking whether $(m_1, m_2) \in R^{\oplus}$ would simply consider m_1 represented as $s_1 \oplus s_2 \oplus \ldots \oplus s_k$ and then scan all the possible permutations of m_2, each represented as $s_1' \oplus s_2' \oplus \ldots \oplus s_k'$, to check that $(s_i, s_i') \in R$ for $i = 1, \ldots, k$. Of course, this naive algorithm is in $O(k!)$.

Example 1. Consider $R = \{(s_1, s_3), (s_1, s_4), (s_2, s_4)\}$, which is not an equivalence relation. Suppose we want to check that $(s_1 \oplus s_2, s_4 \oplus s_3) \in R^{\oplus}$. If we start by matching $(s_1, s_4) \in R$, then we fail because the residual (s_2, s_3) is not in R. However, if we permute the second marking to $s_3 \oplus s_4$, then we succeed because the required pairs (s_1, s_3) and (s_2, s_4) are both in R.

□

Nonetheless, the problem of checking whether $(m_1, m_2) \in R^{\oplus}$ has polynomial time complexity because it can be considered as an instance of the problem of finding a perfect matching in a bipartite graph, where the nodes of the two partitions are the tokens in the two markings, and the edges are defined by the relation R. In fact, the definition of the bipartite graph takes $O(k^2)$ time (where $k = |m_1| = |m_2|$) and, then, the Hopcroft-Karp-Karzanov algorithm [14] for computing the maximum matching has worst-case time complexity $O(h\sqrt{k})$, where h is the number of the edges in the bipartire graph ($h \leq k^2$) and to check whether the maximum matching is perfect can be done simply by checking that the size of the matching equals the number of nodes in each partition, i.e., k. Hence, in evaluating the complexity of the algorithm in Sect. 5, we assume that the complexity of checking whether $(m_1, m_2) \in R^{\oplus}$ is in $O(k^2\sqrt{k})$.

Proposition 2 [10]. *For each place relation $R \subseteq S \times S$, the following hold:*

1. *If R is an equivalence relation, then $R^{\oplus}$ is an equivalence relation.*
2. *If $R_1 \subseteq R_2$, then $R_1^{\oplus} \subseteq R_2^{\oplus}$, i.e., the additive closure is monotone.*
3. *If $(m_1, m_2) \in R^{\oplus}$ and $(m_1', m_2') \in R^{\oplus}$, then $(m_1 \oplus m_1', m_2 \oplus m_2') \in R^{\oplus}$, i.e., the additive closure is additive.* □

Proposition 3 [10]. *For each family of place relations $R_i \subseteq S \times S$, the following hold:*

1. *$\emptyset^{\oplus} = \{(\theta, \theta)\}$, i.e., the additive closure of the empty place relation is a singleton marking relation, relating the empty marking to itself.*
2. *$(\mathscr{I}_S)^{\oplus} = \mathscr{I}_M$, i.e., the additive closure of the identity relation on places $\mathscr{I}_S = \{(s, s) \mid s \in S\}$ is the identity relation on markings $\mathscr{I}_M = \{(m, m) \mid m \in \mathscr{M}(S)\}$.*
3. *$(R^{\oplus})^{-1} = (R^{-1})^{\oplus}$, i.e., the inverse of an additively closed relation R is the additive closure of its inverse R^{-1}.*
4. *$(R_1 \circ R_2)^{\oplus} = (R_1^{\oplus}) \circ (R_2^{\oplus})$, i.e., the additive closure of the composition of two place relations is the compositions of their additive closures.* □

Definition 10 (Place Bisimulation). *Let $N = (S, A, T)$ be a P/T net. A place bisimulation is a relation $R \subseteq S \times S$ such that if $(m_1, m_2) \in R^{\oplus}$ then*

- *$\forall t_1$ such that $m_1[t_1\rangle m_1'$, $\exists t_2$ such that $m_2[t_2\rangle m_2'$ with $(^{\bullet}t_1, {}^{\bullet}t_2) \in R^{\oplus}$, $l(t_1) = l(t_2)$, $(t_1^{\bullet}, t_2^{\bullet}) \in R^{\oplus}$ and $(m_1', m_2') \in R^{\oplus}$,*
- *$\forall t_2$ such that $m_2[t_2\rangle m_2'$, $\exists t_1$ such that $m_1[t_1\rangle m_1'$ with $(^{\bullet}t_1, {}^{\bullet}t_2) \in R^{\oplus}$, $l(t_1) = l(t_2)$, $(t_1^{\bullet}, t_2^{\bullet}) \in R^{\oplus}$ and $(m_1', m_2') \in R^{\oplus}$.*

Two markings m_1 and m_2 are place bisimilar, *denoted by $m_1 \sim_p m_2$, if there exists a place bisimulation R such that $(m_1, m_2) \in R^{\oplus}$.* □

Proposition 4 [1,13]. *For each P/T net $N = (S, A, T)$, relation $\sim_p \subseteq \mathscr{M}(S) \times \mathscr{M}(S)$ is an equivalence relation.* □

By Definition 10, place bisimilarity can be defined as follows:

$\sim_p = \bigcup \{R^\oplus \mid R$ is a place bisimulation$\}$.

By monotonicity of the additive closure (Proposition 2(2)), if $R_1 \subseteq R_2$, then $R_1^\oplus \subseteq R_2^\oplus$. Hence, we can restrict our attention to maximal place bisimulations only:

$\sim_p = \bigcup \{R^\oplus \mid R$ is a *maximal* place bisimulation$\}$.

However, it is not true that

$\sim_p = (\bigcup \{R \mid R$ is a *maximal* place bisimulation$\})^\oplus$

because the union of place bisimulations may not be a place bisimulation, so that its definition is not coinductive. We illustrate this fact by means of the following example.

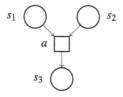

Fig. 1. A simple net

Example 2 Consider the simple P/T net in Fig. 1, with $S = \{s_1, s_2, s_3\}$. It is rather easy to realize that there are only two maximal place bisimulations, namely:

$R_1 = \mathscr{I}_S = \{(s_1, s_1), (s_2, s_2), (s_3, s_3)\}$ and

$R_2 = (R_1 \setminus \mathscr{I}_{\{s_1, s_2\}}) \cup \{(s_1, s_2), (s_2, s_1)\} = \{(s_1, s_2), (s_2, s_1), (s_3, s_3)\}$,

only one of which is an equivalence relation. However, note that their union $R = R_1 \cup R_2$ is not a place bisimulation. In fact, on the one hand $(s_1 \oplus s_1, s_1 \oplus s_2) \in R^\oplus$, but, on the other hand, these two markings do not satisfy the place bisimulation game, because $s_1 \oplus s_1$ is stuck, while $s_1 \oplus s_2$ can fire the a-labeled transition, reaching s_3.   □

4   Branching Place Bisimilarity

Now we define a variant of place bisimulation, which is insensitive, to some extent, to τ-sequential transitions, i.e., τ-labeled transitions whose pre-set and post-set have size one. This relation is inspired to (semi-)branching bisimulation [2,6], a behavioral relation defined over LTSs. In its definition, we use τ-*sequential transition sequences*, usually denoted by σ, which are sequences composed of τ-sequential transitions in $T \cup I(S)$, i.e., τ-sequential net transitions and also idling transitions.

Definition 11 (Branching place bisimulation). *Given a P/T net* $N = (S, A, T)$, *a* branching place bisimulation *is a relation* $R \subseteq S \times S$ *such that if* $(m_1, m_2) \in R^\oplus$

1. $\forall t_1$ such that $m_1[t_1\rangle m_1'$
    (i) either t_1 is τ-sequential and $\exists \sigma, m_2'$ such that σ is τ-sequential, $m_2[\sigma\rangle m_2'$, and $({}^\bullet t_1, {}^\bullet \sigma) \in R$, $({}^\bullet t_1, \sigma^\bullet) \in R$, $(t_1^\bullet, \sigma^\bullet) \in R$ and $(m_1 \ominus {}^\bullet t_1, m_2 \ominus {}^\bullet \sigma) \in R^\oplus$;
    (ii) or there exist σ, t_2, m, m_2' such that σ is τ-sequential, $m_2[\sigma\rangle m[t_2\rangle m_2'$, $l(t_1) = l(t_2)$, $\sigma^\bullet = {}^\bullet t_2$, $({}^\bullet t_1, {}^\bullet \sigma) \in R^\oplus$, $({}^\bullet t_1, {}^\bullet t_2) \in R^\oplus$ $(t_1^\bullet, t_2^\bullet) \in R^\oplus$, and moreover, $(m_1 \ominus {}^\bullet t_1, m_2 \ominus {}^\bullet \sigma) \in R^\oplus$;
2. and, symmetrically, $\forall t_2$ such that $m_2[t_2\rangle m_2'$
    (i) either t_2 is τ-sequential and $\exists \sigma, m_1'$ such that σ is τ-sequential, $m_1[\sigma\rangle m_1'$, and $({}^\bullet \sigma, {}^\bullet t_2) \in R$, $(\sigma^\bullet, {}^\bullet t_2) \in R$, $(\sigma^\bullet, t_2^\bullet) \in R$ and $(m_1 \ominus {}^\bullet \sigma, m_2 \ominus {}^\bullet t_2) \in R^\oplus$;
    (ii) or there exist σ, t_1, m, m_1' such that σ is τ-sequential, $m_1[\sigma\rangle m[t_1\rangle m_1'$, $l(t_1) = l(t_2)$, $\sigma^\bullet = {}^\bullet t_1$, $({}^\bullet \sigma, {}^\bullet t_2) \in R^\oplus$, $({}^\bullet t_1, {}^\bullet t_2) \in R^\oplus$ $(t_1^\bullet, t_2^\bullet) \in R^\oplus$, and moreover, $(m_1 \ominus {}^\bullet \sigma, m_2 \ominus {}^\bullet t_2) \in R^\oplus$.

Two markings m_1 and m_2 are branching place bisimulation equivalent, denoted by $m_1 \approx_p m_2$, if there exists a branching place bisimulation R such that $(m_1, m_2) \in R^\oplus$.   □

Note that, in the either case, by additivity of $R^\oplus$ (cf. Proposition 2(3)), from $(m_1 \ominus {}^\bullet t_1, m_2 \ominus {}^\bullet \sigma) \in R^\oplus$ and $({}^\bullet t_1, \sigma^\bullet) \in R$, we get $(m_1, m_2') \in R^\oplus$, as well as, from $(t_1^\bullet, \sigma^\bullet) \in R$ we get $(m_1', m_2') \in R^\oplus$. Similarly for the or case.

Proposition 5 *For each P/T net $N = (S, A, T)$, the following hold:*

(i) *The identity relation $\mathscr{I}_S$ is a branching place bisimulation.*
(ii) *The inverse relation R^{-1} of a branching place bisimulation R is a branching place bisimulation.*

Proof. Case (i) is obvious. For case (ii), assume $(m_2, m_1) \in (R^{-1})^\oplus$ and $m_2[t_2\rangle m_2'$. By Proposition 3(3), we have that $(m_2, m_1) \in (R^\oplus)^{-1}$ and so $(m_1, m_2) \in R^\oplus$. Since R is a branching place bisimulation, we have that

(i) either t_2 is τ-sequential and $\exists \sigma, m_1'$ such that σ is τ-sequential, $m_1[\sigma\rangle m_1'$, and $({}^\bullet \sigma, {}^\bullet t_2) \in R$, $(\sigma^\bullet, {}^\bullet t_2) \in R$, $(\sigma^\bullet, t_2^\bullet) \in R$ and, moreover, $(m_1 \ominus {}^\bullet \sigma, m_2 \ominus {}^\bullet t_2) \in R^\oplus$;
(ii) or there exist σ, t_1, m, m_1' such that σ is τ-sequential, $m_1[\sigma\rangle m[t_1\rangle m_1'$, $l(t_1) = l(t_2)$, $\sigma^\bullet = {}^\bullet t_1$, $({}^\bullet \sigma, {}^\bullet t_2) \in R^\oplus$, $({}^\bullet t_1, {}^\bullet t_2) \in R^\oplus$ $(t_1^\bullet, t_2^\bullet) \in R^\oplus$, and $(m_1 \ominus {}^\bullet \sigma, m_2 \ominus {}^\bullet t_2) \in R^\oplus$.

Summing up, if $(m_2, m_1) \in (R^{-1})^\oplus$ and $m_2[t_2\rangle m_2'$ (the case when m_1 moves first is symmetric, and so omitted), then

(i) either t_2 is τ-sequential and $\exists \sigma, m_1'$ such that σ is τ-sequential, $m_1[\sigma\rangle m_1'$, and $({}^\bullet t_2, {}^\bullet \sigma) \in R^{-1}$, $({}^\bullet t_2, \sigma^\bullet) \in R^{-1}$, $(t_2^\bullet, \sigma^\bullet) \in R^{-1}$ and $(m_2 \ominus {}^\bullet t_2, m_1 \ominus {}^\bullet \sigma) \in (R^{-1})^\oplus$;
(ii) or there exist σ, t_1, m, m_1' such that σ is τ-sequential, $m_1[\sigma\rangle m[t_1\rangle m_1'$, $l(t_1) = l(t_2)$, $\sigma^\bullet = {}^\bullet t_1$, $({}^\bullet t_2, {}^\bullet \sigma) \in (R^{-1})^\oplus$, $({}^\bullet t_2, {}^\bullet t_1) \in (R^{-1})^\oplus$ $(t_2^\bullet, t_1^\bullet) \in (R^{-1})^\oplus$, and, moreover, $(m_2 \ominus {}^\bullet t_2, m_1 \ominus {}^\bullet \sigma) \in (R^{-1})^\oplus$;

so that R^{-1} is a branching place bisimulation, indeed.   □

Much more challenging is to prove that the relation composition of two branching place bisimulations is a branching place bisimulation. We need a technical lemma first.

Lemma 1 *Let $N = (S, A, T)$ be a P/T net, and R be a place bisimulation.*

1. *For each τ-sequential transition sequence σ_1, for all m_2 such that $({}^\bullet\sigma_1, m_2) \in R^\oplus$, a τ-sequential transition sequence σ_2 exists such that $m_2 = {}^\bullet\sigma_2$ and $(\sigma_1^\bullet, \sigma_2^\bullet) \in R^\oplus$;*
2. *and symmetrically, for each τ-sequential transition sequence σ_2, for all m_1 such that $(m_1, {}^\bullet\sigma_2) \in R^\oplus$, a τ-sequential transition sequence σ_1 exists such that $m_1 = {}^\bullet\sigma_1$ and $(\sigma_1^\bullet, \sigma_2^\bullet) \in R^\oplus$.*

Proof. By symmetry, we prove only case 1, by induction on the length of σ_1.

Base case: $\sigma_1 = \epsilon$. In this trivial case, ${}^\bullet\sigma_1 = \theta$ and so the only possible m_2 is θ as well. We just take $\sigma_2 = \epsilon$ and all the required conditions are trivially satisfied.

Inductive case: $\sigma_1 = \delta_1 t_1$, where $t_1 \in T \cup I(S)$. Hence, by inductive hypothesis, for each m_2 such that $({}^\bullet\delta_1, m_2) \in R^\oplus$, we know that there exists a δ_2 such that $m_2 = {}^\bullet\delta_2$ and $(\delta_1^\bullet, \delta_2^\bullet) \in R^\oplus$.
If $t_1 = i(s)$, then we have to consider two subcases:

- *if $s \in \delta_1^\bullet$, then ${}^\bullet\delta_1 t_1 = {}^\bullet\delta_1$ and $\delta_1 t_1^\bullet = \delta_1^\bullet$. Hence, we can take $\sigma_2 = \delta_2$ and all the required conditions are trivially satisfied;*
- *if $s \notin \delta_1^\bullet$, then ${}^\bullet\delta_1 t_1 = {}^\bullet\delta_1 \oplus s$ and $\delta_1 t_1^\bullet = \delta_1^\bullet \oplus s$. Then, $\forall s'$ such that $(s, s') \in R$, we can take $\sigma_2 = \delta_2 i(s')$, so that $({}^\bullet\delta_1 t_1, {}^\bullet\delta_2 i(s')) \in R^\oplus$, $(\delta_1 t_1^\bullet, \delta_2 i(s')^\bullet) \in R^\oplus$, as required.*

Also if $t_1 \in T$, we have consider two subcases:

- *If $s_1 = {}^\bullet t_1 \in \delta_1^\bullet$, then, since $(\delta_1^\bullet, \delta_2^\bullet) \in R^\oplus$, there exists $s_2 \in \delta_2^\bullet$ such that $(s_1, s_2) \in R$ and $(\delta_1^\bullet \ominus s_1, \delta_2^\bullet \ominus s_2) \in R^\oplus$. Then, by Definition 11, it follows that to the move $s_1 \xrightarrow{\tau} s_1'$:*
  - *(i) Either there exist σ, s_2' such that σ is τ-sequential, $s_2[\sigma\rangle s_2'$, $(s_1, s_2) \in R$ and $(s_1', s_2') \in R$.*
    In this case, we take $\sigma_2 = \delta_2\sigma$, so that $({}^\bullet\delta_1 t_1, {}^\bullet\delta_2\sigma) \in R^\oplus$ (because ${}^\bullet\delta_1 t_1 = {}^\bullet\delta_1$ and ${}^\bullet\delta_2\sigma = {}^\bullet\delta_2$), and $(\delta_1 t_1^\bullet, \delta_2\sigma^\bullet) \in R^\oplus$ (because $\delta_1 t_1^\bullet = (\delta_1^\bullet \ominus s_1) \oplus s_1'$ and $\delta_2\sigma^\bullet = (\delta_2^\bullet \ominus s_2) \oplus s_2'$), as required.
  - *(ii) Or there exist $\sigma, t_2, \bar{s}, s_2'$ such that σt_2 is τ-sequential, $\sigma^\bullet = {}^\bullet t_2$, $s_2[\sigma\rangle\bar{s}[t_2\rangle s_2'$, $(s_1, \bar{s}) \in R$ and $(s_1', s_2') \in R$.*
    In this case, we take $\sigma_2 = \delta_2\sigma t_2$, so that $({}^\bullet\delta_1 t_1, {}^\bullet\delta_2\sigma t_2) \in R^\oplus$, and, moreover, $(\delta_1 t_1^\bullet, \delta_2\sigma t_2^\bullet) \in R^\oplus$, as required.
- *If $s_1 = {}^\bullet t_1 \notin \delta_1^\bullet$, then, for each s_2 such that $(s_1, s_2) \in R$, we follow the same step as above (by Definition 11), and so we omit this part of the proof.*

Proposition 6. *For each P/T net* $N = (S, A, T)$, *the relational composition* $R_1 \circ R_2$ *of two branching place bisimulations* R_1 *and* R_2 *is a branching place bisimulation.*

Proof. Assume $(m_1, m_3) \in (R_1 \circ R_2)^\oplus$ *and* $m_1[t_1\rangle m_1'$. *By Proposition 3(4), we have that* $(m_1, m_3) \in (R_1)^\oplus \circ (R_2)^\oplus$, *and so* m_2 *exists such that* $(m_1, m_2) \in R_1^\oplus$ *and* $(m_2, m_3) \in R_2^\oplus$.

As $(m_1, m_2) \in R_1^\oplus$ *and* R_1 *is a branching place bisimulation, if* $m_1[t_1\rangle m_1'$, *then*

(i) *either* t_1 *is* τ-*sequential and* $\exists \sigma, m_2'$ *such that* σ *is* τ-*sequential,* $m_2[\sigma\rangle m_2'$, *and* $(^\bullet t_1, ^\bullet \sigma) \in R_1$, $(^\bullet t_1, \sigma^\bullet) \in R_1$, $(t_1^\bullet, \sigma^\bullet) \in R_1$ *and* $(m_1 \ominus {}^\bullet t_1, m_2 \ominus {}^\bullet \sigma) \in R_1^\oplus$;

(ii) *or there exist* σ, t_2, m, m_2' *such that* σ *is* τ-*sequential,* $m_2[\sigma\rangle m[t_2\rangle m_2'$, $l(t_1) = l(t_2)$, $\sigma^\bullet = {}^\bullet t_2$, $(^\bullet t_1, ^\bullet \sigma) \in R_1^\oplus$, $(^\bullet t_1, ^\bullet t_2) \in R_1^\oplus$ $(t_1^\bullet, t_2^\bullet) \in R_1^\oplus$, *and moreover,* $(m_1 \ominus {}^\bullet t_1, m_2 \ominus {}^\bullet \sigma) \in R_1^\oplus$.

Let us consider case (i), *i.e., assume that to the move* $m_1[t_1\rangle m_1'$, m_2 *replies with* $m_2[\sigma\rangle m_2'$ *such that* $(^\bullet t_1, ^\bullet \sigma) \in R_1$, $(^\bullet t_1, \sigma^\bullet) \in R_1$, $(t_1^\bullet, \sigma^\bullet) \in R_1$ *and, moreover,* $(m_1 \ominus {}^\bullet t_1, m_2 \ominus {}^\bullet \sigma) \in R_1^\oplus$. *Since* $(m_2, m_3) \in R_2^\oplus$, *there exists a submarking* $\overline{m} \subseteq m_3$ *such that* $(^\bullet \sigma, \overline{m}) \in R_2^\oplus$ *and* $(m_2 \ominus {}^\bullet \sigma, m_3 \ominus \overline{m}) \in R_2^\oplus$. *By Lemma 1, there exists a* τ-*sequential transition sequence* σ' *such that* $\overline{m} = {}^\bullet \sigma'$ *and* $(\sigma^\bullet, \sigma'^\bullet) \in R_2^\oplus$. *Hence,* $m_3[\sigma'\rangle m_3'$, *where* $m_3' = (m_3 \ominus {}^\bullet \sigma') \oplus \sigma'^\bullet$.

Summing up, to the move $m_1[t_1\rangle m_1'$, m_3 *can reply with* $m_3[\sigma'\rangle m_3'$, *in such a way that* $(^\bullet t_1, ^\bullet \sigma') \in R_1 \circ R_2$, $(^\bullet t_1, \sigma'^\bullet) \in R_1 \circ R_2$, $(t_1^\bullet, \sigma'^\bullet) \in R_1 \circ R_2$ *and, moreover,* $(m_1 \ominus {}^\bullet t_1, m_3 \ominus {}^\bullet \sigma') \in (R_1 \circ R_2)^\oplus$, *(by Proposition 3(4)), as required.*

Let us consider case (ii), *i.e., assume that to the move* $m_1[t_1\rangle m_1'$, m_2 *replies with* $m_2[\sigma\rangle m[t_2\rangle m_2'$, *where* σ *is* τ-*sequential,* $l(t_1) = l(t_2)$, $\sigma^\bullet = {}^\bullet t_2$, *and* $(^\bullet t_1, ^\bullet \sigma) \in R_1^\oplus$, $(^\bullet t_1, ^\bullet t_2) \in R_1^\oplus$, $(t_1^\bullet, t_2^\bullet) \in R_1^\oplus$, *and moreover,* $(m_1 \ominus {}^\bullet t_1, m_2 \ominus {}^\bullet \sigma) \in R_1^\oplus$.

Since $(m_2, m_3) \in R_2^\oplus$, *there exists a submarking* $\overline{m} \subseteq m_3$ *such that* $(^\bullet \sigma, \overline{m}) \in R_2^\oplus$ *and* $(m_2 \ominus {}^\bullet \sigma, m_3 \ominus \overline{m}) \in R_2^\oplus$. *By Lemma 1, there exists a* τ-*sequential transition sequence* σ' *such that* $\overline{m} = {}^\bullet \sigma'$ *and* $(\sigma^\bullet, \sigma'^\bullet) \in R_2^\oplus$. *Hence,* $m_3[\sigma'\rangle m'$, *where* $m' = (m_3 \ominus {}^\bullet \sigma') \oplus \sigma'^\bullet$ *and, moreover,* $(m, m') \in R_2^\oplus$.

Since $(m, m') \in R_2^\oplus$, $\sigma^\bullet = {}^\bullet t_2$ *and* $(\sigma^\bullet, \sigma'^\bullet) \in R_2^\oplus$, *there exists* $\underline{m} = \sigma'^\bullet \subseteq m'$ *such that* $(^\bullet t_2, \underline{m}) \in R_2^\oplus$ *and* $(m \ominus {}^\bullet t_2, m' \ominus \underline{m}) \in R_2^\oplus$. *Hence, by Definition 11, to the move* $^\bullet t_2[t_2\rangle t_2^\bullet$, $\underline{m}$ *can reply as follows:*

(a) *Either* t_2 *is* τ-*sequential and* $\exists \overline{\sigma}$ *such that* $\overline{\sigma}$ *is* τ-*sequential,* $\underline{m} = {}^\bullet \overline{\sigma}$, $\underline{m}[\overline{\sigma}\rangle \overline{\sigma}^\bullet$, *and* $(^\bullet t_2, ^\bullet \overline{\sigma}) \in R_2$, $(^\bullet t_2, \overline{\sigma}^\bullet) \in R_2$, $(t_2^\bullet, \overline{\sigma}^\bullet) \in R_2$ *and* $(m \ominus t_2, m' \ominus \overline{\sigma}) \in R_2^\oplus$. *In this case, to the move* $m_1[t_1\rangle m_1'$, m_3 *can reply with* $m_3[\sigma'\rangle m'[\overline{\sigma}\rangle m_3'$, *with* $m_3' = (m' \ominus {}^\bullet \overline{\sigma}) \oplus \overline{\sigma}^\bullet$, *such that* $(^\bullet t_1, ^\bullet \sigma' \overline{\sigma}) \in (R_1 \circ R_2)^\oplus$ *(because* $(^\bullet t_1, ^\bullet \sigma) \in R_1^\oplus$, $\sigma'^\bullet = {}^\bullet \overline{\sigma}$ *and* $(^\bullet \sigma, ^\bullet \sigma') \in R_2^\oplus$), $(^\bullet t_1, \sigma' \overline{\sigma}^\bullet) \in (R_1 \circ R_2)^\oplus$ *(because* $(^\bullet t_1, ^\bullet t_2) \in R_1$, $\sigma'^\bullet = {}^\bullet \overline{\sigma}$ *and* $(^\bullet t_2, \overline{\sigma}^\bullet) \in R_2$), $(t_1^\bullet, \sigma' \overline{\sigma}^\bullet) \in (R_1 \circ R_2)^\oplus$ *(as* $(t_1^\bullet, t_2^\bullet) \in R_1$ *and* $(t_2^\bullet, \overline{\sigma}^\bullet) \in R_2$), *and, moreover,* $(m_1 \ominus {}^\bullet t_1, m_3 \ominus {}^\bullet \sigma' \overline{\sigma}) \in (R_1 \circ R_2)^\oplus$.

(b) *or* $\exists \overline{\sigma}, t_3, \overline{m}$ *such that* $\overline{\sigma}$ *is* τ-*sequential,* $\underline{m} = {}^\bullet \overline{\sigma}$, $\underline{m}[\overline{\sigma}\rangle \overline{m}[t_3\rangle t_3^\bullet$, $l(t_2) = l(t_3)$, $\overline{m} = \overline{\sigma}^\bullet = {}^\bullet t_3$, $(^\bullet t_2, ^\bullet \overline{\sigma}) \in R_2^\oplus$, $(^\bullet t_2, ^\bullet t_3) \in R_2^\oplus$ $(t_2^\bullet, t_3^\bullet) \in R_2^\oplus$, *and* $(m \ominus {}^\bullet t_2, m' \ominus {}^\bullet \overline{\sigma}) \in R_2^\oplus$.

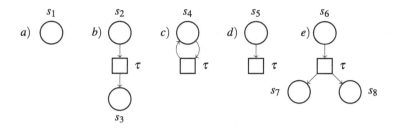

Fig. 2. Some simple nets with silent moves

In this case, to the move $m_2[\sigma\rangle m[t_2\rangle m_2'$, m_3 replies with $m_3[\sigma'\rangle m'[\overline{\sigma}\rangle$ $m''[t_3\rangle m_3'$, with $m_3' = (m' \ominus {}^\bullet\overline{\sigma}) \oplus t_3^\bullet$, such that $\overline{\sigma}$ is τ-sequential, ${}^\bullet\overline{\sigma} = \sigma'{}^\bullet$, and therefore $({}^\bullet\sigma t_2, {}^\bullet\sigma'\overline{\sigma}t_3) \in R_2^\oplus$ (because ${}^\bullet\sigma t_2 = {}^\bullet\sigma$, ${}^\bullet\sigma'\overline{\sigma}t_3 = {}^\bullet\sigma'$ and $({}^\bullet\sigma, {}^\bullet\sigma') \in R_2^\oplus$), and $(\sigma t_2^\bullet, \sigma'\overline{\sigma}t_3^\bullet) \in R_2^\oplus$ (because $\sigma t_2^\bullet = t_2^\bullet$, $\sigma'\overline{\sigma}t_3^\bullet = t_3^\bullet$ and $(t_2^\bullet, t_3^\bullet) \in R_2^\oplus$).

Summing up, to the move $m_1[t_1\rangle m_1'$, m_3 can reply with $m_3[\sigma'\rangle m'[\overline{\sigma}\rangle$ $m''[t_3\rangle m_3'$, such that $({}^\bullet t_1, {}^\bullet\sigma'\overline{\sigma}) \in (R_1 \circ R_2)^\oplus$ (as $({}^\bullet t_1, {}^\bullet\sigma) \in R_1^\oplus$, ${}^\bullet\sigma'\overline{\sigma} = {}^\bullet\sigma'$ and $({}^\bullet\sigma, {}^\bullet\sigma') \in R_2^\oplus$), $({}^\bullet t_1, {}^\bullet t_3) \in (R_1 \circ R_2)^\oplus$ (as $({}^\bullet t_1, {}^\bullet t_2) \in R_1^\oplus$, and $({}^\bullet t_2, {}^\bullet t_3) \in R_2^\oplus$), $(t_1^\bullet, t_3^\bullet) \in (R_1 \circ R_2)^\oplus$ (because $(t_1^\bullet, t_2^\bullet) \in R_1^\oplus$, and $(t_2^\bullet, t_3^\bullet) \in R_2^\oplus$), and $(m_1 \ominus {}^\bullet t_1, m_3 \ominus {}^\bullet\sigma'\overline{\sigma}) \in (R_1 \circ R_2)^\oplus$ (because $(m_1 \ominus {}^\bullet t_1, m_2 \ominus {}^\bullet\sigma) \in R_1^\oplus$ and $(m_2 \ominus {}^\bullet\sigma, m_3 \ominus {}^\bullet\sigma') \in R_2^\oplus$).

The case when m_2 moves first is symmetric, and so omitted. Hence, $R_1 \circ R_2$ is a branching place bisimulation, indeed. □

Theorem 1. *For each P/T net $N = (S, A, T)$, relation $\approx_p \subseteq \mathcal{M}(S) \times \mathcal{M}(S)$ is an equivalence relation.*

Proof. Direct consequence of Propositions 5 and 6. □

Proposition 7 (Branching place bisimilarity is finer than branching interleaving bisimilarity). *For each P/T net $N = (S, A, T)$, $m_1 \approx_p m_2$ implies $m_1 \approx_{bri} m_2$.*

Proof. If $m_1 \approx_p m_2$, then $(m_1, m_2) \in R^\oplus$ for some branching place bisimulation R. Note that $R^\oplus$ is a branching interleaving bisimilarity, so that $m_1 \approx_{bri} m_2$. □

Example 3. Consider the nets in Fig. 2. Of course, $s_1 \approx_p s_2$, as well as $s_1 \approx_p s_4$. However, $s_2 \not\approx_p s_5$, because s_2 cannot respond to the non-τ-sequential move $s_5 \xrightarrow{\tau} \theta$. For the same reason, $s_2 \not\approx_p s_6$. Note that silent transitions that are not τ-sequential are not considered as unobservable. □

By Definition 11, branching place bisimilarity can be defined as follows:
$$\approx_p = \bigcup\{R^\oplus \mid R \text{ is a branching place bisimulation}\}.$$
By monotonicity of the additive closure (Proposition 2(2)), if $R_1 \subseteq R_2$, then $R_1^\oplus \subseteq R_2^\oplus$. Hence, we can restrict our attention to maximal branching place bisimulations only:

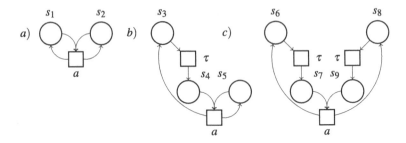

Fig. 3. Some branching place bisimilar nets

$\sim_p = \bigcup \{R^\oplus \mid R \text{ is a } maximal \text{ branching place bisimulation}\}$.
However, it is not true that
$\sim_p = (\bigcup \{R \mid R \text{ is a } maximal \text{ place bisimulation}\})^\oplus$, because the union of
branching place bisimulations may be not a branching place bisimulation.

Example 4. Consider the nets in Fig. 3. It is easy to realize that $s_1 \oplus s_2 \approx_p s_3 \oplus s_5$,
because $R_1 = \{(s_1, s_3), (s_2, s_5), (s_1, s_4)\}$ is a branching place bisimulation. In
fact, to the move $t_1 = s_1 \oplus s_2 \xrightarrow{a} s_1 \oplus s_2$, $s_3 \oplus s_5$ replies with $s_3 \oplus s_5[\sigma\rangle s_4 \oplus$
$s_5[t_2\rangle s_3 \oplus s_5$, where $\sigma = t\, i(s_5)$ (with $t = (s_3, \tau, s_4)$ and $i(s_5) = (s_5, \tau, s_5)$) and
$t_2 = (s_4 \oplus s_5, a, s_3 \oplus s_5)$, such that $(^\bullet t_1, {}^\bullet t_2) \in R_1^\oplus$ and $(t_1^\bullet, t_2^\bullet) \in R_1^\oplus$. Then, to
the move $s_3 \oplus s_5[t\rangle s_4 \oplus s_5$, $s_1 \oplus s_2$ can reply by idling with $s_1 \oplus s_2[\sigma'\rangle s_1 \oplus s_2$,
where $\sigma' = i(s_1)$, and $(^\bullet \sigma', {}^\bullet t) \in R_1^\oplus$, $(\sigma'^\bullet, {}^\bullet t) \in R_1^\oplus$ and $(\sigma'^\bullet, t^\bullet) \in R_1^\oplus$.

Note that also the identity relation $\mathscr{I}_S$, where $S = \{s_1, s_2, s_3, s_4, s_5\}$ is a
branching place bisimulation. However, $R = R_1 \cup \mathscr{I}_S$ is not a branching place
bisimulation, because, for instance, $(s_1 \oplus s_2, s_3 \oplus s_2) \in R^\oplus$, but these two markings are clearly not equivalent, as $s_1 \oplus s_2$ can do a, while $s_3 \oplus s_2$ cannot.

Similarly, one can prove that $s_1 \oplus s_2 \approx_p s_6 \oplus s_8$ because $R_2 = \{(s_1, s_6), (s_2, s_8),$
$(s_1, s_7), (s_2, s_9)\}$ is a branching place bisimulation.   □

5   Branching Place Bisimilarity is Decidable

In order to prove that $\approx_p$ is decidable, we first need a technical lemma which
states that it is decidable to check if a place relation $R \subseteq S \times S$ is a branching
place bisimulation.

Lemma 2. *Given a P/T net $N = (S, A, T)$ and a place relation $R \subseteq S \times S$, it
is decidable if R is a branching place bisimulation.*

*Proof. We want to prove that R is a branching place bisimulation if and only if
the following two conditions are satisfied:*

1. *$\forall t_1 \in T$, $\forall m$ such that $(^\bullet t_1, m) \in R^\oplus$*
   *(a) either t_1 is τ-sequential and there exists an acyclic τ-sequential σ such
   that $m = {}^\bullet \sigma$, $(^\bullet t_1, \sigma^\bullet) \in R$ and $(t_1^\bullet, \sigma^\bullet) \in R$;*

(b) or there exist an acyclic τ-sequential σ and $t_2 \in T$, with $\sigma^\bullet = {}^\bullet t_2$, such that $m = {}^\bullet\sigma$, $l(t_1) = l(t_2)$, $({}^\bullet t_1, {}^\bullet t_2) \in R^\oplus$ and $(t_1^\bullet, t_2^\bullet) \in R^\oplus$.

2. $\forall t_2 \in T$, $\forall m$ such that $(m, {}^\bullet t_2) \in R^\oplus$

(a) either t_2 is τ-sequential and there exists an acyclic τ-sequential σ such that $m = {}^\bullet\sigma$, $(\sigma^\bullet, {}^\bullet t_2) \in R$ and $(\sigma^\bullet, t_2^\bullet) \in R$;

(b) or there exist an acyclic τ-sequential σ and $t_1 \in T$, with $\sigma^\bullet = {}^\bullet t_1$, such that $m = {}^\bullet\sigma$, $l(t_1) = l(t_2)$, $({}^\bullet t_1, {}^\bullet t_2) \in R^\oplus$ and $(t_1^\bullet, t_2^\bullet) \in R^\oplus$.

The implication from left to right is obvious: if R is a branching place bisimulation, then for sure conditions 1 and 2 are satisfied, because, as observed in Remark 1, if there exists a suitable τ-sequential transition sequence σ, then there exists also a suitable acyclic τ-sequential σ' such that ${}^\bullet\sigma = {}^\bullet\sigma'$ and $\sigma^\bullet = \sigma'^\bullet$. For the converse implication, assume that conditions 1 and 2 are satisfied; then we have to prove that the branching place bisimulation game for R holds for all pairs $(m_1, m_2) \in R^\oplus$.

Let $q = \{(s_1, s_1'), (s_2, s_2'), \ldots, (s_k, s_k')\}$ be any multiset of associations that can be used to prove that $(m_1, m_2) \in R^\oplus$. So this means that $m_1 = s_1 \oplus s_2 \oplus \ldots \oplus s_k$, $m_2 = s_1' \oplus s_2' \oplus \ldots \oplus s_k'$ and that $(s_i, s_i') \in R$ for $i = 1, \ldots, k$. If $m_1[t_1\rangle m_1'$, then $m_1' = m_1 \ominus {}^\bullet t_1 \oplus t_1^\bullet$. Consider the multiset of associations $p = \{(\bar{s}_1, \bar{s}_1'), \ldots, (\bar{s}_h, \bar{s}_h')\} \subseteq q$, with $\bar{s}_1 \oplus \ldots \oplus \bar{s}_h = {}^\bullet t_1$. Note that $({}^\bullet t_1, \bar{s}_1' \oplus \ldots \oplus \bar{s}_h') \in R^\oplus$. Therefore, by condition 1,

(a) either t_1 is τ-sequential and there exists an acyclic τ-sequential σ such that $m = {}^\bullet\sigma$, $({}^\bullet t_1, \sigma^\bullet) \in R$ and $(t_1^\bullet, \sigma^\bullet) \in R$;

(b) or there exist an acyclic τ-sequential σ and $t_2 \in T$, with $\sigma^\bullet = {}^\bullet t_2$, such that $m = {}^\bullet\sigma$, $l(t_1) = l(t_2)$, $({}^\bullet t_1, {}^\bullet t_2) \in R^\oplus$ and $(t_1^\bullet, t_2^\bullet) \in R^\oplus$.

In case (a), since ${}^\bullet\sigma \subseteq m_2$, also $m_2[\sigma\rangle m_2'$ is firable, where $m_2' = m_2 \ominus {}^\bullet\sigma \oplus \sigma^\bullet$, so that $({}^\bullet t_1, \sigma^\bullet) \in R$, $(t_1^\bullet, \sigma^\bullet) \in R$ and, finally, $(m_1 \ominus {}^\bullet t_1, m_2 \ominus {}^\bullet\sigma) \in R^\oplus$, as required. Note that the last condition holds because, from the multiset q of matching pairs for m_1 and m_2, we have removed those in p. In case (b), since ${}^\bullet\sigma \subseteq m_2$, also $m_2[\sigma\rangle m[t_2\rangle m_2'$ is firable, where $m_2' = m_2 \ominus {}^\bullet\sigma \oplus t_2^\bullet$, so that $l(t_1) = l(t_2)$, $({}^\bullet t_1, {}^\bullet t_2) \in R^\oplus$, $(t_1^\bullet, t_2^\bullet) \in R^\oplus$ and, finally, $(m_1 \ominus {}^\bullet t_1, m_2 \ominus {}^\bullet\sigma) \in R^\oplus$, as required.

If $m_2[t_2\rangle m_2'$, then we have to use an argument symmetric to the above, where condition 2 is used instead. Hence, we have proved that conditions 1 and 2 are enough to prove that R is a branching place bisimulation.

Finally, observe that the set T is finite and, for each $t_1 \in T$, the number of markings m such that $({}^\bullet t_1, m) \in R^\oplus$ and $(m, {}^\bullet t_1) \in R^\oplus$ is finite as well. More precisely, this part of the procedure takes $O(q \cdot \frac{(n+p-1)!}{(n-1)! \cdot p!} \cdot (p^2 \sqrt{p}))$ time where $q = |T|$, $n = |S|$ and p is the least number such that $|{}^\bullet t| \leq p$ for all $t \in T$, because the distribution of p tokens over n places is given by the binomial coefficient $\binom{n+p-1}{p} = \frac{(n+p-1)!}{(n-1)! \cdot p!}$ and checking if such a marking of size p is related to ${}^\bullet t_1$ takes $O(p^2 \sqrt{p})$ time.

Moreover, for each pair (t_1, m) satisfying the condition $(^\bullet t_1, m) \in R^\oplus$, we have to check conditions (a) and (b), each checkable in a finite amount of time. In fact, for case (a), we have to check if there exists a place s such that $(^\bullet t_1, s) \in R$ and $(t_1^\bullet, s) \in R$, which is reachable from m by means of an acyclic τ-1-sequential transition sequence σ; this condition is decidable because we have at most n places to examine and for each candidate place s, we can check whether a suitable acyclic τ-1-sequential σ exists. Similarly, in case (b) we have to consider all the transitions t_2 such that $(^\bullet t_1, ^\bullet t_2) \in R^\oplus$ and $(t_1^\bullet, t_2^\bullet) \in R^\oplus$ and check if at least one of these is reachable from m by means of an acyclic τ-sequential transition sequence σ such that $^\bullet \sigma = m$ and $\sigma^\bullet = ^\bullet t_2$ and the existence of such a σ is decidable. Therefore, in a finite amount of time we can decide if a given place relation R is actually a branching place bisimulation. □

Theorem 2 (Branching place bisimilarity is decidable). *Given a P/T net $N = (S, A, T)$, for each pair of markings m_1 and m_2, it is decidable whether $m_1 \approx_p m_2$.*

Proof. If $|m_1| \neq |m_2|$, then $m_1 \not\approx_p m_2$ by Proposition 1. Otherwise, let $|m_1| = k = |m_2|$. As $|S| = n$, the set of all the place relations over S is of size 2^n. Let us list such relations as: $R_1, R_2, \ldots, R_{2^n}$. Hence, for $i = 1, \ldots, 2^n$, by Lemma 2 we can decide whether R_i is a branching place bisimulation and, in such a case, we can check whether $(m_1, m_2) \in R_i^\oplus$ in $O(k^2 \sqrt{k})$ time. As soon as we found a branching place bisimulation R_i such that $(m_1, m_2) \in R_i^\oplus$, we stop concluding that $m_1 \approx_p m_2$. If none of the R_i is a branching place bisimulation such that $(m_1, m_2) \in R_i^\oplus$, then we can conclude that $m_1 \not\approx_p m_2$. □

6 Conclusion and Future Research

Place bisimilarity [1] is the only decidable [13] behavioral equivalence for P/T nets which respects the expected causal behavior, as it is slightly finer than *structure preserving bisimilarity* [7], in turn slightly finer than *fully-concurrent bisimilarity* [3]. Thus, it is the only equivalence for which it is possible (at least, in principle) to verify algorithmically the (causality-preserving) correctness of an implementation by exhibiting a place bisimulation between its specification and implementation.

It is sometimes argued that place bisimilarity is too discriminating. In particular, [1] and [7] argue that a *sensible* equivalence should not distinguish markings whose behaviors are patently the same, such as marked Petri nets that differ only in their unreachable parts. As an example, consider the net in Fig. 4, discussed in [1]. Clearly, markings s_1 and s_4 are equivalent, also according to all the behavioral equivalences discussed in [7], except for place bisimilarity. As a matter of fact, a place bisimulation R containing the pair (s_1, s_4) would require also the pairs (s_2, s_5) and (s_3, s_6), but then this place relation R cannot be a place bisimulation because $(s_2 \oplus s_3, s_5 \oplus s_6) \in R^\oplus$, but $s_2 \oplus s_3$ can perform c, while this is not possible for $s_5 \oplus s_6$. Nonetheless, we would like to argue in favor of place bisimilarity, despite this apparent paradoxical example.

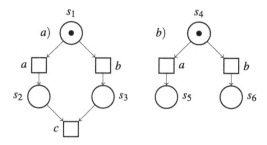

Fig. 4. Two non-place bisimilar nets

As a matter of fact, our interpretation of place bisimilarity is that this equivalence is an attempt of giving semantics to *unmarked* nets, rather than to marked nets, so that the focus shifts from the common (but usually undecidable) question *When are two markings equivalent?* to the more restrictive (but decidable) question *When are two places equivalent?* A possible (preliminary, but not accurate enough) answer to the latter question may be: two places are equivalent if, whenever the same number of tokens are put on these two places, the behavior of the marked nets is the same. If we reinterpret the example of Fig. 4 in this perspective, we clearly see that place s_1 and place s_4 cannot be considered as equivalent because, even if the markings s_1 and s_4 are equivalent, nonetheless the marking $2 \cdot s_1$ is not equivalent to the marking $2 \cdot s_4$, as only the former can perform the trace abc.

A place bisimulation R considers two places s_1 and s_2 as equivalent if $(s_1, s_2) \in R$, as, by definition of place bisimulation, they must behave the same in any R-related context. Back to our example in Fig. 4, if (s_1, s_4) would belong to R, then also $(2 \cdot s_1, 2 \cdot s_4)$ should belong to $R^{\oplus}$, but then we discover that the place bisimulation game does not hold for this pair of markings, so that R cannot be a place bisimulation.

Moreover, if we consider the duality between the process algebra FNM (a dialect of CCS, extended with multi-party interaction) and P/T nets, proposed in [9], we may find further arguments supporting this more restrictive interpretation of net behavior. In fact, an *unmarked* P/T net N can be described by an FNM system of equations, where each equation defines a constant C_i (whose body is a sequential process term t_i), representing place s_i. Going back to the nets in Fig. 4, according to this duality, the constant C_1 for place s_1 is not equivalent (in any reasonable sense) to the constant C_4 for place s_4 because these two constants describe all the potential behaviors of these two places, which are clearly different! Then, the marked net $N(m_0)$ is described by a parallel term composed of as many instances of C_i as the tokens that are present in s_i for m_0, encapsulated by a suitably defined restriction operator $(\nu L)-$. Continuing the example, it turns out that $(\nu L)C_1$ is equivalent to $(\nu L)C_4$ because the markings s_1 and s_4 are equivalent, but $(\nu L)(C_1|C_1)$ is not equivalent to $(\nu L)(C_4|C_4)$ because the markings $2 \cdot s_1$ is not equivalent to the marking $2 \cdot s_4$, as discussed above.

Furthermore, on the subclass of BPP nets (i.e., nets whose transitions have singleton pre-set), place bisimilarity specializes to *team bisimilarity* [10], which is unquestionably the most appropriate behavioral equivalence for BPP nets, as it coincides with *structure-preserving bisimilarity* [7], hence matching all the relevant criteria expressed in [7] for a sensible behavioral equivalence.

Finally, there are at least the following three important technical differences between place bisimilarity and other coarser, causality-respecting equivalences, such as fully-concurrent bisimilarity [3].

1. A fully-concurrent bisimulation is a complex relation – composed of cumbersome triples of the form (process, bijection, process) – that must contain infinitely many triples if the net system offers never-ending behavior. (Indeed, not even one single case study of a system with never-ending behavior has been developed for this equivalence.) On the contrary, a place bisimulation is always a very simple finite relation over the finite set of places. (And a simple case study is described in [13].)

2. A fully-concurrent bisimulation proving that m_1 and m_2 are equivalent is a relation specifically designed for the initial markings m_1 and m_2. If we want to prove that, e.g., $n \cdot m_1$ and $n \cdot m_2$ are fully-concurrent bisimilar (which may not hold!), we have to construct a new fully-concurrent bisimulation to this aim. Instead, a place bisimulation R relates those places which are considered equivalent under all the possible R-related contexts. Hence, if R justifies that $m_1 \sim_p m_2$ as $(m_1, m_2) \in R^\oplus$, then for sure the same R justifies that $n \cdot m_1$ and $n \cdot m_2$ are place bisimilar, as also $(n \cdot m_1, n \cdot m_2) \in R^\oplus$.

3. Finally, while place bisimilarity is decidable [13], fully-concurrent bisimilarity is undecidable on finite P/T nets [5].

The newly defined *branching place bisimilarity* is the only extension of the place bisimilarity idea to P/T nets with silent moves that has been proved decidable, even if the time complexity of the decision procedure we have proposed is exponential in the size of the net.

Of course, this behavioral relation may be subject to the same criticisms raised to place bisimilarity and also its restrictive assumption that only τ-sequential transitions can be abstracted away can be criticized, as its applicability to real case studies may appear rather limited. In the following, we try to defend our point of view.

First, on the subclass of BPP nets, branching place bisimilarity coincides with *branching team bisimilarity* [12], a very satisfactory equivalence which is actually coinductive and, for this reason, also very efficiently decidable in polynomial time. Moreover, on the subclass of *finite-state machines* (i.e., nets whose transitions have singleton pre-set and singleton, or empty, post-set), branching team bisimilarity has been axiomatized [11] on the process algebra CFM [9], which can represent all (and only) the finite-state machines, up to net isomorphism.

Second, we conjecture that branching place bisimilarity does respect the causal behavior of P/T nets. In particular, we conjecture that *branching fully-*

concurrent bisimilarity [12, 22] (which is undecidable) is strictly coarser than $\approx_p$, because it may equate nets whose silent transitions are not τ-sequential (and also may relate markings of different size). For instance, consider the net in Fig. 5. Of course, the markings $s_1 \oplus s_3$ and $s_5 \oplus s_6$ are branching fully-concurrent bisimilar: to the move $s_1 \oplus s_3[t_1\rangle s_2 \oplus s_3$, where $t_1 = (s_1, \tau, s_2)$, $s_5 \oplus s_6$ can reply with $s_5 \oplus s_6[t_2\rangle s_7 \oplus s_8$, where $t_2 = (s_5 \oplus s_6, \tau, s_7 \oplus s_8)$ and the reached markings are clearly equivalent. However, $s_1 \oplus s_3 \not\approx_p s_5 \oplus s_6$ because $s_1 \oplus s_3$ cannot reply to the move $s_5 \oplus s_6[t_2\rangle s_7 \oplus s_8$, as t_2 is not τ-sequential (i.e., it can be seen as the result of a synchronization), while t_1 is τ-sequential.

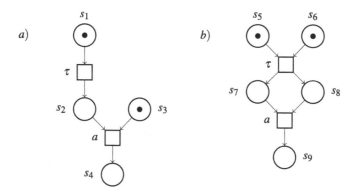

Fig. 5. Two branching fully-concurrent P/T nets

We already argued in the introduction that it is very much questionable whether a synchronization can be considered as unobservable, even if this idea is rooted in the theory of concurrency from the very beginning. As a matter of fact, in CCS [17] and in the π-calculus [18, 24], the result of a synchronization is a silent, τ-labeled (hence unobservable) transition. However, the silent label τ is used in these process algebras for two different purposes:

- First, to ensure that a synchronization is strictly binary: since the label τ cannot be used for synchronization, by labeling a synchronization transition by τ any further synchronization of the two partners with other parallel components is prevented (i.e., multi-party synchronization is disabled).
- Second, to describe that the visible effect of the transition is null: a τ-labeled transition can be considered unobservable and can be abstracted away, to some extent.

Nonetheless, it is possible to modify slightly these process algebras by introducing two different actions for these different purposes. In fact, the result of a binary synchronization can be some *observable* label, say λ (or even $\lambda(a)$, if the name of the channel a is considered as visible), for which no co-label exists, so that further synchronization is impossible. While the action τ, that can be used as a prefix, is used to denote some local, internal (hence unobservable) computation. In this way, a net semantics for these process algebras (in the style of, e.g., [9]) would generate τ-sequential P/T nets, that are amenable to be compared by means of branching place bisimilarity.

As a final comment, we want to discuss an apparently insurmountable limitation of our approach. In fact, the extension of the place bisimulation idea to nets with silent transitions that are not τ-sequential seems very hard, or even impossible. Consider again the two P/T nets in Fig. 5. If we want that $s_1 \oplus s_3$ be related to $s_5 \oplus s_6$, we need to include the pairs (s_1, s_5) and (s_3, s_6). If the marking $s_5 \oplus s_6$ silently reaches $s_7 \oplus s_8$, then $s_1 \oplus s_3$ can respond by idling (and in such a case we have to include the pairs (s_1, s_7) and (s_3, s_8)) or by performing the transition $s_1 \overset{\tau}{\longrightarrow} s_2$ (and in such a case we have to include the pairs (s_2, s_7) and (s_3, s_8)). In any case, the candidate place relation R should be of the form $\{(s_1, s_5), (s_3, s_6), (s_3, s_8), \ldots\}$. However, this place relation cannot be a place bisimulation of any sort because, on the one hand, $(s_1 \oplus s_3, s_5 \oplus s_8) \in R^{\oplus}$ but, on the other hand, $s_1 \oplus s_3$ can eventually perform a, while $s_5 \oplus s_8$ is stuck.

Nonetheless, this negative observation is coherent with our intuitive interpretation of (branching) place bisimilarity as a way to give semantics to *unmarked* nets. In the light of the duality between P/T nets and the FNM process algebra discussed above [9], a place is interpreted as a sequential process type (and each token in this place as an instance of a sequential process of that type, subject to some restriction); hence, a (branching) place bisimulation essentially states which kinds of sequential processes (composing the distributed system represented by the Petri net) are to be considered equivalent. In our example above, it makes no sense to consider place s_1 and place s_5 as equivalent, because the corresponding FNM constants C_1 and C_5 have completely different behavior: C_5 can interact (with C_6), while C_1 can only perform some internal, local transition.

Future work will be devoted to find more efficient algorithms for checking branching place bisimilarity. One idea could be to build directly the set of maximal branching place bisimulations, rather than to scan all the place relations to check whether they are branching place bisimulations, as we did in the proof of Theorem 2.

Acknowledgements. The anonymous referees are thanked for their useful comments and suggestions.

References

1. Autant, C., Belmesk, Z., Schnoebelen, P.: Strong bisimilarity on nets revisited. In: Aarts, E.H.L., van Leeuwen, J., Rem, M. (eds.) PARLE 1991. LNCS, vol. 506, pp. 295–312. Springer, Heidelberg (1991). https://doi.org/10.1007/3-540-54152-7_71
2. Basten, T.: Branching bisimilarity is an equivalence indeed!. Inf. Process. Lett. **58**(3), 141–147 (1996)
3. Best, E., Devillers, R., Kiehn, A., Pomello, L.: Concurrent bisimulations in Petri nets. Acta Inf. **28**(3), 231–264 (1991)

4. Desel, J., Reisig, W.: Place/transition petri nets. In: Reisig, W., Rozenberg, G. (eds.) ACPN 1996. LNCS, vol. 1491, pp. 122–173. Springer, Heidelberg (1998). https://doi.org/10.1007/3-540-65306-6_15
5. Esparza, J.: Decidability and complexity of Petri net problems — an introduction. In: Reisig, W., Rozenberg, G. (eds.) ACPN 1996. LNCS, vol. 1491, pp. 374–428. Springer, Heidelberg (1998). https://doi.org/10.1007/3-540-65306-6_20
6. van Glabbeek, R.J., Weijland, W.P.: Branching time and abstraction in bisimulation semantics. J. ACM **43**(3), 555–600 (1996)
7. Glabbeek, R.J.: Structure preserving bisimilarity, supporting an operational petri net semantics of CCSP. In: Meyer, R., Platzer, A., Wehrheim, H. (eds.) Correct System Design. LNCS, vol. 9360, pp. 99–130. Springer, Cham (2015). https://doi.org/10.1007/978-3-319-23506-6_9
8. Gorrieri, R., Versari, C.: Introduction to Concurrency Theory. TTCSAES. Springer, Cham (2015). https://doi.org/10.1007/978-3-319-21491-7
9. Gorrieri, R.: Process Algebras for Petri Nets: The Alphabetization of Distributed Systems. EATCS Monographs in Computer Science, Springer, Heidelberg (2017). https://doi.org/10.1007/978-3-319-55559-1
10. Gorrieri, R.: Team bisimilarity, and its associated modal logic, for BPP nets. Acta Informatica 1–41 (2020). https://doi.org/10.1007/s00236-020-00377-4
11. Gorrieri, R.: Team equivalences for finite-state machines with silent moves. Inf. Comput. **275** (2020). https://doi.org/10.1016/j.ic.2020.104603
12. Gorrieri, R.: Causal semantics for BPP nets with silent moves. Fundam. Inform. (2020, to appear)
13. Gorrieri, R.: Place bisimilarity is decidable, indeed!, arXiv:2104.01392, April 2021
14. Hopcroft, J.E., Karp, R.M.: An $n^{5/2}$ algorithm for maximum matchings in bipartite graphs. SIAM J. Comput. **2**(4), 225–231 (1973)
15. Jančar, P.: Undecidability of bisimilarity for Petri nets and some related problems. Theoret. Comput. Sci. **148**(2), 281–301 (1995)
16. Keller, R.: Formal verification of parallel programs. Commun. ACM **19**(7), 561–572 (1976)
17. Milner, R.: Communication and Concurrency. Prentice-Hall (1989)
18. Milner, R., Parrow, J., Walker, D.: A calculus of mobile processes. Inf. Comput. **100**(1), 1–77 (1992)
19. Olderog, E.R.: Nets, Terms and Formulas, Cambridge Tracts in Theoretical Computer Science 23. Cambridge University Press (1991)
20. Park, D.: Concurrency and automata on infinite sequences. In: Deussen, P. (ed.) GI-TCS 1981. LNCS, vol. 104, pp. 167–183. Springer, Heidelberg (1981). https://doi.org/10.1007/BFb0017309
21. Peterson, J.L.: Petri Net Theory and the Modeling of Systems. Prentice-Hall (1981)
22. Pinchinat, S.: Des bisimulations pour la sémantique des systèmes réactifs, Génie logiciel [cs.SE]. Ph.D. thesis, Institut National Polytechnique de Grenoble - INPG (1993)
23. Reisig, W.: Petri Nets: An Introduction, EATCS Monographs in Theoretical Computer Science, Springer, Heidelberg (1985). https://doi.org/10.1007/978-3-642-69968-9
24. Sangiorgi, D., Walker, D.: The π-calculus: A Theory of Mobile Processes. Cambridge University Press (2001)

Prioritise the Best Variation

Wen Kokke[1]([⊠]) and Ornela Dardha[2]

[1] University of Edinburgh, Edinburgh, UK
wen.kokke@ed.ac.uk
[2] University of Glasgow, Glasgow, UK
ornela.dardha@glasgow.ac.uk

Abstract. Binary session types guarantee communication safety and
session fidelity, but *alone* they cannot rule out deadlocks arising from
the interleaving of different sessions. In Classical Processes (CP) [53]—a
process calculus based on classical linear logic—deadlock freedom is guaranteed by combining channel creation and parallel composition under the
same logical cut rule. Similarly, in Good Variation (GV) [39,54]—a linear concurrent λ-calculus—deadlock freedom is guaranteed by combining
channel creation and thread spawning under the same operation, called
fork. In both CP and GV, deadlock freedom is achieved at the expense
of expressivity, as the only processes allowed are tree-structured. Dardha
and Gay [19] define Priority CP (PCP), which allows cyclic-structured
processes and restores deadlock freedom by using *priorities*, in line with
Kobayashi and Padovani [34,44]. Following PCP, we present Priority GV
(PGV), a variant of GV which decouples channel creation from thread
spawning. Consequently, we type cyclic-structured processes and restore
deadlock freedom by using priorities. We show that our type system is
sound by proving subject reduction and progress. We define an encoding
from PCP to PGV and prove that the encoding preserves typing and is
sound and complete with respect to the operational semantics.

Keywords: Session types · π-calculus · Functional programming ·
Deadlock freedom · GV · CP

1 Introduction

Session types [29,30,47] are types for protocols. Regular types ensure functions
are used according to their specification. Session types ensure *communication
channels* are used according to their protocols. Session types have been studied in
many settings. For instance, in the π-calculus [29,30,47], a foundational calculus
for communication and concurrency, and in concurrent λ-calculi [26], including
the focus of our paper: Good Variation [39,54, GV].

Supported by the EU HORIZON 2020 MSCA RISE project 778233 "Behavioural Application Program Interfaces" (BehAPI).

© IFIP International Federation for Information Processing 2021
Published by Springer Nature Switzerland AG 2021
K. Peters and T. A. C. Willemse (Eds.): FORTE 2021, LNCS 12719, pp. 100–119, 2021.
https://doi.org/10.1007/978-3-030-78089-0_6

GV is a concurrent λ-calculus with *binary* session types, where each channel is shared between exactly two processes. Binary session types guarantee two crucial properties *communication safety*—*e.g.*, if the protocol says to transmit an integer, you transmit an integer—and *session fidelity*—*e.g.*, if the protocol says send, you send. A third crucial property is *deadlock freedom*, which ensures that processes do not have cyclic dependencies—*e.g.*, when two processes wait for each other to send a value. Binary session types *alone* are insufficient to rule out deadlocks arising from interleaved sessions, but several additional techniques have been developed to guarantee deadlock freedom in session-typed π-calculus and concurrent λ-calculus.

In the π-calculus literature, there have been several attempts at developing Curry-Howard correspondences between session-typed π-calculus and linear logic [27]: Caires and Pfenning's πDILL [9] corresponds to dual intuitionistic linear logic [4], and Wadler's Classical Processes [53, CP] corresponds to classical linear logic [27, CLL]. Both calculi guarantee deadlock freedom, which they achieve by restricting structure of processes and shared channels to *trees*, by combing name restriction and parallel composition into a single construct, corresponding to the logical cut. This ensures that two processes can only communicate via exactly one series of channels, which rules out interleavings of sessions, and guarantees deadlock freedom. There are many downsides to combining name restriction and parallel composition, such as lack of modularity, difficulty typing structural congruence and formulating label-transition semantics, which have led to various approaches to decoupling these constructs. Hypersequent CP [37,38,41] and Linear Compositional Choreographies [14] decouple them, but maintain the correspondence to CLL and allow only tree-structured processes. Priority CP [20, PCP] weakens the correspondence to CLL in exchange for a more expressive language which allows cyclic-structured processes. PCP decouples CP's cut rule into two separate constructs: one for parallel composition via a mix rule, and one for name restriction via a cycle rule. To restore deadlock freedom, PCP uses *priorities* [34,44]. Priorities encode the *order of actions* and rule out bad cyclic interleavings. Dardha and Gay [20] prove cycle-elimination for PCP, adapting the cut-elimination proof for classical linear logic, and deadlock freedom follows as a corollary.

CP and GV are related via a pair of translations which satisfy simulation [40], and which can be tweaked to satisfy reflection. The two calculi share the same strong guarantees. GV achieves deadlock freedom via a similar syntactic restriction: it combines channel creation and thread spawning into a single operation, called "fork", which is related to the cut construct in CP. Unfortunately, as with CP, this syntactic restriction has its downsides.

Our aim is to develop a more expressive version of GV while maintaining deadlock freedom. While process calculi have their advantages, *e.g.*, their succinctness, we chose to work with GV for several reasons. In general, concurrent λ-calculi support higher-order functions, and have a capability for abstraction not usually present in process calculi. Within a concurrent λ-calculus, one can derive extensions of the communication capabilities of the language via well-understood extensions of the functional fragment, *e.g.*, we can derive internal/external choice from sum types. Concurrent λ-calculi maintain a clear separation between the program which the user writes and the configurations which represent the state of the system as it evaluates the program. However, our main motivation is that results obtained for λ-calculi transfer more easily to real-world functional programming languages. Case in point: we easily adapted the type system of PGV to Linear Haskell [6], which gives us a library for deadlock-free session-typed programming [36]. The benefit of working specifically with GV, as opposed to other concurrent λ-calculi, is its relation to CP [53], and its formal properties, including deadlock freedom. We thus pose our research question for GV:

RQ: Can we design a more expressive GV which guarantees deadlock freedom for cyclic-structured processes?

We follow the line of work from CP to Priority CP, and present Priority GV (PGV), a variant of GV which decouples channel creation from thread spawning, thus allowing cyclic-structured processes, but which nonetheless guarantees deadlock freedom via priorities. This closes the circle of the connection between CP and GV [53], and their priority-based versions, PCP [20] and PGV. We make the following main contributions:

(Sect. 2) **Priority GV.** We present Priority GV (Sect. 2, PGV), a session-typed functional language with priorities, and prove subject reduction (Theorem 1) and progress (Theorem 2). We addresses several problems in the original GV language, most notably: (a) PGV does not require the pseudo-type $S^{\sharp}$; and (b) its structural congruence is type preserving. PGV answers our research question positively as it allows cyclic-structured binary session-typed processes that are deadlock free.

(Sect. 3) **Translation from PCP to PGV.** We present a *sound and complete encoding* of Priority CP [20] in PGV (Sect. 3). We prove the encoding preserves typing (Theorem 4) and satisfies operational correspondence (Theorems 5 and 6). To obtain a tight correspondence, we update PCP, moving away from commuting conversions and reduction as cut elimination towards reduction based on structural congruence, as it is standard in process calculi.

2   Priority GV

We present Priority GV (PGV), a session-typed functional language based on GV [39,54] which uses priorities à la Kobayashi and Padovani [34,45] to enforce deadlock freedom. Priority GV offers a more fine-grained analysis of communication structures, and by separating channel creation form thread spawning it allows cyclic structures. We illustrate this with two programs in PGV, examples 1 and 2. Each program contains two processes—the main process, and the child process created by **spawn**—which communicate using *two* channels. The child process receives a unit over the channel x/x', and then sends a unit over the channel y/y'. The main process does one of two things: (a) in example 1, it sends a unit over the channel x/x', and then waits to receive a unit over the channel y/y'; (b) in Example 2, it does these in the opposite order, which results in a deadlock. PGV is more expressive than GV: Example 1 is typeable and guaranteed deadlock-free in PGV, but is not typeable in GV [53] and not guaranteed deadlock-free in GV's predecessor [26]. We believe PGV is a non-conservative *extension* of GV, as CP can be embedded in a Kobayashi-style system [22].

Example 1 (Cyclic Structure).   *Example 2 (Deadlock).*

$$
\begin{aligned}
&\textbf{let } (x, x') = \textbf{new in} \\
&\textbf{let } (y, y') = \textbf{new in} \\
&\textbf{spawn} \left(\begin{array}{l} \textbf{let } ((), x') = \textbf{recv } x' \textbf{ in} \\ \textbf{let } y = \textbf{send } ((), y) \textbf{ in} \\ \textbf{wait } x'; \textbf{close } y \end{array} \right) ; \\
&\textbf{let } x = \textbf{send } ((), x) \textbf{ in} \\
&\underline{\textbf{let } ((), y') = \textbf{recv } y' \textbf{ in}} \\
&\textbf{close } x; \textbf{wait } y'
\end{aligned}
\qquad
\begin{aligned}
&\textbf{let } (x, x') = \textbf{new in} \\
&\textbf{let } (y, y') = \textbf{new in} \\
&\textbf{spawn} \left(\begin{array}{l} \textbf{let } ((), x') = \textbf{recv } x' \textbf{ in} \\ \textbf{let } y = \textbf{send } ((), y) \textbf{ in} \\ \textbf{wait } x' \textbf{ close } y \end{array} \right) ; \\
&\textbf{let } ((), y') = \textbf{recv } y' \textbf{ in} \\
&\underline{\textbf{let } x = \textbf{send } ((), x) \textbf{ in}} \\
&\textbf{close } x \textbf{ wait } y'
\end{aligned}
$$

Session Types. Session types (S) are defined by the following grammar:

$$ S ::= \, !^o T.S \mid \, ?^o T.S \mid \textbf{end}_!^o \mid \textbf{end}_?^o $$

Session types $!^o T.S$ and $?^o T.S$ describe the endpoints of a channel over which we send or receive a value of type T, and then proceed as S. Types $\textbf{end}_!^o$ and $\textbf{end}_?^o$ describe endpoints of a channel whose communication has finished, and over which we must synchronise before closing the channel. Each connective in a session type is annotated with a *priority* $o \in \mathbb{N}$.

Types. Types (T, U) are defined by the following grammar:

$$ T, U ::= T \times U \mid \textbf{1} \mid T + U \mid \textbf{0} \mid T \multimap^{p,q} U \mid S $$

Types $T \times U$, $\textbf{1}$, $T + U$, and $\textbf{0}$ are the standard linear λ-calculus product type, unit type, sum type, and empty type. Type $T \multimap^{p,q} U$ is the standard linear function type, annotated with *priority bounds* $p, q \in \mathbb{N} \cup \{\bot, \top\}$. Every session type is also a type. Given a function with type $T \multimap^{p,q} U$, p is a *lower bound* on the priorities of the endpoints captured by the body of the function, and q is an *upper bound* on the priority of the communications that take place as a result of applying the function. The type of *pure functions* $T \multimap U$, *i.e.*, those which perform no communications, is syntactic sugar for $T \multimap^{\top, \bot} U$.

Environments. Typing environments Γ, Δ associate types to names. Environments are linear, so two environments can only be combined as Γ, Δ if their names are distinct, *i.e.*, $\text{fv}(\Gamma) \cap \text{fv}(\Delta) = \varnothing$.

$$ \Gamma, \Delta ::= \varnothing \mid \Gamma, x : T $$

Duality. Duality plays a crucial role in session types. The two endpoints of a channel are assigned dual types, ensuring that, for instance, whenever one program *sends* a value on a channel, the program on the other end is waiting to *receive*. Each session type S has a dual, written $\overline{S}$. Duality is an involutive function which *preserves priorities*:

$$\overline{!^oT.S} = ?^oT.\overline{S} \qquad \overline{?^oT.S} = !^oT.\overline{S} \qquad \overline{\mathbf{end}_!^o} = \mathbf{end}_?^o \qquad \overline{\mathbf{end}_?^o} = \mathbf{end}_!^o$$

Priorities. Function $\mathrm{pr}(\cdot)$ returns the smallest priority of a session type. The type system guarantees that the top-most connective always holds the smallest priority, so we simply return the priority of the top-most connective:

$$\mathrm{pr}(!^oT.S) = o \qquad \mathrm{pr}(?^oT.S) = o \qquad \mathrm{pr}(\mathbf{end}_!^o) = o \qquad \mathrm{pr}(\mathbf{end}_?^o) = o$$

We extend the function $\mathrm{pr}(\cdot)$ to types and typing contexts by returning the smallest priority in the type or context, or $\top$ if there is no priority. We use $\sqcap$ and $\sqcup$ to denote the minimum and maximum:

$$
\begin{aligned}
\min_{\mathrm{pr}}(T \times U) &= \min_{\mathrm{pr}}(T) \sqcap \min_{\mathrm{pr}}(U) & \min_{\mathrm{pr}}(\mathbf{1}) &= \top \\
\min_{\mathrm{pr}}(T + U) &= \min_{\mathrm{pr}}(T) \sqcap \min_{\mathrm{pr}}(U) & \min_{\mathrm{pr}}(\mathbf{0}) &= \top \\
\min_{\mathrm{pr}}(T \multimap^{p,q} U) &= p & \min_{\mathrm{pr}}(S) &= \mathrm{pr}(S) \\
\min_{\mathrm{pr}}(\Gamma, x : A) &= \min_{\mathrm{pr}}(\Gamma) \sqcap \min_{\mathrm{pr}}(A) & \min_{\mathrm{pr}}(\varnothing) &= \top
\end{aligned}
$$

Terms. Terms (L, M, N) are defined by the following grammar:

$$
\begin{aligned}
L, M, N ::=\ & x \mid K \mid \lambda x.M \mid M\,N \\
& \mid\ () \mid M; N \mid (M, N) \mid \mathbf{let}\ (x, y) = M\ \mathbf{in}\ N \\
& \mid\ \mathbf{inl}\ M \mid \mathbf{inr}\ M \mid \mathbf{case}\ L\ \{\mathbf{inl}\ x \mapsto M;\ \mathbf{inr}\ y \mapsto N\} \mid \mathbf{absurd}\ M \\
K \quad ::=\ & \mathbf{link} \mid \mathbf{new} \mid \mathbf{spawn} \mid \mathbf{send} \mid \mathbf{recv} \mid \mathbf{close} \mid \mathbf{wait}
\end{aligned}
$$

Let x, y, z, and w range over variable names. Occasionally, we use a, b, c, and d. The term language is the standard linear λ-calculus with products, sums, and their units, extended with constants K for the communication primitives.

Constants are best understood in conjunction with their typing and reduction rules in Figs. 1 and 2. Briefly, **link** links two endpoints together, forwarding messages from one to the other, **new** creates a new channel and returns a pair of its endpoints, and **spawn** spawns off its argument as a new thread. The **send** and **recv** functions send and receive values on a channel. However, since the typing rules for PGV ensure the linear usage of endpoints, they also return a new copy of the endpoint to continue the session. The **close** and **wait** functions close a channel. We use syntactic sugar to make terms more readable: we write $\mathbf{let}\ x = M\ \mathbf{in}\ N$ in place of $(\lambda x.N)\,M$, $\lambda().M$ in place of $\lambda z.z; M$, and $\lambda(x, y).M$ in place of $\lambda z.\mathbf{let}\ (x, y) = z\ \mathbf{in}\ M$. We recover **fork** as $\lambda x.\mathbf{let}\ (y, z) = \mathbf{new}\ ()\ \mathbf{in}\ \mathbf{spawn}\ (\lambda().x\,y); z$.

Internal and External Choice. Typically, session-typed languages feature constructs for internal and external choice. In GV, these can be defined in terms of the core language, by sending or receiving a value of a sum type [39]. We use the following syntactic sugar for internal ($S \oplus^o S'$) and external ($S \mathbin{\&}^o S'$) choice and their units:

$$S \oplus^o S' \triangleq {!}^o(\overline{S} + \overline{S'}).\mathbf{end}_!^{o+1} \qquad \oplus^o\{\} \triangleq {!}^o\mathbf{0}.\mathbf{end}_!^{o+1}$$
$$S \mathbin{\&}^o S' \triangleq {?}^o(S + S').\mathbf{end}_?^{o+1} \qquad \mathbin{\&}^o\{\} \triangleq {?}^o\mathbf{0}.\mathbf{end}_?^{o+1}$$

As the syntax for units suggests, these are the binary and nullary forms of the more common n-ary choice constructs $\oplus^o\{l_i : S_i\}_{i \in I}$ and $\mathbin{\&}^o\{l_i : S_i\}_{i \in I}$, which one may obtain generalising the sum types to variant types. For simplicity, we present only the binary and nullary forms.

Similarly, we use syntactic sugar for the term forms of choice, which combine sending and receiving with the introduction and elimination forms for the sum and empty types. There are two constructs for binary internal choice, expressed using the meta-variable ℓ which ranges over $\{\mathbf{inl}, \mathbf{inr}\}$. As there is no introduction for the empty type, there is no construct for nullary internal choice:

$$\mathbf{select}\ \ell \quad \triangleq \lambda x.\mathbf{let}\ (y, z) = \mathbf{new}\ \mathbf{in}\ \mathbf{close}\ (\mathbf{send}\ (\ell\ y, x)); z$$
$$\mathbf{offer}\ L\ \{\mathbf{inl}\ x \mapsto M; \mathbf{inr}\ y \mapsto N\} \triangleq$$
$$\quad \mathbf{let}\ (z, w) = \mathbf{recv}\ L\ \mathbf{in}\ \mathbf{wait}\ w; \mathbf{case}\ z\ \{\mathbf{inl}\ x \mapsto M;\ \mathbf{inr}\ y \mapsto N\}$$
$$\mathbf{offer}\ L\ \{\} \triangleq \mathbf{let}\ (z, w) = \mathbf{recv}\ L\ \mathbf{in}\ \mathbf{wait}\ w; \mathbf{absurd}\ z$$

Operational Semantics. Priority GV terms are evaluated as part of a configuration of processes. Configurations are defined by the following grammar:

$$\phi ::= \bullet \mid \circ \qquad \mathcal{C}, \mathcal{D}, \mathcal{E} ::= \phi\ M \mid \mathcal{C} \parallel \mathcal{D} \mid (\nu x x')\mathcal{C}$$

Configurations ($\mathcal{C}$, $\mathcal{D}$, $\mathcal{E}$) consist of threads $\phi\ M$, parallel compositions $\mathcal{C} \parallel \mathcal{D}$, and name restrictions $(\nu x x')\mathcal{C}$. To preserve the functional nature of PGV, where programs return a single value, we use flags (ϕ) to differentiate between the main thread, marked $\bullet$, and child threads created by **spawn**, marked $\circ$. Only the main thread returns a value. We determine the flag of a configuration by combining the flags of all threads in that configuration:

$$\bullet + \circ = \bullet \qquad \circ + \bullet = \bullet \qquad \circ + \circ = \circ \qquad (\bullet + \bullet\ \text{is undefined})$$

The use of $\circ$ for child threads [39] overlaps with the use of the meta-variable o for priorities [20]. Both are used to annotate sequents: flags appear on the sequent in configuration typing, and priorities in term typing. To distinguish the two symbols, they are typeset in a different font and a different colour.

Term reduction.

$$
\begin{array}{lll}
\text{E-Lam} & (\lambda x.M)\,V & \longrightarrow_M M\{V/x\} \\
\text{E-Unit} & \text{let } () = () \text{ in } M & \longrightarrow_M M \\
\text{E-Pair} & \text{let } (x,y) = (V,W) \text{ in } M & \longrightarrow_M M\{V/x\}\{W/y\} \\
\text{E-Inl} & \text{case inl } V \ \{\text{inl } x \mapsto M;\ \text{inr } y \mapsto N\} \longrightarrow_M M\{V/x\} \\
\text{E-Inr} & \text{case inr } V \ \{\text{inl } x \mapsto M;\ \text{inr } y \mapsto N\} \longrightarrow_M N\{V/y\}
\end{array}
$$

$$
\frac{\text{E-Lift}}{\begin{array}{c} M \longrightarrow_M M' \\ \hline E[M] \longrightarrow_M E[M'] \end{array}}
$$

Structural congruence.

$$
\begin{array}{llll}
\text{SC-LinkSwap} & \mathcal{F}[\text{link } (x,y)] & \equiv \mathcal{F}[\text{link } (y,x)] \\
\text{SC-ResLink} & (\nu xy)(\phi \text{ link } (x,y)) & \equiv \phi\,() \\
\text{SC-ResSwap} & (\nu xy)\mathcal{C} & \equiv (\nu yx)\mathcal{C} \\
\text{SC-ResComm} & (\nu xy)(\nu zw)\mathcal{C} & \equiv (\nu zw)(\nu xy)\mathcal{C}, \text{ if } \{x,y\} \cap \{z,w\} = \varnothing \\
\text{SC-ResExt} & (\nu xy)(\mathcal{C} \parallel \mathcal{D}) & \equiv \mathcal{C} \parallel (\nu xy)\mathcal{D}, \text{ if } x,y \notin \text{fv}(\mathcal{C}) \\
\text{SC-ParNil} & \mathcal{C} \parallel \circ() & \equiv \mathcal{C} \\
\text{SC-ParComm} & \mathcal{C} \parallel \mathcal{D} & \equiv \mathcal{D} \parallel \mathcal{C} \\
\text{SC-ParAssoc} & \mathcal{C} \parallel (\mathcal{D} \parallel \mathcal{E}) & \equiv (\mathcal{C} \parallel \mathcal{D}) \parallel \mathcal{E}
\end{array}
$$

Configuration reduction.

$$
\begin{array}{lll}
\text{E-Link} & (\nu xy)(\mathcal{F}[\text{link } (w,x)] \parallel \mathcal{C}) \longrightarrow_c \mathcal{F}[()] \parallel \mathcal{C}\{w/y\} \\
\text{E-New} & \mathcal{F}[\text{new } ()] \longrightarrow_c (\nu xy)(\mathcal{F}[(x,y)]), \text{ if } x,y \notin \text{fv}(\mathcal{F}) \\
\text{E-Spawn} & \mathcal{F}[(\text{spawn } V)] \longrightarrow_c \mathcal{F}[()] \parallel \circ V\,() \\
\text{E-Send} & (\nu xy)(\mathcal{F}[\text{send } (V,x)] \parallel \mathcal{F}'[\text{recv } y]) \longrightarrow_c (\nu xy)(\mathcal{F}[x] \parallel \mathcal{F}'[(V,y)]) \\
\text{E-Close} & (\nu xy)(\mathcal{F}[\text{wait } x] \parallel \mathcal{F}'[\text{close } y]) \longrightarrow_c \mathcal{F}[()] \parallel \mathcal{F}'[()]
\end{array}
$$

$$
\frac{\text{E-LiftC}}{\begin{array}{c} \mathcal{C} \longrightarrow_c \mathcal{C}' \\ \hline \mathcal{G}[\mathcal{C}] \longrightarrow_c \mathcal{G}[\mathcal{C}'] \end{array}}
\qquad
\frac{\text{E-LiftM}}{\begin{array}{c} M \longrightarrow_M M' \\ \hline \mathcal{F}[M] \longrightarrow_M \mathcal{F}[M'] \end{array}}
\qquad
\frac{\text{E-LiftSC}}{\begin{array}{c} \mathcal{C} \equiv \mathcal{C}' \quad \mathcal{C}' \longrightarrow_c \mathcal{D}' \quad \mathcal{D}' \equiv \mathcal{D} \\ \hline \mathcal{C} \longrightarrow_c \mathcal{D} \end{array}}
$$

Fig. 1. Operational semantics for PGV.

Values (V, W), evaluation contexts (E), thread evaluation contexts $(\mathcal{F})$, and configuration contexts $(\mathcal{G})$ are defined by the following grammars:

$$
\begin{array}{lll}
V, W & ::= & x \mid K \mid \lambda x.M \mid () \mid (V,W) \mid \text{inl } V \mid \text{inr } V \\
E & ::= & \Box \mid E\,M \mid V\,E \\
& & \mid E; N \mid (E,M) \mid (V,E) \mid \text{let } (x,y) = E \text{ in } M \\
& & \mid \text{inl } E \mid \text{inr } E \mid \text{case } E \ \{\text{inl } x \mapsto M;\ \text{inr } y \mapsto N\} \mid \text{absurd } E \\
\mathcal{F} & ::= & \phi\,E \\
\mathcal{G} & ::= & \Box \mid \mathcal{G} \parallel \mathcal{C} \mid (\nu xy)\mathcal{G}
\end{array}
$$

We factor the reduction relation of PGV into a deterministic reduction on terms $(\longrightarrow_M)$ and a non-deterministic reduction on configurations $(\longrightarrow_c)$, see

Fig. 1. We write $\longrightarrow^+_M$ and $\longrightarrow^+_C$ for the transitive closures, and $\longrightarrow^\star_M$ and $\longrightarrow^\star_C$ for the reflexive-transitive closures.

Term reduction is the standard call-by-value, left-to-right evaluation for GV, and only deviates from reduction for the linear λ-calculus in that it reduces terms to values *or* ready terms waiting to perform a communication action.

Configuration reduction resembles evaluation for a process calculus: E-LINK, E-SEND, and E-CLOSE perform communications, E-LIFTC allows reduction under configuration contexts, and E-LIFTSC embeds a structural congruence $\equiv$. The remaining rules mediate between the process calculus and the functional language: E-NEW and E-SPAWN evaluate the **new** and **spawn** constructs, creating the equivalent configuration constructs, and E-LIFTM embeds term reduction.

Structural congruence satisfies the following axioms: SC-LINKSWAP allows swapping channels in the link process. SC-RESLINK allows restriction to applied to link which is structurally equivalent to the terminated process, thus allowing elimination of unnecessary restrictions. SC-RESSWAP allows swapping channels and SC-RESCOMM states that restriction is commutative. SC-RESEXT is the standard scope extrusion rule. Rules SC-PARNIL, SC-PARCOMM and SC-PARASSOC state that parallel composition uses the terminated process as the neutral element; it is commutative and associative.

While our configuration reduction is based on the standard evaluation for GV, the increased expressiveness of PGV allows us to simplify the relation on two counts. (a) *We decompose the* **fork** *construct.* In GV, **fork** creates a new channel, spawns a child thread, and, when the child thread finishes, it closes the channel to its parent. In PGV, these are three separate operations: **new**, **spawn**, and **close**. We no longer require that every child thread finishes by returning a terminated channel. Consequently, we also simplify the evaluation of the **link** construct. Intuitively, evaluating **link** causes a substitution: if we have a channel bound as (νxy), then **link** (w, x) replaces all occurrences of y by w. However, in GV, **link** is required to return a terminated channel, which means that the semantics for *link* must *create* a fresh channel of type $\mathbf{end}_!/\mathbf{end}_?$. The endpoint of type $\mathbf{end}_!$ is returned by the *link* construct, and a **wait** on the other endpoint guards the *actual* substitution. In PGV, evaluating **link** simply causes a substitution. (b) *Our structural congruence is type preserving.* Consequently, we can embed it directly into the reduction relation. In GV, this is not the case, and subject reduction relies on proving that if $\equiv \longrightarrow_C$ ends up in an ill-typed configuration, we can rewrite it to a well-typed configuration using $\equiv$.

Typing. Figure 2 gives the typing rules for PGV. Typing rules for terms are at the top of Fig. 2. Terms are typed by a judgement $\Gamma \vdash^p M : T$ stating that "a term M has type T and an upper bound on its priority p under the typing environment Γ". Typing for the linear λ-calculus is standard. Linearity is ensured by splitting environments on branching rules, requiring that the environment in the variable rule consists of just the variable, and the environment in the constant and unit rules are empty. Constants K are typed using type schemas, and embedded using T-CONST (mid of Fig. 2). The typing rules treat *all variables*

Static Typing Rules.

$$\textsc{T-Var}\quad \frac{}{x : T \vdash^{\bot} x : T} \qquad \textsc{T-Const}\quad \frac{}{\varnothing \vdash^{\bot} K : T} \qquad \textsc{T-Lam}\quad \frac{\Gamma, x : T \vdash^{q} M : U}{\Gamma \vdash^{\bot} \lambda x.M : T \multimap^{\min_{\mathrm{pr}}(\Gamma), q} U}$$

$$\textsc{T-App}\quad \frac{\Gamma \vdash^{p} M : T \multimap^{p', q'} U \qquad \Delta \vdash^{q} N : T \qquad p < \mathbf{min}_{\mathbf{pr}}(\Delta) \qquad q < p'}{\Gamma, \Delta \vdash^{p \sqcup q \sqcup q'} M\ N : U}$$

$$\textsc{T-Unit}\quad \frac{}{\varnothing \vdash^{\bot} () : \mathbf{1}} \qquad \textsc{T-LetUnit}\quad \frac{\Gamma \vdash^{p} M : \mathbf{1} \qquad \Delta \vdash^{q} N : T \qquad p < \mathbf{min}_{\mathbf{pr}}(\Delta)}{\Gamma, \Delta \vdash^{p \sqcup q} M; N : T}$$

$$\textsc{T-Pair}\quad \frac{\Gamma \vdash^{p} M : T \qquad \Delta \vdash^{q} N : U \qquad p < \mathbf{min}_{\mathbf{pr}}(\Delta)}{\Gamma, \Delta \vdash^{p \sqcup q} (M, N) : T \times U}$$

$$\textsc{T-LetPair}\quad \frac{\Gamma \vdash^{p} M : T \times T' \qquad \Delta, x : T, y : T' \vdash^{q} N : U \qquad p < \mathbf{min}_{\mathbf{pr}}(\Delta, T, T')}{\Gamma, \Delta \vdash^{p \sqcup q} \mathbf{let}\ (x, y) = M\ \mathbf{in}\ N : U}$$

$$\textsc{T-Inl}\quad \frac{\Gamma \vdash^{p} M : T \qquad \mathbf{min}_{\mathbf{pr}}(T) = \mathbf{min}_{\mathbf{pr}}(U)}{\Gamma \vdash^{p} \mathbf{inl}\ M : T + U} \qquad \textsc{T-Inr}\quad \frac{\Gamma \vdash^{p} M : U \qquad \mathbf{min}_{\mathbf{pr}}(T) = \mathbf{min}_{\mathbf{pr}}(U)}{\Gamma \vdash^{p} \mathbf{inr}\ M : T + U}$$

$$\textsc{T-CaseSum}\quad \frac{\Gamma \vdash^{p} L : T + T' \qquad \Delta, x : T \vdash^{q} M : U \qquad \Delta, y : T' \vdash^{q} N : U \qquad p < \mathbf{min}_{\mathbf{pr}}(\Delta)}{\Gamma, \Delta \vdash^{p \sqcup q} \mathbf{case}\ L\ \{\mathbf{inl}\ x \mapsto M;\ \mathbf{inr}\ y \mapsto N\} : U}$$

$$\textsc{T-Absurd}\quad \frac{\Gamma \vdash^{p} M : \mathbf{0}}{\Gamma, \Delta \vdash^{p} \mathbf{absurd}\ M : T}$$

Type Schemas for Constants.

$$\mathbf{link} : S \times \overline{S} \multimap \mathbf{1} \qquad \mathbf{new} : \mathbf{1} \multimap S \times \overline{S} \qquad \mathbf{spawn} : (\mathbf{1} \multimap^{p,q} \mathbf{1}) \multimap \mathbf{1}$$

$$\mathbf{send} : T \times !^{o}T.S \multimap^{\top, o} S \qquad \mathbf{recv} : ?^{o}T.S \multimap^{\top, o} T \times S$$

$$\mathbf{close} : \mathbf{end}_{!}^{o} \multimap^{\top, o} \mathbf{1} \qquad \mathbf{wait} : \mathbf{end}_{?}^{o} \multimap^{\top, o} \mathbf{1}$$

Runtime Typing Rules.

$$\textsc{T-Main}\quad \frac{\Gamma \vdash^{p} M : T}{\Gamma \vdash^{\bullet} \bullet M} \qquad \textsc{T-Child}\quad \frac{\Gamma \vdash^{p} M : \mathbf{1}}{\Gamma \vdash^{\circ} \circ M} \qquad \textsc{T-Res}\quad \frac{\Gamma, x : S, y : \overline{S} \vdash^{\phi} \mathcal{C}}{\Gamma \vdash^{\phi} (\nu x y)\mathcal{C}} \qquad \textsc{T-Par}\quad \frac{\Gamma \vdash^{\phi} \mathcal{C} \qquad \Delta \vdash^{\phi'} \mathcal{D}}{\Gamma, \Delta \vdash^{\phi + \phi'} \mathcal{C} \parallel \mathcal{D}}$$

Fig. 2. Typing rules for PGV.

as linear resources, even those of non-linear types such as **1**. However, the rules can easily be extended to allow values with unrestricted usage [53].

The only non-standard feature of the typing rules is the priority annotations. Priorities are based on *obligations/capabilities* used by Kobayashi [34], and simplified to single priorities following Padovani [44]. The integration of priorities into GV is adapted from Padovani and Novara [45]. Paraphrasing Dardha and Gay [20], priorities obey the following two laws: (i) an action with lower priority happens before an action with higher priority; and (ii) communication requires *equal* priorities for dual actions.

In PGV, we keep track of a lower and upper bound on the priorities of a term, *i.e.*, while evaluating the term, when does it start communicating, and when does it finish. The upper bound is written on the sequent, whereas the lower bound is approximated from the typing environment. Typing rules for sequential constructs enforce sequentially, *e.g.*, the typing for $M; N$ has a side condition which requires that the upper bound of M is smaller than the lower bound of N, *i.e.*, M finishes before N starts. The typing rule for **new** ensures that both endpoints of a channel share the same priorities. Together, these two constraints guarantee deadlock freedom.

To illustrate this, let's go back to the deadlocked program in Example 2. Crucially, it composes the terms below in parallel. While each of these terms itself is well-typed, they impose opposite conditions on the priorities, so their composition is ill-typed. (We omit the priorities on **end**$_!$ and **end**$_?$.)

$$\frac{y' : ?^{o'}\,\mathbf{1}.\mathbf{end}_? \vdash^{o'} \mathbf{recv}\;y' : \mathbf{1} \times \mathbf{end}_?}{x : !^o\mathbf{1}.\mathbf{end}_!, y' : \mathbf{end}_? \vdash^p \mathbf{let}\;x = \mathbf{send}\;((),x)\;\mathbf{in}\;\dots : \mathbf{1}} \quad o' < o$$

$$x : !^o\mathbf{1}.\mathbf{end}_!, y' : ?^{o'}\,\mathbf{1}.\mathbf{end}_? \vdash^p \mathbf{let}\;((),y') = \mathbf{recv}\;y'\;\mathbf{in}\;\mathbf{let}\;x = \mathbf{send}\;((),x)\;\mathbf{in}\;\dots : \mathbf{1}$$

$$\frac{x' : ?^{o}\mathbf{1}.\mathbf{end}_? \vdash^o \mathbf{recv}\;x' : \mathbf{1} \times \mathbf{end}_?}{y : !^{o'}\,\mathbf{1}.\mathbf{end}_!, x' : \mathbf{end}_? \vdash^q \mathbf{let}\;y = \mathbf{send}\;((),y)\;\mathbf{in}\;\dots : \mathbf{1}} \quad o < o'$$

$$y : !^{o'}\,\mathbf{1}.\mathbf{end}_!, x' : ?^o\mathbf{1}.\mathbf{end}_? \vdash^q \mathbf{let}\;((),x') = \mathbf{recv}\;x'\;\mathbf{in}\;\mathbf{let}\;y = \mathbf{send}\;((),y)\;\mathbf{in}\;\dots : \mathbf{1}$$

Closures suspend communication, so T-LAM stores the priority bounds of the function body on the function type, and T-APP restores them. For instance, $\lambda x.\mathbf{send}\;(x,y)$ is assigned the type $A \multimap^{o,o} S$, *i.e.*, a function which, when applied, starts and finishes communicating at priority o.

$$\frac{\dfrac{}{\mathbf{send} : A \times !^o A.S \multimap^{\top,o} S} \quad \dfrac{\dfrac{}{x : A \vdash^{\perp} x : A} \quad \dfrac{}{x : A, y : !^o A.S \vdash^{\perp} y : !^o A.S}}{x : A, y : !^o A.S \vdash^{\perp} (x,y) : A \times !^o A.S}}{\dfrac{x : A, y : !^o A.S \vdash^o \mathbf{send}\;(x,y) : S}{y : !^o A.S \vdash^{\perp} \lambda x.\mathbf{send}\;(x,y) : A \multimap^{o,o} S}}$$

Typing rules for configurations are at the bottom of Fig. 2. Configurations are typed by a judgement $\Gamma \vdash^\phi \mathcal{C}$ stating that "a configuration $\mathcal{C}$ with flag ϕ is well typed under typing environment Γ". Configuration typing is based on the standard typing for GV. Terms are embedded either as main or as child threads. The priority bound from the term typing is discarded, as configurations contain no further blocking actions. Main threads are allowed to return a value, whereas child threads are required to return the unit value. Sequents are annotated with a flag ϕ, which ensures that there is at most one main thread.

While our configuration typing is based on the standard typing for GV, it differs on two counts: (i) *we require that child threads return the unit value*, as opposed to a terminated channel; and (ii) *we simplify typing for parallel composition*. In order to guarantee deadlock freedom, in GV each parallel composition must split *exactly one* channel of the channel pseudo-type $S^\sharp$ into two endpoints of type S and $\bar{S}$. Consequently, associativity of parallel composition does not preserve typing. In PGV, we guarantee deadlock freedom using priorities, which removes the need for the channel pseudo-type $S^\sharp$, and simplifies typing for parallel composition, while restoring type preservation for the structural congruence.

Subject Reduction. Unlike with previous versions of GV, structural congruence, term reduction, and configuration reduction are all type preserving.

We must show that substitution preserves priority constraints. For this, we prove Lemma 1, which shows that values have finished all their communication, and that any priorities in the type of the value come from the typing environment.

Lemma 1. *If $\Gamma \vdash^p V : T$, then $p = \bot$, and $min_{\mathrm{pr}}(\Gamma) = min_{\mathrm{pr}}(T)$.*

Lemma 2 (Substitution).
If $\Gamma, x : U' \vdash^p M : T$ and $\Theta \vdash^q V : U'$, then $\Gamma, \Theta \vdash^p M\{V/x\} : T$.

Lemma 3 (Subject Reduction, $\longrightarrow_M$).
If $\Gamma \vdash^p M : T$ and $M \longrightarrow_M M'$, then $\Gamma \vdash^p M' : T$.

Lemma 4 (Subject Congruence, $\equiv$).
If $\Gamma \vdash^\phi \mathcal{C}$ and $\mathcal{C} \equiv \mathcal{C}'$, then $\Gamma \vdash^\phi \mathcal{C}'$.

Theorem 1 (Subject Reduction, $\longrightarrow_\mathcal{C}$).
If $\Gamma \vdash^\phi \mathcal{C}$ and $\mathcal{C} \longrightarrow_\mathcal{C} \mathcal{C}'$, then $\Gamma \vdash^\phi \mathcal{C}'$.

Progress and Deadlock Freedom. PGV satisfies progress, as PGV configurations either reduce or are in normal form. However, the normal forms may seem surprising at first, as evaluating a well-typed PGV term does not necessarily produce *just* a value. If a term returns an endpoint, then its normal form contains a thread which is ready to communicate on the dual of that endpoint. This behaviour is not new to PGV. Let us consider an example, adapted from Lindley and Morris [39], in which a term returns an endpoint linked to an echo server. The echo server receives a value and sends it back unchanged. Consider the program which creates a new channel, with endpoints x and x', spawns off an echo server listening on x, and then returns x':

- **let** $(x, x') = $ **new** **in** $\mathrm{echo}_x \triangleq$ **let** $(y, x) = $ **recv** x **in**
  spawn $(\lambda().\mathrm{echo}_x); x'$ **let** $x = $ **send** (y, x) **in** **close** x

If we reduce the above program, we get $(\nu x x')(\circ\ \mathbf{echo}_x\ \|\ \bullet\ x')$. Clearly, no more evaluation is possible, even though the configuration contains the thread $\circ\ \mathbf{echo}_x$, which is blocked on x. In Corollary 1 we will show that if a term does not return an endpoint, it must produce *only* a value.

Actions are terms which perform communication actions and which synchronise between two threads. *Ready terms* are terms which perform communication actions, either by themselves, *e.g.*, creating a new channel or thread, or with another thread, *e.g.*, sending or receiving. Progress for the term language is standard for GV, and deviates from progress for linear λ-calculus only in that terms may reduce to values or *ready terms*:

Definition 1 (Actions). *A term acts on an endpoint x if it is* **send** (V, x), **recv** x, **close** x, *or* **wait** x. *A term is an* action *if it acts on some endpoint x.*

Definition 2 (Ready Terms). *A term L is* ready *if it is of the form $E[M]$, where M is of the form* **new**, **spawn** N, **link** (x, y), *or M acts on x. In the latter case, we say that L is* ready to act on x.

Lemma 5 (Progress, $\longrightarrow_M$). *If $\Gamma \vdash^p M : T$ and Γ contains only session types, then: (a) M is a value; (b) $M \longrightarrow_M N$ for some N; or (c) M is ready.*

Canonical forms deviate from those for GV, in that we opt to move all ν-binders to the top. The standard GV canonical form, alternating ν-binders and their corresponding parallel compositions, does not work for PGV, since multiple channels may be split across a single parallel composition.

A configuration either reduces, or it is equivalent to configuration in normal form. Crucial to the normal form is that each term M_i is blocked on the corresponding channel x_i, and hence no two terms act on dual endpoints. Furthermore, no term M_i can perform a communication action by itself, since those are excluded by the definition of actions. Finally, as a corollary, we get that well-typed terms which do not return endpoints return *just* a value:

Definition 3 (Canonical Forms). *A configuration $\mathcal{C}$ is in canonical form if it is of the form $(\nu x_1 x'_1) \ldots (\nu x_n x'_n)(\circ M_1 \| \cdots \| \circ M_m \| \bullet N)$ where no term M_i is a value.*

Lemma 6 (Canonical Forms). *If $\Gamma \vdash^\bullet \mathcal{C}$, there exists some $\mathcal{D}$ such that $\mathcal{C} \equiv \mathcal{D}$ and $\mathcal{D}$ is in canonical form.*

Definition 4 (Normal Forms). *A configuration $\mathcal{C}$ is in normal form if it is of the form $(\nu x_1 x'_1) \ldots (\nu x_n x'_n)(\circ M_1 \| \cdots \| \circ M_m \| \bullet V)$ where each M_i is ready to act on x_i.*

Theorem 2 (Progress, $\longrightarrow_\mathcal{C}$). *If $\varnothing \vdash^\bullet \mathcal{C}$ and $\mathcal{C}$ is in canonical form, then either $\mathcal{C} \longrightarrow_\mathcal{C} \mathcal{D}$ for some $\mathcal{D}$; or $\mathcal{C} \equiv \mathcal{D}$ for some $\mathcal{D}$ in normal form.*

Proof (Sketch). Our proof follows that of Kobayashi [34, theorem 2]. We apply Lemma 5 to each thread. Either we obtain a reduction, or each child thread is ready and the main thread ready or a value. We pick the ready term L with the smallest priority bound. If L contains new, spawn, or a link, we apply E-NEW, E-SPAWN, or E-LINK. Otherwise, L must be ready on some x_i. Linearity guarantees there is some thread L' which acts on x'_i. If L' is ready, priority typing guarantees it is ready on x'_i, and we apply E-SEND or E-CLOSE. If L' is not ready, it must be the main thread returning a value. We move L into the ith position and repeat until we either find a reduction or reach normal form.

Corollary 1. *If* $\varnothing \vdash^\phi \mathcal{C}$, $\mathcal{C} \longrightarrow\!\!\!\!/_{\mathcal{C}}$, *and* $\mathcal{C}$ *contains no endpoints, then* $\mathcal{C} \equiv \phi\, V$ *for some value* V.

It follows immediately from Theorem 2 and Corollary 1 that a term which does not return an endpoint will complete all its communication actions, thus satisfying deadlock freedom.

3   Relation to Priority CP

We present a correspondence between Priority GV and an updated version of Priority CP [20, PCP], which is Wadler's CP [53] with priorities. This correspondence connects PGV to (a relaxed variant of) classical linear logic.

3.1   Revisiting Priority CP

Types. (A, B) in PCP correspond to linear logic connectives annotated with priorities $o \in \mathbb{N}$. Typing environments, duality, and the priority function $\mathrm{pr}(\cdot)$ are defined as expected.

$$A, B ::= A \otimes^o B \mid A \,\mathfrak{N}^o\, B \mid \mathbf{1}^o \mid \perp^o \mid A \oplus^o B \mid A \,\&^o\, B \mid \mathbf{0}^o \mid \top^o$$

Processes. (P, Q) in PCP are defined by the following grammar.

$$
\begin{aligned}
P, Q ::= &\; x \leftrightarrow y \mid (\nu xy)P \mid (P \parallel Q) \mid \mathbf{0} \\
&\mid x[y].P \mid x[].P \mid x(y).P \mid x().P \\
&\mid x \triangleleft \mathrm{inl}.P \mid x \triangleleft \mathrm{inr}.P \mid x \triangleright \{\mathrm{inl} : P; \mathrm{inr} : Q\} \mid x \triangleright \{\}
\end{aligned}
$$

Processes are typed by sequents $P \vdash \Gamma$, which correspond to the one-sided sequents in classical linear logic. Differently from PGV, in PCP we do not need to store the greatest priority on the sequent, as, due to the absence of higher-order functions, we cannot compose processes *sequentially*.

PCP decomposes cut into T-Res and T-Par rules—corresponding to cycle and mix rules, respectively—and guarantees deadlock freedom by using priority constraints, $e.g.,$, as in T-Send.

$$\frac{\text{T-Res}}{P \vdash \Gamma, x : A, y : A^{\perp}} \qquad \frac{\text{T-Par}}{P \vdash \Gamma \quad Q \vdash \Delta} \qquad \frac{\text{T-Send}}{P \vdash \Gamma, y : A, x : B \quad o < \min_{\mathrm{pr}}(\Gamma, A, B)}{x[y].P \vdash \Gamma, x : A \otimes^{o} B}$$

The main change we make to PCP is *removing commuting conversions* and defining an operational semantics based on structural congruence. Commuting conversions are necessary if we want our reduction strategy to correspond *exactly* to cut elimination. However, from the perspective of process calculi, commuting conversions behave strangely: they allow an input/output action to be moved to the top of a process, thus potentially blocking actions which were previously possible. This makes CP, and Dardha and Gay's PCP [20], non-confluent. As Lindley and Morris [39] show, all communications that can be performed *with* the use of commuting conversions, can also be performed *without* them, using structural congruence.

In particular for PCP, commuting conversions break our intuition that an action with lower priority *occurs before* an action with higher priority. To cite Dardha and Gay [20] *"if a prefix on a channel endpoint x with priority o is pulled out at top level, then to preserve priority constraints in the typing rules [..], it is necessary to increase priorities of all actions after the prefix on x"* by $o + 1$. One benefit of removing commuting conversions is that we no longer need to dynamically change the priorities during reduction, which means that the intuition for priorities holds true in our updated version of PCP. Furthermore, we can safely define reduction on untyped processes, which means that type and priority information is erasable!

We prove closed progress for our updated PCP.

Theorem 3 (Progress, $\Longrightarrow$). *If $P \vdash \varnothing$, then either $P = \mathbf{0}$ or there exists a Q such that $P \Longrightarrow Q$.*

3.2 Correspondence Between PGV and PCP

We illustrate the relation between PCP and PGV by defining a translation from PCP to PGV. The translation on types is defined as follows:

$$(\!|A \otimes^{o} B|\!) = !^{o}\overline{(\!|A|\!)}.(\!|B|\!) \quad (\!|\mathbf{1}^{o}|\!) = \mathbf{end}_{!}^{o} \quad (\!|A \,\mathbin{⅋}^{o} B|\!) = ?^{o}(\!|A|\!).(\!|B|\!) \quad (\!|\perp^{o}|\!) = \mathbf{end}_{?}^{o}$$
$$(\!|A \oplus^{o} B|\!) = (\!|A|\!) \oplus^{o} (\!|B|\!) \quad (\!|\mathbf{0}^{o}|\!) = \oplus^{o}\{\} \quad (\!|A \,\&^{o} B|\!) = (\!|A|\!) \,\&^{o} (\!|B|\!) \quad (\!|\top^{o}|\!) = \&^{o}\{\}$$

There are two separate translations on processes. The main translation, $(\!|\cdot|\!)_M$, translates processes to *terms*:

$$
\begin{aligned}
(\!|x \leftrightarrow y|\!)_M &= \mathbf{link}\ (x, y) \\
(\!|(\nu xy)P|\!)_M &= \mathbf{let}\ (x, y) = \mathbf{new}\ \mathbf{in}\ (\!|P|\!)_M \\
(\!|P \parallel Q|\!)_M &= \mathbf{spawn}\ (\lambda().(\!|P|\!)_M);\ (\!|Q|\!)_M \\
(\!|\mathbf{0}|\!)_M &= () \\
(\!|x[].P|\!)_M &= \mathbf{close}\ x;\ (\!|P|\!)_M \\
(\!|x().P|\!)_M &= \mathbf{wait}\ x;\ (\!|P|\!)_M \\
(\!|x[y].P|\!)_M &= \mathbf{let}\ (y, z) = \mathbf{new}\ \mathbf{in}\ \mathbf{let}\ x = \mathbf{send}\ (z, x)\ \mathbf{in}\ (\!|P|\!)_M \\
(\!|x(y).P|\!)_M &= \mathbf{let}\ (y, x) = \mathbf{recv}\ x\ \mathbf{in}\ (\!|P|\!)_M \\
(\!|x \triangleleft \mathrm{inl}.P|\!)_M &= \mathbf{let}\ x = \mathbf{select}\ \mathrm{inl}\ x\ \mathbf{in}\ (\!|P|\!)_M \\
(\!|x \triangleleft \mathrm{inr}.P|\!)_M &= \mathbf{let}\ x = \mathbf{select}\ \mathrm{inr}\ x\ \mathbf{in}\ (\!|P|\!)_M \\
(\!|x \triangleright \{\mathrm{inl} : P; \mathrm{inr} : Q\}|\!)_M &= \mathbf{offer}\ x\ \{\mathrm{inl}\ x \mapsto (\!|P|\!)_M; \mathrm{inr}\ x \mapsto (\!|Q|\!)_M\} \\
(\!|x \triangleright \{\}|\!)_M &= \mathbf{offer}\ x\ \{\}
\end{aligned}
$$

Unfortunately, the operational correspondence along $(\!|\cdot|\!)_M$ is unsound, as it translates ν-binders and parallel compositions to **new** and **spawn**, which can reduce to their equivalent configuration constructs using E-NEW and E-SPAWN. The same goes for ν-binders which are inserted when translating bound send to unbound send. For instance, the process $x[y].P$ is blocked, but its translation uses **new** and can reduce. To address this issue, we use a second translation, $(\!|\cdot|\!)_C$, which is equivalent to $(\!|\cdot|\!)_M$ followed by reductions using E-NEW and E-SPAWN:

$$
\begin{aligned}
(\!|(\nu xy)P|\!)_C &= (\nu xy)(\!|P|\!)_C \\
(\!|P \parallel Q|\!)_C &= (\!|P|\!)_C \parallel (\!|Q|\!)_C \\
(\!|x[y].P|\!)_C &= (\nu yz)(\circ\ \mathbf{let}\ x = \mathbf{send}\ (z, x)\ \mathbf{in}\ (\!|P|\!)_M) \\
(\!|x \triangleleft \mathrm{inl}.P|\!)_C &= (\nu yz)(\circ\ \mathbf{let}\ x = \mathbf{close}\ (\mathbf{send}\ (\mathrm{inl}\ y, x));\ z\ \mathbf{in}\ (\!|P|\!)_M) \\
(\!|x \triangleleft \mathrm{inr}.P|\!)_C &= (\nu yz)(\circ\ \mathbf{let}\ x = \mathbf{close}\ (\mathbf{send}\ (\mathrm{inr}\ y, x));\ z\ \mathbf{in}\ (\!|P|\!)_M) \\
(\!|P|\!)_C &= \circ(\!|P|\!)_M, \quad \text{if none of the above apply}
\end{aligned}
$$

Typing environments are translated pointwise, and sequents $P \vdash \Gamma$ are translated as $(\!|\Gamma|\!) \vdash^\circ (\!|P|\!)_C$, where $\circ$ indicates a child thread. Translated processes do not have a main thread. The translations $(\!|\cdot|\!)_M$ and $(\!|\cdot|\!)_C$ preserve typing, and the latter induces a sound and complete operational correspondence.

Lemma 7 (Preservation, $(\!|\cdot|\!)_M$). *If $P \vdash \Gamma$, then $(\!|\Gamma|\!) \vdash^p (\!|P|\!)_M : \mathbf{1}$.*

Theorem 4 (Preservation, $(\!|\cdot|\!)_C$). *If $P \vdash \Gamma$, then $(\!|\Gamma|\!) \vdash^\circ (\!|P|\!)_C$.*

Lemma 8. *For any P, either:*

- $\circ\,(\!|P|\!)_M = (\!|P|\!)_C$; *or*
- $\circ\,(\!|P|\!)_M \longrightarrow_C^+ (\!|P|\!)_C$, *and for any $\mathcal{C}$, if $\circ\,(\!|P|\!)_M \longrightarrow_C \mathcal{C}$, then $\mathcal{C} \longrightarrow_C^\star (\!|P|\!)_C$.*

Theorem 5 (Operational Correspondence, Soundness, $(\![\cdot]\!)_c$). *If* $P \vdash \Gamma$ *and* $(\![P]\!)_c \longrightarrow_C \mathcal{C}$, *there exists a* Q *such that* $P \Longrightarrow^+ Q$ *and* $\mathcal{C} \longrightarrow_C^\star (\![Q]\!)_c$.

Theorem 6 (Operational Correspondence, Completeness, $(\![\cdot]\!)_c$). *If* $P \vdash \Gamma$ *and* $P \Longrightarrow Q$, *then* $(\![P]\!)_c \longrightarrow_C^+ (\![Q]\!)_c$.

4 Related Work and Discussion

Deadlock Freedom and Progress. Deadlock freedom and progress are well studied properties in the π-calculus. For the 'standard' typed π-calculus, an important line of work starts from Kobayashi's approach to deadlock freedom [33], where priorities are values from an abstract poset. Kobayashi [34] simplifies the abstract poset to pairs of naturals, called *obligations* and *capabilities*. Padovani simplifies these further to a single natural, called a *priority* [44], and adapts obligations/capabilities to session types [43].

For the session-typed π-calculus, Dezani *et al.* [25] guarantee progress by allowing only one active session at a time. Dezani [24] introduces a partial order on channels, similar to Kobayashi [33]. Carbone and Debois [11] define progress for session typed π-calculus in terms of a *catalyser* which provides the missing counterpart to a process. Carbone *et al.* [10] use catalysers to show that progress is a compositional form of lock-freedom and can be lifted to session types via the encoding of session types to linear types [18,21,35]. Vieira and Vasconcelos [51] use single priorities and an abstract partial order to guarantee deadlock freedom in a binary session-typed π-calculus and building on conservation types.

While our work focuses on *binary* session types, it is worth to discuss related work on Multiparty Session Types (MPST). The line of work on MPST starts with Honda *et al.* [31], which guarantees deadlock freedom *within a single session*, but not for session interleaving. Bettini *et al.* [7] follow a technique similar to Kobayashi's for MPST. The main difference with our work is that we associate priorities with communication actions, where Bettini *et al.* [7] associate them with channels. Carbone and Montesi [13] combine MPST with choreographies and obtain a formalism that satisfies deadlock freedom. Deniélou and Yoshida [23] introduce *multiparty compatibility* which generalises duality in binary session types. They synthesise safe and deadlock-free global types from local types leveraging LTSs and communicating automata. Castellani *et al.* [16] guarantee lock freedom, a stronger property than deadlock freedom, for MPST with *internal delegation*, where participants in the same session are allowed to delegate tasks to each other, and internal delegation is captured by the global type. Scalas and Yoshida [46] provide a revision of the foundations for MPST, and offer a less complicated and more general theory, by removing duality/consistency. The type systems is parametric and type checking is decidable, but allows for a novel integration of model checking techniques. More protocols and processes can be typed and are guaranteed to be free of deadlocks.

Neubauer and Thiemann [42] and Vasconcelos *et al.* [49,50] introduce the first functional language with session types. Such works did not guarantee deadlock freedom until GV [39,53]. Toninho *et al.* [48] present a translation of simply-typed λ-calculus into session-typed π-calculus, but their focus is not on deadlock freedom.

Ties with Logic. The correspondence between logic and types lays the foundation for functional programming [54]. Since its inception by Girard [27], linear logic has been a candidate for a foundational correspondence for concurrent programs. A correspondence with linear π-calculus was established early on by Abramsky [1] and Bellin and Scott [5]. Many years later, several correspondences between linear logic and the π-calculus with binary session types were proposed. Caires and Pfenning [9] propose a correspondence with dual intuitionistic linear logic, while Wadler [53] proposes a correspondence with classical linear logic. Both guarantee deadlock freedom as a consequence of cut elimination. Dardha and Gay [20] integrate Kobayashi and Padovani's work on priorities [34,44] with CP, loosening its ties to linear logic in exchange for expressivity. Dardha and Pérez [22] compare priorities à la Kobayashi with tree restrictions à la CP, and show that the latter is a subsystem of the former. Balzer *et al.* [2] introduce sharing at the cost of deadlock freedom, which they restore using an approach similar to priorities [3]. Carbone *et al.* [12,15] give a logical view of MPST with a generalised duality. Caires and Pérez [8] give a presentation of MPST in terms of binary session types and the use of a *medium process* which guarantee protocol fidelity and deadlock freedom. Their binary session types are rooted in linear logic. Ciobanu and Horne [17] give the first Curry-Howard correspondence between MPST and BV [28], a conservative extension of linear logic with a noncommutative operator for sequencing. Horne [32] give a system for subtyping and multiparty compatibility where compatible processes are race free and deadlock free using a Curry-Howard correspondence, similar to the approach in [17].

Conclusion. We answered our research question by presenting Priority GV, a session-typed functional language which allows cyclic communication structures and uses priorities to ensure deadlock freedom. We showed its relation to Priority CP [20] via an operational correspondence.

Future Work. Our formalism so far only captures the core of GV. In future work, we plan to explore recursion, following Lindley and Morris [40] and Padovani and Novara [45], and sharing, following Balzer and Pfenning [2] or Voinea *et al.* [52].

Acknowledgements. The authors would like to thank Simon Fowler, April Gonçalves, and Philip Wadler for their comments on the manuscript.

References

1. Abramsky, S.: Proofs as processes. Theor. Comput. Sci. **135**(1), 5–9 (1994). https://doi.org/10.1016/0304-3975(94)00103-0
2. Balzer, S., Pfenning, F.: Manifest sharing with session types. Proc. ACM Program. Lang. **1**(ICFP), 37:1–37:29 (2017). https://doi.org/10.1145/3110281

3. Balzer, S., Toninho, B., Pfenning, F.: Manifest deadlock-freedom for shared session types. In: Caires, L. (ed.) ESOP 2019. LNCS, vol. 11423, pp. 611–639. Springer, Cham (2019). https://doi.org/10.1007/978-3-030-17184-1_22

4. Barber, A.: Dual intuitionistic linear logic (1996). https://www.lfcs.inf.ed.ac.uk/reports/96/ECS-LFCS-96-347/ECS-LFCS-96-347.pdf

5. Bellin, G., Scott, P.J.: On the π-calculus and linear logic. Theor. Comput. Sci. **135**(1), 11–65 (1994). https://doi.org/10.1016/0304-3975(94)00104-9

6. Bernardy, J.P., Boespflug, M., Newton, R.R., Jones, S.P., Spiwack, A.: Linear Haskell: practical linearity in a higher-order polymorphic language. In: Proceedings of POPL, vol. 2, pp. 1–29 (2018). https://doi.org/10.1145/3158093

7. Bettini, L., Coppo, M., D'Antoni, L., De Luca, M., Dezani-Ciancaglini, M., Yoshida, N.: Global progress in dynamically interleaved multiparty sessions. In: van Breugel, F., Chechik, M. (eds.) CONCUR 2008. LNCS, vol. 5201, pp. 418–433. Springer, Heidelberg (2008). https://doi.org/10.1007/978-3-540-85361-9_33

8. Caires, L., Pérez, J.A.: Multiparty session types within a canonical binary theory, and beyond. In: Albert, E., Lanese, I. (eds.) FORTE 2016. LNCS, vol. 9688, pp. 74–95. Springer, Cham (2016). https://doi.org/10.1007/978-3-319-39570-8_6

9. Caires, L., Pfenning, F.: Session types as intuitionistic linear propositions. In: Gastin, P., Laroussinie, F. (eds.) CONCUR 2010. LNCS, vol. 6269, pp. 222–236. Springer, Heidelberg (2010). https://doi.org/10.1007/978-3-642-15375-4_16

10. Carbone, M., Dardha, O., Montesi, F.: Progress as compositional lock-freedom. In: Kühn, E., Pugliese, R. (eds.) COORDINATION 2014. LNCS, vol. 8459, pp. 49–64. Springer, Heidelberg (2014). https://doi.org/10.1007/978-3-662-43376-8_4

11. Carbone, M., Debois, S.: A graphical approach to progress for structured communication in web services. In: Proceedings of ICE. Electronic Proceedings in Theoretical Computer Science, vol. 38, pp. 13–27 (2010). https://doi.org/10.4204/EPTCS.38.4

12. Carbone, M., Lindley, S., Montesi, F., Schürmann, C., Wadler, P.: Coherence generalises duality: a logical explanation of multiparty session types. In: Proceedings of of CONCUR. LIPIcs, vol. 59, pp. 33:1–33:15. Leibniz-Zentrum für Informatik (2016). https://doi.org/10.4230/LIPIcs.CONCUR.2016.33

13. Carbone, M., Montesi, F.: Deadlock-freedom-by-design: multiparty asynchronous global programming. In: Proceedings of POPL, pp. 263–274 (2013). https://doi.org/10.1145/2480359.2429101

14. Carbone, M., Montesi, F., Schürmann, C.: Choreographies, logically. Distrib. Comput. **31**(1), 51–67 (2018). https://doi.org/10.1007/978-3-662-44584-6_5

15. Carbone, M., Montesi, F., Schürmann, C., Yoshida, N.: Multiparty session types as coherence proofs. In: Proceedings of CONCUR. LIPIcs, vol. 42, pp. 412–426. Leibniz-Zentrum für Informatik (2015). https://doi.org/10.1007/s00236-016-0285-y

16. Castellani, I., Dezani-Ciancaglini, M., Giannini, P., Horne, R.: Global types with internal delegation. Theor. Comput. Sci. **807**, 128–153 (2020). https://doi.org/10.1016/j.tcs.2019.09.027

17. Ciobanu, G., Horne, R.: Behavioural analysis of sessions using the calculus of structures. In: Mazzara, M., Voronkov, A. (eds.) PSI 2015. LNCS, vol. 9609, pp. 91–106. Springer, Cham (2016). https://doi.org/10.1007/978-3-319-41579-6_8

18. Dardha, O.: Recursive session types revisited. In: Proceedings of BEAT. Electronic Proceedings in Theoretical Computer Science, vol. 162, pp. 27–34 (2014). https://doi.org/10.4204/EPTCS.162.4

19. Dardha, O., Gay, S.J.: A new linear logic for deadlock-free session-typed processes. In: Baier, C., Dal Lago, U. (eds.) FoSSaCS 2018. LNCS, vol. 10803, pp. 91–109. Springer, Cham (2018). https://doi.org/10.1007/978-3-319-89366-2_5

20. Dardha, O., Gay, S.J.: A new linear logic for deadlock-free session typed processes. In: Proceedings of FoSSaCS (2018). http://www.dcs.gla.ac.uk/~ornela/publications/DG18-Extended.pdf

21. Dardha, O., Giachino, E., Sangiorgi, D.: Session types revisited. In: Proceedings of PPDP, pp. 139–150. ACM (2012). https://doi.org/10.1145/2370776.2370794

22. Dardha, O., Pérez, J.A.: Comparing type systems for deadlock-freedom (2018). https://arxiv.org/abs/1810.00635

23. Deniélou, P.-M., Yoshida, N.: Multiparty compatibility in communicating automata: characterisation and synthesis of global session types. In: Fomin, F.V., Freivalds, R., Kwiatkowska, M., Peleg, D. (eds.) ICALP 2013. LNCS, vol. 7966, pp. 174–186. Springer, Heidelberg (2013). https://doi.org/10.1007/978-3-642-39212-2_18

24. Dezani-Ciancaglini, M., de'Liguoro, U., Yoshida, N.: On progress for structured communications. In: Barthe, G., Fournet, C. (eds.) TGC 2007. LNCS, vol. 4912, pp. 257–275. Springer, Heidelberg (2008). https://doi.org/10.1007/978-3-540-78663-4_18

25. Dezani-Ciancaglini, M., Mostrous, D., Yoshida, N., Drossopoulou, S.: Session types for object-oriented languages. In: Thomas, D. (ed.) ECOOP 2006. LNCS, vol. 4067, pp. 328–352. Springer, Heidelberg (2006). https://doi.org/10.1007/11785477_20

26. Gay, S.J., Vasconcelos, V.T.: Linear type theory for asynchronous session types. J. Funct. Program. **20**(1), 19–50 (2010). https://doi.org/10.1017/S0956796809990268

27. Girard, J.Y.: Linear logic. Theor. Comput. Sci. **50**, 1–102 (1987). https://doi.org/10.1016/0304-3975(87)90045-4

28. Guglielmi, A.: A system of interaction and structure. ACM Trans. Comput. Log. **8**(1), 1 (2007). https://doi.org/10.1145/1182613.1182614

29. Honda, K.: Types for dyadic interaction. In: Best, E. (ed.) CONCUR 1993. LNCS, vol. 715, pp. 509–523. Springer, Heidelberg (1993). https://doi.org/10.1007/3-540-57208-2_35

30. Honda, K., Vasconcelos, V.T., Kubo, M.: Language primitives and type discipline for structured communication-based programming. In: Hankin, C. (ed.) ESOP 1998. LNCS, vol. 1381, pp. 122–138. Springer, Heidelberg (1998). https://doi.org/10.1007/BFb0053567

31. Honda, K., Yoshida, N., Carbone, M.: Multiparty asynchronous session types. In: Proceedings of POPL, vol. 43, no. 1, pp. 273–284. ACM (2008). https://doi.org/10.1145/2827695

32. Horne, R.: Session subtyping and multiparty compatibility using circular sequents. In: Proceedings of CONCUR. LIPIcs, vol. 171, pp. 12:1–12:22. Leibniz-Zentrum für Informatik (2020). https://doi.org/10.4230/LIPIcs.CONCUR.2020.12

33. Kobayashi, N.: A partially deadlock-free typed process calculus. ACM Trans. Program. Lang. Syst. **20**(2), 436–482 (1998). https://doi.org/10.1145/276393.278524

34. Kobayashi, N.: A new type system for deadlock-free processes. In: Baier, C., Hermanns, H. (eds.) CONCUR 2006. LNCS, vol. 4137, pp. 233–247. Springer, Heidelberg (2006). https://doi.org/10.1007/11817949_16

35. Kobayashi, N.: Type systems for concurrent programs (2007)

36. Kokke, W., Dardha, O.: Deadlock-free session types in Linear Haskell (2021). https://arxiv.org/abs/2103.14481

37. Kokke, W., Montesi, F., Peressotti, M.: Better late than never: a fully-abstract semantics for classical processes. Proc. ACM Program. Lang. **3**(POPL) (2019). https://doi.org/10.1145/3290337

38. Kokke, W., Montesi, F., Peressotti, M.: Taking linear logic apart. In: Proceedings of Linearity & TLLA. Electronic Proceedings in Theoretical Computer Science, vol. 292, pp. 90–103. Open Publishing Association (2019). https://doi.org/10.4204/EPTCS.292.5

39. Lindley, S., Morris, J.G.: A semantics for propositions as sessions. In: Proceedings of ESOP, pp. 560–584 (2015). https://doi.org/10.1007/978-3-662-46669-8_23

40. Lindley, S., Morris, J.G.: Talking bananas: structural recursion for session types. In: Proceedings of ICFP. ACM (2016). https://doi.org/10.1145/2951913.2951921

41. Montesi, F., Peressotti, M.: Classical transitions (2018). https://arxiv.org/abs/1803.01049

42. Neubauer, M., Thiemann, P.: An implementation of session types. In: Jayaraman, B. (ed.) PADL 2004. LNCS, vol. 3057, pp. 56–70. Springer, Heidelberg (2004). https://doi.org/10.1007/978-3-540-24836-1_5

43. Padovani, L.: From lock freedom to progress using session types. In: Proceedings of PLACES. vol. 137, pp. 3–19. Electronic Proceedings in Theoretical Computer Science (2013). https://doi.org/10.4204/EPTCS.137.2

44. Padovani, L.: Deadlock and lock freedom in the linear π-calculus. In: Proceedings of CSL-LICS, pp. 72:1–72:10. ACM (2014). https://doi.org/10.1145/2603088.2603116

45. Padovani, L., Novara, L.: Types for deadlock-free higher-order programs. In: Graf, S., Viswanathan, M. (eds.) FORTE 2015. LNCS, vol. 9039, pp. 3–18. Springer, Cham (2015). https://doi.org/10.1007/978-3-319-19195-9_1

46. Scalas, A., Yoshida, N.: Less is more: multiparty session types revisited. Proc. ACM Program. Lang. **3**(POPL) (2019). https://doi.org/10.1145/3290343

47. Takeuchi, K., Honda, K., Kubo, M.: An interaction-based language and its typing system. In: Halatsis, C., Maritsas, D., Philokyprou, G., Theodoridis, S. (eds.) PARLE 1994. LNCS, vol. 817, pp. 398–413. Springer, Heidelberg (1994). https://doi.org/10.1007/3-540-58184-7_118

48. Toninho, B., Caires, L., Pfenning, F.: Functions as session-typed processes. In: Birkedal, L. (ed.) FoSSaCS 2012. LNCS, vol. 7213, pp. 346–360. Springer, Heidelberg (2012). https://doi.org/10.1007/978-3-642-28729-9_23

49. Vasconcelos, V., Ravara, A., Gay, S.: Session types for functional multithreading. In: Gardner, P., Yoshida, N. (eds.) CONCUR 2004. LNCS, vol. 3170, pp. 497–511. Springer, Heidelberg (2004). https://doi.org/10.1007/978-3-540-28644-8_32

50. Vasconcelos, V.T., Gay, S.J., Ravara, A.: Type checking a multithreaded functional language with session types. Theor. Comput. Sci. **368**(1–2), 64–87 (2006). https://doi.org/10.1016/j.tcs.2006.06.028

51. Torres Vieira, H., Thudichum Vasconcelos, V.: Typing progress in communication-centred systems. In: De Nicola, R., Julien, C. (eds.) COORDINATION 2013. LNCS, vol. 7890, pp. 236–250. Springer, Heidelberg (2013). https://doi.org/10.1007/978-3-642-38493-6_17

52. Voinea, A.L., Dardha, O., Gay, S.J.: Resource sharing via capability-based multiparty session types. In: Ahrendt, W., Tapia Tarifa, S.L. (eds.) IFM 2019. LNCS, vol. 11918, pp. 437–455. Springer, Cham (2019). https://doi.org/10.1007/978-3-030-34968-4_24

53. Wadler, P.: Propositions as sessions. J. Funct. Program. **24**(2–3), 384–418 (2014). https://doi.org/10.1017/S095679681400001X

54. Wadler, P.: Propositions as types. Commun. ACM **58**(12), 75–84 (2015). https://doi.org/10.1145/2699407

Towards Multi-layered Temporal Models:
A Proposal to Integrate Instant Refinement in CCSL

Mathieu Montin[1,2]([⊠]) [ID] and Marc Pantel[2] [ID]

[1] INRIA, Team Veridis, University of Lorraine, LORIA, Team MOSEL,
Nancy, France
mathieu.montin@loria.fr
[2] IRIT, ENSEEIHT, INPT, Team ACADIE, Toulouse, France

Abstract. For the past 50 years, temporal constraints have been a key driver in the development of critical systems, as ensuring their safety requires their behaviour to meet stringent temporal requirements. A well established and promising approach to express and verify such temporal constraints is to rely on formal modelling languages. One such language is CCSL, first introduced as part of the MARTE UML profile, which allows the developer, through entities called clocks, to abstract any system into events on which constraints can be expressed, and then assessed using TimeSquare, a tool which implements its operational semantics. By nature, CCSL handles horizontal separation (component based design at one step in the system development) of concerns through the notion of clocks, but does not yet take into account the other major separation of concerns used in modern system development: vertical separation, also called refinement in the literature (relations between the various steps of the system development). This paper proposes an approach to extend CCSL with a notion of refinement in order to handle temporal models relying on both vertical and horizontal parts. Our proposal relies on the notion of multi-layered time to provide two new CCSL relations expressing two different yet complementary notions of refinement. Their integration with the other CCSL constructs is discussed and their use is illustrated while the relevance and future impacts of this extended version of CCSL is detailed.

Keywords: CCSL · Refinement · Temporal constraints

1 Introduction

1.1 Ubiquity of Complex Systems

Software engineering as a discipline has been developed since the middle of the 20th century. The overall consensus is that the 1960s, and especially the 1968 convention held in Marktoberdorf, Germany [2] paved the way to where software engineering stands now: present in every single field and company, as well as in

K. Peters and T. A. C. Willemse (Eds.): FORTE 2021, LNCS 12719, pp. 120–137, 2021.
https://doi.org/10.1007/978-3-030-78089-0_7

our everyday life. While we believe that the future induced by this evolution will be bright, it can only be so provided we have the ability to validate and verify the software being made, in other words, to assess that it satisfies its user requirements.

This need is especially potent when considering critical systems, the failure of which could have significant negative impacts on human lives. These critical systems usually rely on various interactions with other entities: other software through logical interfaces, human beings through user interfaces and the outside world through both sensors, which provide the software with data, and actuators, which control various physical mechanisms. The software from the last category are called cyber physical systems (CPS) [6]. The actuators in these systems have their output constantly adjusted using the data received from their sensors, and such adjustments must usually be done in real time, thus inducing various hard temporal constraints. These critical systems are especially present in fields such as aeronautics, cars, railway, space exploration, energy transformation, health-care and factory automation, which means they have the potential to be of huge size and complexity, and their modelling usually needs to be heterogeneous.

1.2 Heterogeneous Modelling

Complexity of description is a major issue in the development and especially in the verification of such software. This complexity, as described in the theory of complex systems [15] is wide and encompasses a large number of possible cases. It was first called algorithmic complexity [7] by its inventors Kolmogorov and Chaitin and was originally related to the size of the simplest algorithm in a given language that can implement the system. This notion is nowadays wider, as the system in question can either be huge in size, and provide a sophisticated service, composed of numerous small size elements which are required to interact with one another in a correct manner, or simply display a very complex behaviour which translates into a deep conceptual challenge for the developers. Such complexity is very concrete, especially in CPS where it can often be found in its various incarnations.

While developing and assessing such complex systems, the notion of separation of concerns becomes mandatory, as first introduced by Dijsktra himself in [5]. A concern is a specific element of a system that must be handled during its development. As for separation of concerns, while the name being somewhat self-explanatory, it hides a higher level of complexity depending on the nature of the concerns that are separated from others. Indeed, these concerns can be of various nature: functional, physical, logical, abstraction, human, sociological, ergonomic, psychological, economical, ethical... and can refer to various macroscopic elements: provided service, provided quality of service... including the process, methods and tools for the development itself.

There are three main kinds of separation of concerns that are usually called horizontal, transversal and vertical. Horizontal separation of concerns (also called Component Based Design) consists in applying a divide and conquer strategy that splits a problem in sub-problems recursively until reaching problems that are small enough to be solved efficiently. The complexity of the remaining components is reduced as their size is small, however they usually mix different issues to be solved: security, functionality, performance... The second kind of separation, called transversal, corresponds to Aspect Oriented Engineering. It proposes to isolate each of the previous issues to handle them separately with the most appropriate tools. The third kind is called vertical separation: it consists in separating a given development into different steps from an abstract specification to a concrete implementation through a process usually called refinement.

1.3   Handling Vertical and Horizontal Separation over Temporal Constraints

Our work revolves around the handling of the behavioural aspect of a system development, that is, the specific aspect-oriented subpart of a system concerning temporal constraints over its behaviour. These temporal constraints are not absolute in our work, which means there will be no mention of worst-case execution time (WCET) in this work, but rather they are relative from one component to another. The word "component" is not used lightly: this indeed corresponds to the various components horizontal separation has revealed and defined, and the way these components interact time-wise with one another. This horizontal description is usually expressed using languages such as CCSL, which will be described in more details in Sect. 2.3. This language relies on a notion of time and time structure that will be summarized in Sect. 2.1. However, this kind of description do not handle vertical separation, which means temporal constraints between layers of refinement cannot be expressed. In this paper, we propose a way to express such constraints, and we motivate the need for such expressiveness. A first step in that direction consists in using a multi-layered time structure that will be described in more details in Sect. 2.2. Then, new CCSL relations can be defined between several layers of refinement, thus allowing relations to be cross-specifications. We call the first new relation 1-N refinement, which is described in Sect. 5, while the second relation, called 1-1 refinement, is depicted in Sect. 6. Both these relations will be formally defined, illustrated by examples, and integrated to CCSL through properties of preservation with existing CCSL relations. An example of system which could benefit from such a multi-layered temporal description is depicted in Sect. 3.

2   Theoretical Ground

This section presents the state-of-the-art elements that we rely on in this work. They consist of common notions about time, as well as one of our previous works regarding multi-layered temporal structures and basic notions about CCSL.

2.1   Time, Partial Orders and Time Structures

In order to model the temporal relations between instants in the execution of complex systems, the usual approach is to use strict partial orders which allow some instants to be coincident and some others to be unrelated, as opposed to total orders where all instants must be related one way or another. A strict partial order is a mathematical structure composed of four entities, which are the following:

- A set of instants, I.
- A first relation over the elements of I, $\_\approx\_$, called coincidence.
- A second relation over the elements of I, $\_<\_$, called precedence.
- A proof that $\_\approx\_$ and $\_<\_$ satisfy the properties of strict partial ordering.

The last element of this structure ties all the others together. By giving the right properties to the relations, it also gives them the intended semantics, that is: $\_<\_$ is a strict precedence between elements of I and $\_\approx\_$ is a relation of equivalence between its elements. These properties are as follows:

1. $\_\approx\_$ is an equivalence relation	
• $\_\approx\_$ is reflexive	$\in I : i \approx$
• $\_\approx\_$ is transitive	$\forall(i,j,k) \in I^3 : i \approx j \land j \approx k \Rightarrow j \approx$
• $\_\approx\_$ is symmetrical	$\forall(i,j) \in I^2 : i \approx j \Rightarrow j \approx i$
2. $\_<\_$ is irreflexive towards $\_\approx\_$	$\forall(i,j) \in I^2 : i < j \Rightarrow \neg i \approx j$
3. $\_<\_$ is transitive	$\forall(i,j,k) \in I^3 : i < j \land j < k \Rightarrow j < k$
4. $\_<\_$ respects $\_\approx\_$	
• on the left	$\forall(i,j,k) \in I^3 : i \approx j \land i < k \Rightarrow j < k$
• on the right	$\forall(i,j,k) \in I^3 : i \approx j \land k < i \Rightarrow k < j$

As an example of modelling with such structure, let us consider Alice and her usual morning routine: Alice gets up, after which she either takes a bath first followed by eating or vice versa. She always sings while in the bath. After that, she takes off for work. The two possible traces depicting her behaviour over a single day are shown in Fig. 1. They consist of the following possible events: getting up (u), bathing (b), singing (s), eating (e) and taking off (o).

(a) A first possible behaviour   (b) A second possible behaviour

Fig. 1. Both possible behaviours

These possible behaviours can be merged using a time structure [16], with an underlying partial order, that is depicted on Fig. 2. The events b and e are concurrent and are not linked by any of the two relations composing the strict partial order. The blue vertical dashed line represents coincidence (when events occur simultaneously) while the red arrows represent precedence (one occurs strictly before the other). Note that, while we arbitrary chose to represent e before b, it does not mean that e precedes b, and e could equally have been placed somewhere else between u and o.

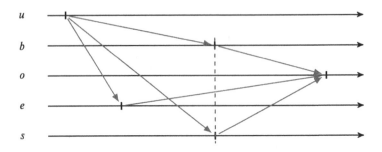

Fig. 2. Alice's morning routine time structure

2.2 Multi-layered Time Structures

Such time structures naturally embed a notion of horizontal separation. Indeed, each of the events it contains is, by definition, separated from the others, regardless of the origin of said events. In other words, whether these events come from the same system or from different systems that are being coordinated is not relevant. However, all these events are from the same step in the development, in other words, they are from the same layer of refinement. This level of observation is fundamental when dealing with such systems because a trace or a time structure only takes into account observable events. For instance, in Alice's morning routine, one did not chose to represent details about her different activities, such as the way she showers (from hair to toes) or the sequence of songs she sings while doing so. These more precise events could have been depicted in another time structure. In a previous work [10], we showed that, should we assign a partial order to any of these depictions (any of these levels of refinement), then these partial orders $((<_c, \approx_c)$ and $(<_a, \approx_a))$ would be related with one another in a certain formal manner, as follows:

Let Ω be the set of all sets: $\forall I \in \Omega, \forall (<_c, <_a, \approx_c, \approx_a) \in (I \times I)^4 :$

$(<_c, \approx_c) <_r (<_a, \approx_a) \overset{d}{\Longleftrightarrow}$

$$\forall (i,j) \in I : \begin{pmatrix} i <_c j \Rightarrow i <_a j \vee i \approx_a j & (1) \\ i <_a j \Rightarrow i <_c j & (2) \\ i \approx_c j \Rightarrow i \approx_a j & (3) \\ i \approx_a j \Rightarrow i \approx_c j \vee i <_c j \vee j <_c i & (4) \end{pmatrix}$$

In this definition, the level annotated by the index c is the lowest (the more concrete) level of observation and a is the highest (the more abstract). We state what it means for a pair of relations to refine another pair of relations. We can only compare pairs of relations that are bound to the same underlying set of instants. This relation is composed of four predicates, each of which indicates how one of the four relations is translated into the other level of observation.

- *Precedence abstraction:* If a strictly precedes b in the lower level, then it can either be coincident to it in the higher level or still precede it. This means that a distinction which is visible at a lower level can either disappear at a higher one or remain visible, depending on the behaviour of the refinement for these instants – *Equation (1)*
- *Precedence embodiment:* If a strictly precedes b in the higher level, then it can only precede it in the lower level. This means that the distinction between these instants already existed in the higher level and thus cannot be lost when refining. Looking closer at a system preserves precedence between instants – *Equation (2)*
- *Coincidence abstraction:* If a is coincident to b in the lower level, they stay coincident in the higher level. Looking at the system from a higher point of view cannot reveal temporal distinction between events – *Equation (3)*
- *Coincidence embodiment:* If a is coincident to b in the higher level then these instants cannot be independent in the lower level; they will still be related but nothing can be said on the nature of this relation – *Equation (4)*

2.3   Horizontal Constraints with CCSL

A time structure, such as the one depicted in Fig. 1 displays relations between the occurrences of events. A time structure can either be seen as a specification, depicting how occurrences of events should be related, or as a temporal implementation of an informal specification depicting how the instants are bound to one another to respect a certain set of relations. Said relations can be expressed informally, but it is usually preferred to use a formal modelling language dedicated to this purpose. CCSL [1] is such a language, that was defined in the MARTE [13] UML [12] profile. It proposes various relations and expressions to bind events (that are named clocks in the language) with one another. Considering two clocks c_1 and c_1, here are the CCSL relations that will be used in the rest of the paper to illustrate our contribution:

1. *Subclocking:* $\qquad\qquad\qquad\qquad c_1 \sqsubseteq c_2 \Longleftrightarrow \forall i \in c_1, \exists j \in c_2, i \approx j$

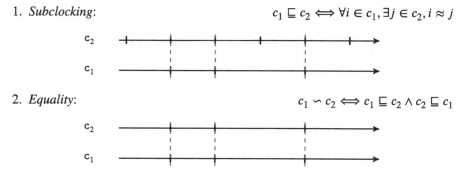

2. *Equality:* $\qquad\qquad\qquad\qquad c_1 \backsim c_2 \Longleftrightarrow c_1 \sqsubseteq c_2 \wedge c_2 \sqsubseteq c_1$

3. *Exclusion:* $c_1 \sharp c_2 \iff \forall (i,j) \in (c_1 \times c_2), \neg i \approx j$

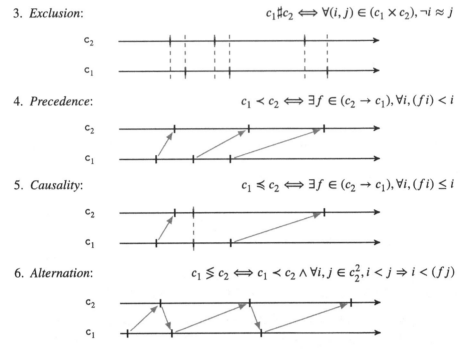

4. *Precedence:* $c_1 \prec c_2 \iff \exists f \in (c_2 \to c_1), \forall i, (fi) < i$

5. *Causality:* $c_1 \preccurlyeq c_2 \iff \exists f \in (c_2 \to c_1), \forall i, (fi) \leq i$

6. *Alternation:* $c_1 \lessapprox c_2 \iff c_1 \prec c_2 \wedge \forall i, j \in c_2^2, i < j \Rightarrow i < (fj)$

These definitions were first introduced in [3] and were mechanized in a previous work [9]. Note that the precedence and causality definitions have been simplified for the purpose of this paper, as there are more constraints to the binding function f (it has to be bijective for instance). These additional constraints are naturally taken into account in the formal mechanized counterpart of this work. All details can be found in [8].

3   An Example of Multi-layered Modelling

This theoretical ground provides us with the following expressiveness: it is possible to specify the behaviour of several sub-parts of a given system, as well as several parts from different systems. It is also possible to compare orders with one another towards the notion of refinement, but it is not yet possible to express constraints and properties between events coming from different layers of refinement. This section proposes a simple example as to why this is relevant, after which solutions will be proposed.

3.1   The *Deadlock* Petrinet

Petri nets [14] are considered by many as the assembler of concurrent system, and their semantics is well known in the field. Let us consider the Petrinet on Fig. 3a, usually used as a toy example for teaching purposes.

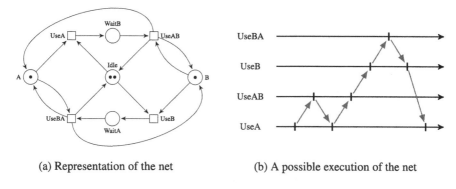

(a) Representation of the net   (b) A possible execution of the net

Fig. 3. The *Deadlock* Petrinet

This net, however simple, is very interesting because it allows a state of deadlock to be reached. It consists of two processes, initially in the *Idle* state, that can execute two tasks: the first one requires the use of the resource A, then the resource B after which, both resources are freed, and the second task is similar, although the resources are required in the opposite order. Both these tasks can be done by any of the two processes, a possibly infinite number of time, as long as both resources are freed when the second process begins the other task. If both processes require the use of their first resource concurrently, a deadlock state is reached because both processes will endlessly wait for the resource that is currently retained by the other process. A possible execution of this net resulting in a deadlock is depicted in Fig. 3b. After two cycles of the first task and one cycle of the second one, resource B is required by a process while resource A is required by the second one, hence the deadlock.

3.2 A Functional View of the System

As surprising as it may seem, writing constraints on this system to forbid the deadlock case is not trivial. In CCSL for instance, it is not easy to forbid inter-twining of events. In this specific example, although it is possible, expressing that an occurrence of *UseB* should never occur between an occurrence of *UseA* and *UseAB* – and vice versa – requires a long list of constraints.

An easier way of expressing such constraints arises when considering the abstract function the deadlock-free runs of this systems fulfil. In this case, such a function is simple: the net should execute both tasks an arbitrary number of times, in an arbitrary order, regardless of the resources actually used in these tasks. Such a behaviour can be depicted as an automaton, shown on Fig. 4a, with a possible execution of the system shown on Fig. 4b.

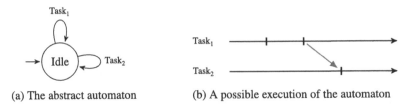

(a) The abstract automaton   (b) A possible execution of the automaton

Fig. 4. A functional representation of the *Deadlock* net

3.3   Binding the Two Levels of Description

The interesting aspect of this second system is that it does not allow any faulty behaviour, which the first one does. Should we synchronize these systems, we would be able to forbid such deadlocked runs. The usual manner to coordinate systems with one another is to compose them horizontally, as seen in Sect. 2.3. However, this implicitly means that both systems are part of a global system whose behaviour has to be specified. In this case, this is not true because they do not live in the same level of observation. The second one can be seen as a specification while the first one can be seen as a possible implementation of said specification. In other words, expressing the fact that the *Deadlock* net should behave as an instantiation – a more concrete implementation – of the automaton should ensure the correctness of its behaviour. This means that constraints between layers of refinement need to be expressed in such cases. Both levels should have their own layer of time, and both these layers should satisfy the relation depicted in Sect. 2.2. In each of these layers, constraints specific to each system can be written and expressed using CCSL. The remaining step is to allow the definition of inter-layers constraints, binding each abstract clock with concrete clocks that refine the event it represents. In our example, this would mean expressing that both *UseA* and *UseAB* should refine *Task₁* while *UseB* and *UseBA* should refine *Task₂*, which the following sections will make possible.

4   Stakes of the Approach

On the Combination of CCSL *and Refinement.* In Sect. 2.2 we summarized a previous work on the modelling of refinement which proposes to handle the different layers of refinement by assigning each of them a separate partial order and ordering these orders with a specific relation. CCSL is itself based on partial orders which means both our notion of refinement and CCSL share the same formalism, which allows us to mix them together to assess how CCSL would behave when combined with refinement.

On Preservation over Instants. Before starting the investigation on this conjunction, it is important to note that the notion of refinement depicted in Sect. 2.2 is fundamentally between partial orders. In that regard, it preserves by nature any required property over these, because it has been defined in that purpose. For instance, the strict precedence between instants is preserved through embodiment using the second equation, while the coincidence between instants is preserved through abstraction as depicted by the third equation. In other words, there is no need to prove that our refinement relation preserves the right properties in terms of instants ordering, because it does so by definition.

On Preservation over Clocks. However, these preservations are natural in terms of relations between instants, yet not for relations between clocks. This means that it would be a mistake to assume that, by nature, relations between clocks should be preserved by the use of our relation of refinement. This makes the following question relevant: are there some properties between clocks which, when specified at a given level of refinement, could be transferred into another level of refinement without additional requirements? Sects. 5 and 6 aim at answering this question after introducing refinement relations between clocks.

An Example of What to Expect. Let us take an example of what to expect here: at a given level of refinement, we know that subclocking is transitive. This means that, given three clocks c_1, c_2 and c_3, if we know that $c_1 \sqsubseteq c_2$ and $c_2 \sqsubseteq c_3$ then we can deduce $c_1 \sqsubseteq c_3$. In other words, the property $c_1 \sqsubseteq c_3$ does not need to be given as an additional constraint, because it is deducible from the rest of the context. In the following sections, we try to assess such assumptions, but in various levels of refinement.

Refinement Between Clocks. In order to establish such properties, we need to express what it means for clocks to refine one another. In that purpose, we propose two different relations of refinement between clocks, both of which bind clocks from different levels of refinement. The first one, depicted in Sect. 5 considers a refinement with an arbitrary number of refined ticks for a given abstract event, whereas Sect. 6 depicts a more constrained form of refinement, where a single concrete tick is allowed. These two notions are motivated both by their own expressiveness and by the various CCSL relations they preserve.

Temporal Context. We consider, for the remaining of this paper, two layers of refinement characterized by two partial orders $(<_c, \approx_c)$ and $(<_a, \approx_a)$. We assume that these partial orders satisfy $(<_c, \approx_c) <_r (<_a, \approx_a)$. In that context, whenever a CCSL relation will be mentioned, it will be prefixed with the layer of refinement in which it is defined. For instance, if a clock c_1 precedes a clock c_2 in the abstract layer of refinement, it will be written $c_1 \prec_a c_2$. Usually, the level of abstraction of each clock will also be written, in which case the previous relation becomes $c_{a_1} \prec_a c_{a_2}$. In this context, preservation properties toward other existing CCSL constructs can and will be discussed. As a side node, every result that is given afterwards has been mechanized and proved using the AGDA proof assistant [11], even though said mechanization will not be detailed in this paper. The full development is provided in the first author PhD report [8].

5   A First Generic CCSL Relation of Refinement

5.1   Definition of 1-N Refinement

Intuition. 1-N refinement aims at modelling the most common, unconstrained relation of refinement. In that purpose, each abstract clock can be refined by several concrete clocks, while each concrete event must be abstracted by a single abstract clock. In addition, each tick of the abstract clock can be refined by a strictly positive number of ticks for each of its concrete clocks. These number of ticks can vary throughout the execution of the system. The following example and definition will emphasize and capture these informal requirements.

Example. Let us consider the following situation: a worker is driving nails in a plank of wood. In the abstract level, we consider driving the nail as an atomic action, while in the more concrete level we consider hitting the nail with the hammer the atomic action. Depending on how strong the worker is, it might take him a few strikes for each nail to be driven in the plank, and each nail could be driven in a different number of strikes. In this case, the clock $Striking_c$ is a 1-N refinement of the clock $Driving_a$, as shown on Fig. 5, where the blue rectangles are the equivalence classes from the abstract point of view. Note that, although this picture is very similar to the ones in Sect. 2.3, both clocks are here coming from different levels of abstraction.

Fig. 5. An example of 1-N refinement

Definition. As a formalization of the previous example and intuition, we define a relation of 1-N refinement between clocks, with C being the set of all clocks:

$$\forall (c_c, c_a) \in C^2, c_c \precsim_{1-n} c_a \Leftrightarrow (\forall i \in c_a, \exists j \in c_c, i \approx_a j) \wedge (\forall j \in c_c, \exists i \in c_a, i \approx_a j)$$

This definition is composed of two parts as follows:

- Any tick of the abstract clock is refined by a tick of the concrete clock. These two ticks are coincident from the abstract point of view.
- Any tick of the concrete clock is a refinement of a tick of the abstract clock. These two ticks are coincident in terms of the abstract partial order.

The required unicity of the abstraction in terms of ticks is a direct consequence of this definition. Indeed, should we take two abstract ticks of the same concrete ticks, they are coincident from the abstract point of view by transitivity of the abstract coincidence, since both are coincident with the concrete tick.

Moreover, clocks are subsets of totally ordered instants which means that two coincident instants of the same clock are in fact propositionally equal, which ensures the required unicity. Such a property, along with others, is part of the formal counterpart of this work.

5.2 1-N Refinement and Coincidence-Based CCSL Relations

We experiment how 1-N refinement behaves when combined with CCSL notions related to coincidence, that is subclocking, equality, exclusion and union.

Subclocking. As coincidence between instants is preserved through abstraction, one would expect subclocking to be preserved as well. This has been proven to hold, which lead to the following theorem:

$$\forall (c_{c_1}, c_{c_2}, c_{a_1}, c_{a_2}) \in C^4, (c_{c_1} \lesssim_{1-n} c_{a_1}) \wedge (c_{c_2} \lesssim_{1-n} c_{a_2}) \wedge (c_{c_1} \sqsubseteq_c c_{c_2}) \Rightarrow (c_{a_1} \sqsubseteq_a c_{a_2})$$

Equality. Since equality between clocks is a case of double subclocking, equality is also preserved through abstraction, which leads to the following theorem:

$$\forall (c_{c_1}, c_{c_2}, c_{a_1}, c_{a_2}) \in C^4, (c_{c_1} \lesssim_{1-n} c_{a_1}) \wedge (c_{c_2} \lesssim_{1-n} c_{a_2}) \wedge (c_{c_1} \frown_c c_{c_2}) \Rightarrow (c_{a_1} \frown_a c_{a_2})$$

Exclusion. Refining excluded clocks cannot create coincident instants, which means the refined clocks are excluded as well. This makes sense because the abstract excluded clocks have ticks that are all in different equivalence classes regarding abstract coincidence and the refined instants are still in these classes and thus cannot be coincident even from the lower point of view. This leads to the following theorem:

$$\forall (c_{c_1}, c_{c_2}, c_{a_1}, c_{a_2}) \in C^4, (c_{c_1} \lesssim_{1-n} c_{a_1}) \wedge (c_{c_2} \lesssim_{1-n} c_{a_2}) \wedge (c_{a_1} \# _a c_{a_2}) \Rightarrow (c_{c_1} \# _c c_{c_2})$$

Multiple Concrete Clocks. When two clocks refine the same one, it means that these two clocks track events that are part of the events tracked by the clock being refined. Thus, it is natural to assume that the union of these clocks is still a refinement of the abstract clock. In CCSL, the union of two clocks is simply the union of their ticks, which leads to the following theorem:

$$\forall (c_c, c_{c_1}, c_{c_2}, c_a) \in C^4, (c_{c_1} \lesssim_{1-n} c_a) \wedge (c_{c_2} \lesssim_{1-n} c_a) \wedge (c_c = c_{c_1} \cup c_{c_2}) \Rightarrow (c_c \lesssim_{1-n} c_a)$$

Multiple Abstract Clocks. On the other hand, when a clock is a refinement of two clocks, this means that the event it tracks is deduced from two entities, leading us to assume that these entities are in fact the same. And indeed, our formalism implies such clock equality, showing that our clock refinement is not symmetrical, as expected. Here is the related theorem:

$$\forall (c_c, c_{a_1}, c_{a_2}) \in C^3, \qquad (c_c \lesssim_{1-n} c_{a_1}) \wedge (c_c \lesssim_{1-n} c_{a_2}) \Rightarrow (c_{a_1} \frown_a c_{a_2})$$

5.3   1-N Refinement and Precedence-Based CCSL Relations

While 1-N refinement preserves CCSL notions related to coincidence, as depicted
in Sect. 5.2, investigating its impact on notions based on precedence (precedence,
causality and alternation) is not as fruitful since proving the refinement is not
sufficient for these relations to be transported. Figure 6 shows two examples
where these relations are not preserved. On these pictures, the two levels of
abstraction are represented, as well as two clocks per level. The brown arrows
represent the abstraction of an event while the orange ones show an embodiment
of such event, following the relation of refinement between these clocks. The
functions f_a and f_c are the binding functions of the precedences from both levels.
Note that the instants i, j, k, l are not suffixed because they do not belong to
any specific level of abstraction, or rather, they belong to both.

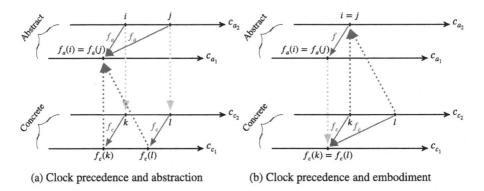

(a) Clock precedence and abstraction   (b) Clock precedence and embodiment

Fig. 6. 1-N refinement and precedence

Abstraction of Precedence. Figure 6a shows an example where a concrete prece-
dence ($c_{c_1} \prec_c c_{c_2}$) is not preserved through abstraction ($\neg(c_{a_1} \prec_a c_{a_2})$). This
happens because two instants that precede one another in the concrete level (in
this case $f_c(k)$ and $f_c(l)$) might be two instances of the same abstract event.
In this case two events that precede one another in this abstract level (i and j)
might actually be mapped to the same abstract event by the precedence function
f_a we are trying to build, which should be injective.

Embodiment of Precedence. Figure 6b shows an example where an abstract prece-
dence ($c_{a_1} \prec_a c_{a_2}$) is not preserved through embodiment ($\neg(c_{c_1} \prec_c c_{c_2})$). This
happens for a similar reason that invalidates the injectivity of the function f_c.

Causality and Alternation. Since alternation is a specific case of precedence
and causality is a less constrained case of precedence, both of these relations
are necessarily not preserved through abstraction nor embodiment as well. The
reason is that a tick can be refined by several ticks of the same clock, thus
possibly compromising the injectivity of the binding function. Should we forbid
such behaviour, more relations could be preserved, which leads to 1-1 refinement.

6   A Second Specific CCSL Relation of Refinement

6.1   Definition of 1-1 Refinement

Intuition. 1-N refinement does not provide an immediate preservation of coincidence-based CCSL relations, because it depicts a situation where such relations simply are not propagated from one level to the next. As stated before, the underlying reason of such a limitation does not lie in our ability to model refinement, but rather in the arbitrary number of ticks each concrete clock can have bound to a single abstract tick. By changing the number of ticks, relations around precedence are naturally not preserved because they rely on the bijectivity of the binding function. 1-1 refinement is meant to preserve the number of ticks through refinement. Each abstract clock can still be refined by an arbitrary number of concrete clocks, although each abstract tick can now only be refined by a single tick of each concrete clocks.

Example. Going back at our example from Sect. 5.1, we can imagine that the worker can now drive nails more efficiently: he manages to do so in only one strike but it requires an increased accuracy, which is now considered a new concrete step for the purposes of this example. This new situation is depicted on Fig. 7.

Fig. 7. An example of 1-1 refinement

Definition. This example leads to the definition of a relation of 1-1 refinement, which only differs from 1-N refinement by the uniqueness of the refined ticks:

$$\forall (c_c, c_a) \in_{c_c} C^2_a, \underset{c_c}{\precapprox}_{1-1} c_a \Leftrightarrow (\forall i \in c_a, \exists! j \in c_c, i \approx_a j) \land (\forall j \in c_c, \exists i \in c_a, i \approx_a j)$$

6.2   1-1 Refinement and Coincidence-Based CCSL Relations

Since unique existence implies existence, 1-1 refinement is trivially a specific case of 1-N refinement, which means that any theorem regarding 1-N refinement and coincidence-based CCSL relation shown in Sect. 5.2 still holds for 1-1 refinement.

6.3   1-1 Refinement and Precedence-Based CCSL Relations

The main goal of the 1-1 refinement is to provide CCSL with a notion of clock refinement which naturally preserves precedence-related relations. In this section we investigate to what extent this preservation is guaranteed.

Embodiment of Causality. Causality is not preserved through embodiment, even with 1-1 refinement, and for a very concrete reason, which is that abstract coincidence can basically be transformed into any relation in the concrete level. Let us take four clocks c_{c_1}, c_{c_2}, c_{a_1} and c_{a_2} such that $c_{c_1} \lesssim_{1-1} c_{a_1}$ and $c_{c_2} \lesssim_{1-1} c_{a_2}$. If we have $c_{a_1} \preccurlyeq_a c_{a_2}$ this means that for a tick i of c_{a_2} we have a tick j of c_{a_1} such that $f(i) = j$ and $i \leq_a j$. It is thus possible that $i \approx_a j$ which, in the concrete level, can be transformed into $j \prec_c i$ which invalidates preservation of causality through embodiment.

Embodiment of Precedence. Precedence, however, does not exhibit this kind of possibility, and is preserved directly through embodiment, leading to the following theorem:

$$\forall (c_{c_1} c_{c_2} c_{a_1} c_{a_2}) \in \mathcal{C}_{c_1}^4 \, (c_{c_1} \lesssim_{1-1} c_{a_1}) \wedge (c_{c_2} \lesssim_{1-1} c_{a_2}) \wedge (c_{a_1} \prec_a c_{a_2}) \Rightarrow (c_{c_1} \prec_c c_{c_2})$$

Abstraction of Precedence and Causality. Causality and precedence are both preserved in term of causality through abstraction. In other words, causality is fully preserved while precedence becomes causality, leading to the following theorem:

$$\forall (c_{c_1} c_{c_2} c_{a_1} c_{a_2}) \in \mathcal{C}_{c_2}^4 \, (c_{c_2} \lesssim_{1-1} c_{a_2}) \wedge ((c_{c_1} \prec_c c_{c_2}) \vee (c_{c_1} \preccurlyeq_c c_{c_2})) \Rightarrow (c_{a_1} \preccurlyeq_a c_{a_2})$$

1-1 Refinement and Alternation. Since precedence is transformed into causality through abstraction, we cannot expect alternation to be preserved in the process. As for embodiment, Fig. 8 gives us an example as to why it does not hold either. Indeed, in the concrete level we can see that both $f_c(k)$ and $f_c(l)$ precede k, invalidating alternation.

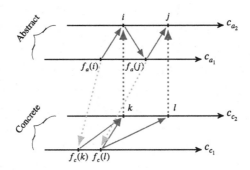

Fig. 8. Alternation and embodiment

7 Additional Relations of Refinement

An interesting question would be to assess whether other refinement relations would be relevant both in terms of expressiveness and in terms of direct preservation of properties, as even 1-1 refinement does not ensure the preservation of all CCSL constructs. This section briefly (and informally) explores this question.

1-X Refinement. 1-X refinement could mean both of the following ideas: either X is set and each abstract tick is refined by X ticks of the concrete clock, or X is not set in which case each abstract tick must be refined by the same number of concrete ticks. In the second case, the first ticks of the abstract clocks fixes X, after which the clocks behaves as in the first case, which makes both of them very similar. Both cases are a special case of 1-N refinement and could be modelled as such, although they would not improve the number of direct properties that are preserved, since they both change the overall number of ticks of the clocks. Thus 1-X refinement seems moderately relevant.

N-N Refinement. The notions of 1-1 and 1-N refinement revolves around the number of possible ticks rather than on the number of possible clocks. In both cases, these refinements are actually 1-N in terms of clocks. Although a N-N refinement in terms of clocks is easy to imagine (and is actually currently supported by considering the union of the N abstract clocks) it is hard to picture what a N-N refinement in terms of ticks would mean. This could mean that any number of ticks of the concrete clock would stand for any number of abstract ticks, which does not seem reasonable nor relevant when thinking in terms of concrete vs abstract. This could require further inspection but 1-1 and 1-N refinement as presented seem to capture the essence of behavioural refinement.

8 Conclusion

8.1 Assessments

We propose an extension of the CCSL modelling language in order to enable refinement checking between models. We provide two new CCSL relations expressing the expected refinement between two clocks: 1-1 refinement and 1-N refinement. These relations are different from the others CCSL relations because they build a bridge between clocks coming from two different specifications rather than a bridge between clocks from the same specification. Each of these different specifications must correspond to a given level of refinement in the sense that their partial order should comply with the relation of refinement which was defined in Sect. 2.2. This extends the spectrum of CCSL through the expression of relations between models at the various development steps.

This extended CCSL can be used in multiple ways. It allows the developer to compare CCSL models in terms of refinement. It also provides a wide area of experimentation around refinement, since 1-1 refinement is more constrained but directly preserves most CCSL relation while 1-N refinement is more permissive but only preserves the main ones. Inside this range, our formal framework enables the definition and proof of preservation of other intermediate refinement relations less constraining than the 1-1 and preserving more properties than the 1-N. Moreover, this context can directly be used throughout the development of time critical systems using the classical top down design where a concrete model must refine an abstract one, in other words the concrete implementation must comply with the abstract specification. Finally, this work can be used to make explicit a common behaviour between various concrete models in the same way inheritance captures common aspects between different classes in object oriented design, ultimately enabling their synchronisation through this common behavioural interface.

In the process of adding refinement to CCSL, we investigated how CCSL operators behave when coupled with the new relations of clock refinement. These preservation properties have been mathematically proven in our formal framework that is thoroughly depicted in [8]. This investigation was fruitful when dealing with CCSL notions which are bound to coincidences using 1-N refinement, in the sense that refinement is fairly regular towards coincidence. It was even more fruitful when considering 1-1 refinement, which was designed to preserve properties of precedence, and which succeeded at doing so. These preservation properties are fundamental regarding the confidence we have in capturing the essence of refinement between clocks in a multi-layered temporal context.

8.2 Perspectives

TimeSquare [4] is an operational semantics for CCSL that provides us with a modelling environment with associated simulation and verification tools. It is based on a single partial order on which clocks are built conforming to a given specification. The main perspective for our work would be to enrich both the official version of CCSL and TimeSquare with our notions of refinement, so that actual engineers could design their multi-layered models in CCSL and verify them using TimeSquare. This requires to extend the core of the tool to handle multiple partial orders instead of one.

Our work could be used to specify relations between systems whose behaviours are similar, in the sense that they refine the same specification. We would like to emphasize and validate this aspect of our work through examples in that direction.

Our approach could be used on a wider range of case studies including more complex industrial size ones that would probably emphasize the relevance of our approach in modelling the behaviour of complex systems.

References

1. André, C., Mallet, F.: Clock Constraints in UML/MARTE CCSL. Research Report RR-6540, INRIA (2008)

2. Buxton, J.N., Randell, B.: Software Engineering Techniques: Report of a Conference Sponsored by the NATO Science Committee, Rome, Italy, 27–31 October 1969, Brussels, Scientific Affairs Division, NATO (1970)
3. Deantoni, J., André, C., Gascon, R.: CCSL denotational semantics. Research Report RR-8628 (2014)
4. DeAntoni, J., Mallet, F.: TimeSquare: treat your models with logical time. In: Furia, C.A., Nanz, S. (eds.) TOOLS 2012. LNCS, vol. 7304, pp. 34–41. Springer, Heidelberg (2012). https://doi.org/10.1007/978-3-642-30561-0_4
5. Dijkstra, E.W.: On the Role of Scientific Thought, pp. 60–66. Springer, New York (1982). https://doi.org/10.1007/978-1-4612-5695-3_12
6. Lee, E.A.: Cyber physical systems: design challenges. In: 11th IEEE International Symposium on Object-Oriented Real-Time Distributed Computing (ISORC 2008), Orlando, Florida, USA, 5–7 May 2008, pp. 363–369. IEEE Computer Society (2008). https://doi.org/10.1109/ISORC.2008.25
7. Li, M., Vitányi, P.M.B.: An introduction to Kolmogorov Complexity and Its Applications. Texts and Monographs in Computer Science. Springer, New York (1993). https://doi.org/10.1007/978-1-4757-3860-5
8. Montin, M.: A formal framework for heterogeneous systems semantics. Ph.D. thesis (2020). http://montin.perso.enseeiht.fr/these.pdf
9. Montin, M., Pantel, M.: Mechanizing the denotational semantics of the clock constraint specification language. In: Abdelwahed, E.H., Bellatreche, L., Golfarelli, M., Méry, D., Ordonez, C. (eds.) MEDI 2018. LNCS, vol. 11163, pp. 385–400. Springer, Cham (2018). https://doi.org/10.1007/978-3-030-00856-7_26
10. Montin, M., Pantel, M.: Ordering strict partial orders to model behavioral refinement. In: Derrick, J., Dongol, B., Reeves, S. (eds.) Proceedings 18th Refinement Workshop, Refine@FM 2018. EPTCS, vol. 282, Oxford, UK, 18 July 2018, pp. 23–38 (2018). https://doi.org/10.4204/EPTCS.282.3
11. Norell, U.: Towards a practical programming language based on dependent type theory. Ph.D. thesis, Department of Computer Science and Engineering, Chalmers University of Technology, SE-412 96 Göteborg, Sweden, September 2007
12. (OMG), O.M.G.: Unified modeling language, December 2017. https://www.omg.org/spec/UML/About-UML/
13. (OMG), O.M.G.: UML profile for MARTE, April 2019. https://www.omg.org/spec/MARTE/About-MARTE/
14. Petri, C.A.: Kommunikation mit Automaten. Dissertation, Schriften des IIM 2, Rheinisch-Westfälisches Institut für Instrumentelle Mathematik an der Universität Bonn, Bonn (1962)
15. Thurner, S., Hanel, R., Klimek, P.: Introduction to the Theory of Complex Systems. Oxford University Press, Oxford (2018)
16. Winskel, G.: Event structures. In: Brauer, W., Reisig, W., Rozenberg, G. (eds.) ACPN 1986. LNCS, vol. 255, pp. 325–392. Springer, Heidelberg (1987). https://doi.org/10.1007/3-540-17906-2_31

A Case Study on Parametric Verification of Failure Detectors

Thanh-Hai Tran[1]([✉]), Igor Konnov[2], and Josef Widder[2]

[1] TU Wien, Vienna, Austria
tran@forsyte.at
[2] Informal Systems, Vienna, Austria

Abstract. Partial synchrony is a model of computation in many distributed algorithms and modern blockchains. Correctness of these algorithms requires the existence of bounds on message delays and on the relative speed of processes after reaching Global Stabilization Time (GST). This makes partially synchronous algorithms parametric in time bounds, which renders automated verification of partially synchronous algorithms challenging. In this paper, we present a case study on formal verification of both safety and liveness of a Chandra and Toueg failure detector that is based on partial synchrony. To this end, we specify the algorithm and the partial synchrony assumptions in three frameworks: TLA$^+$, Ivy, and counter automata. Importantly, we tune our modeling to use the strength of each method: (1) We are using counters to encode message buffers with counter automata, (2) we are using first-order relations to encode message buffers in Ivy, and (3) we are using both approaches in TLA$^+$. By running the tools for TLA$^+$ (TLC and APALACHE) and counter automata (FAST), we demonstrate safety for fixed time bounds. This helped us to find the inductive invariants for fixed parameters, which we used as a starting point for the proofs with Ivy. By running Ivy, we prove safety for arbitrary time bounds. Moreover, we show how to verify liveness of the failure detector by reducing the verification problem to safety verification. Thus, both properties are verified by developing inductive invariants with Ivy. We conjecture that correctness of other partially synchronous algorithms may be proven by following the presented methodology.

Keywords: Failure detectors · TLA$^+$ · Counter automata · FAST · Ivy

1 Introduction

Distributed algorithms play a crucial role in modern infrastructure, but they are notoriously difficult to understand and to get right. Network topologies, message delays, faulty processes, the relative speed of processes, and fairness conditions might lead to behaviors that were neither intended nor anticipated by algorithm

© IFIP International Federation for Information Processing 2021
Published by Springer Nature Switzerland AG 2021
K. Peters and T. A. C. Willemse (Eds.): FORTE 2021, LNCS 12719, pp. 138–156, 2021.
https://doi.org/10.1007/978-3-030-78089-0_8

designers. Hence, many specification and verification techniques for distributed algorithms [14,22,27,29] have been developed.

Verification techniques for distributed algorithms usually focus on two models of computation: synchrony [31] and asynchrony [19,20]. Synchrony is hard to implement in real systems, while many basic problems in fault-tolerant distributed computing are unsolvable in asynchrony.

Partial synchrony lies between synchrony and asynchrony, and escapes their shortcomings. To guarantee liveness properties, proof-of-stake blockchains [9,35] and distributed algorithms [8,11] assume time constraints under partial synchrony. That is the existence of bounds Δ on message delay, and Φ on the relative speed of processes after some time point. This combination makes partially synchronous algorithms parametric in time bounds. While partial synchrony is important for system designers, it is challenging for verification.

We thus investigate verification of distributed algorithms under partial synchrony, and start with the specific class of failure detectors: a Chandra and Toueg failure detector [11]. This is a well-known algorithm under partial synchrony that provides a service that can be used to solve many problems in fault-tolerant distributed computing.

Contributions. In this paper, we do parametric verification of both safety and liveness of the Chandra and Toueg failure detector in case of unknown bounds Δ and Φ. In this case, both Δ and Φ are arbitrary, and the constraints on message delay and the relative speeds hold in every execution from the start.

1. We extend the cutoff results in [34] for partial synchrony. In a nutshell, a cutoff for a parameterized algorithm $\mathcal{A}$ and a property ϕ is a number k such that ϕ holds for every instance of $\mathcal{A}$ if and only if ϕ holds for instances with k processes [7,16]. While the cutoff results [34] are for synchrony or asynchrony, our results are for partial synchrony. Hence, we verify the Chandra and Toueg failure detector under partial synchrony by checking instances with two processes.
2. We introduce the encoding techniques to efficiently specify the failure detector based on our cutoff results. These techniques can tune our modeling to use the strength of the tools: FAST, Ivy, and model checkers for TLA$^+$.
3. We demonstrate how to reduce the liveness properties Eventually Strong Accuracy, and Strong Completeness to safety properties.
4. We check the safety property Strong Accuracy, and the mentioned liveness properties on instances with fixed parameters by using FAST, and model checkers for TLA$^+$.
5. To verify cases of arbitrary bounds Δ and Φ, we find and prove inductive invariants of the failure detector with the interactive theorem prover Ivy. We reduce the liveness properties to safety properties by applying the mentioned techniques. While our specifications are not in the decidable theories that Ivy supports, Ivy requires no additional user assistance to prove most of our inductive invariants.

Related work. Research papers about partially synchronous algorithms, including papers about failure detectors [1,2,24] contain manual proofs and no formal specifications. Without these details, proving those distributed algorithms with interactive theorem provers [13,29] is impossible. To test a candidate I for an inductive invariant with fixed parameters, system designers can apply probabilistic random checking with TLA$^+$ and TLC [23]. However, this approach randomly explores a subset of the execution space. Hence, it can show a counterexample to induction, but cannot prove that I is an inductive invariant. We prove inductive invariants in small cases with the model checker APALACHE [18]. System designers can use timed automata [3] and parametric verification frameworks [4,25,26] to specify and verify timed systems. In the context of timed systems, we are aware of only one paper about verification of failure detectors [5]. In this paper, the authors used three tools, namely UPPAAL [25], mCRL2 [10], and FDR2 [30] to verify small instances of a failure detector based on a logical ring arrangement of processes. Their verification approach required that message buffers were bounded, and had restricted behaviors in the specifications. Moreover, they did not consider the bound Φ on the relative speed of processes. In contrast, there are no restrictions on message buffers, and no ring topology in the Chandra and Toueg failure detector. Moreover, our work is to verify the Chandra and Toueg failure detector in case of arbitrary bounds. In recent years, automatic parameterized verification techniques [14,19,31] have been introduced for distributed systems, but they are designed for synchronous and/or asynchronous models. Interactive theorem provers have been used to prove correctness of distributed algorithms recently. For example, researchers proved safety of Tendermint consensus with Ivy [17].

Structure. In Sect. 2, we summarize the Chandra and Toueg failure detector, and the cutoff results in [34]. In Sect. 3, we extend the cutoff results in [34] for partial synchrony. Our encoding technique is presented in Sect. 4. In Sect. 5, we present how to reduce the mentioned liveness properties to safety ones. Experiments for small Δ and Φ are described in Sect. 6. Ivy proofs for parametric Δ and Φ are discussed in Sect. 7.

2   Preliminaries

This section describes the Chandra and Toueg failure detector [11], and the cutoff results [34] that we can extend to this failure detector under partial synchrony.

A failure detector can be seen as an oracle to get information about crash failures in the distributed system. A failure detector usually guarantees some of the following properties [11] (numbers $1..N$ denote the process identifiers):

- Strong Accuracy: No process is suspected before it crashes.
     $\mathbf{G}(\forall p, q \in 1..N : (Correct(p) \land Correct(q)) \Rightarrow \neg Suspected(p, q))$
- Eventual Strong Accuracy: There is a time after which correct processes are not suspected by any correct process.
     $\mathbf{FG}(\forall p, q \in 1..N : (Correct(p) \land Correct(q)) \Rightarrow \neg Suspected(p, q))$
- Strong Completeness: Eventually every crashed process is permanently suspected by every correct process.
     $\mathbf{FG}(\forall p, q \in 1..N : (Correct(p) \land \neg Correct(q)) \Rightarrow Suspected(p, q))$

Algorithm 1. The eventually perfect failure detector algorithm in [11]

1: *Every process $p \in 1..N$ executes the following*:
2: **for all** $q \in 1..N$ **do**   ▷ Initalization step
3:    $timeout\,[p, q] :=$ default-value
4:    $suspected\,[p, q] := \bot$

5: Send "alive" to all $q \in 1..N$ ▷ Task 1: repeat periodically
6: **for all** $q \in 1..N$ **do**   ▷ Task 2: repeat periodically
7:    **if** $suspected\,[p, q] = \bot$ **and not hear** q during last $timeout\,[p, q]$ ticks **then**
8:        $suspected\,[p, q] := \top$

9: **if** $suspected\,[p, q]$ **then**   ▷ Task 3: when receive "alive" from q
10:    $timeout\,[p, q] := timeout\,[p, q] + 1$
11:    $suspected\,[p, q] := \bot$

where **F** and **G** are operators in LTL (linear temporal logic), predicate $Suspect(p, q)$ refers to whether process p suspects process q in crashing, and predicate $Correct(p)$ refers to whether process p is correct. However, process p might crash later (and not recover). A crashed process p satisfies $\neg Correct(p)$.

Algorithm 1 presents the pseudo-code of the failure detector of [11]. A system instance has N processes that communicate with each other by sending-to-all and receiving messages through unbounded N^2 point-to-point communication channels. A process performs local computation based on received messages (we assume that a process also receives the messages that it sends to itself). In one system step, all processes may take up to one step. Some processes may crash, i.e., stop operating. Correct processes follow Algorithm 1 to detect crashes in the system. Initially, every correct process sets a default value for a timeout of each other, i.e. how long it should wait for others and assumes that no processes have crashed (Line 4). Every correct process p has three tasks: (i) repeatedly sends an "alive" message to all (Line 5), and (ii) repeatedly produces predictions about crashes of other processes based on timeouts (Line 6), and (iii) increases a timeout for process q if p has learned that its suspicion on q is wrong (Line 9). Notice that process p raises suspicion on the operation of process q (Line 6) by considering only information related to q: $timeout\,[p, q]$, $suspected\,[p, q]$, and messages that p has received from q recently.

Algorithm 1 does not satisfy Eventually Strong Accuracy under asynchrony since there exists no bound on message delay, and messages sent by correct processes might always arrive after the timeout expired. Liveness of the failure detector is based on the existence of bounds Δ on the message delay, and Φ on the relative speed of processes after reaching the global stabilization at some time point T_0 [11]. There are many models of partial synchrony [11,15]. In this paper, we focus only on the case of unknown bounds Δ and Φ because other models might call for abstractions. In this case, $T_0 = 1$, and both parameters Δ and Φ are arbitrary. Moreover, the following constraints hold in every execution:

– Constraint 1: If message m is placed in the message buffer from process q to process p by some $Send(m, p)$ at a time $s_1 \geq 1$, and if process p executes a

Receive(*p*) at a time s_2 with $s_2 \geq s_1 + \Delta$, then message *m* must be delivered to *p* at time s_2 or earlier.

– Constraint 2: In every contiguous time interval $[t, t + \Phi]$ with $t \geq 1$, every correct process must take at least one step.

These constraints make the failure detector parametric in Δ and Φ.

Moreover, Algorithm 1 is parameterized by the initial value of the timeout. If a default value of the timeout is too small, there exists a case in which sent messages are delivered after the timeout expired. It violates Strong Accuracy.

In [34], Thanh-Hai et al. defined a class of symmetric point-to-point distributed algorithms that contains the failure detector [11], and proved cutoffs on the number of processes for this class under asynchrony. These cutoff results guarantee that analyzing instances with two processes is sufficient to reason about the correctness of all instances of the Chandra and Toueg failure detector under asynchrony. In the following section, we will generalize this result to partial synchrony, which allows us to verify the mentioned properties on the failure detector by checking instances with only two processes.

3   Cutoffs of the Failure Detector

In this section, we extend the cutoffs of symmetric point-to-point distributed algorithms in [34] for partial synchrony.

Notice that time parameters in partial synchrony only reduce the execution space compared to asynchrony. Hence, we can formalize the system behaviors under partial synchrony by extending the formalization of the system behaviors under asynchrony in [34] with the notion of time, message ages, time constraints under partial synchrony. (Our formalization is left for the full report [32].)

In a nutshell, our cutoff results for the symmetric point-to-point class allow us to verify the mentioned properties on the failure detector under partial synchrony by checking small instances with one and/or two processes. Intuitively, the proofs of our cutoffs are based on the following observations:

– The global transition system and the desired property are symmetric [34].
– Let $\mathcal{G}_2$ and $\mathcal{G}_N$ be two instances of a symmetric point-to-point algorithm with 2 and *N* processes, respectively. By [34], two instances $\mathcal{G}_2$ and $\mathcal{G}_N$ are trace equivalent under a set of predicates in the desired property.
– We will now discuss that the constraints maintain partial synchrony. Let π_N be an execution in $\mathcal{G}_N$. We construct an execution π_2 in $\mathcal{G}_2$ by applying the index projection to π_N (formally defined in [34]). Intuitively, the index projection discards processes 3..*N* as well as their corresponding messages and buffers. Moreover, for every $k, \ell \in \{1, 2\}$, the index projection preserves (i) at which point in time process *k* takes a step, and (ii) what action process *k* takes at a time $t \geq 0$, and (iii) messages from process *k* to process ℓ. Figure 1 demonstrates an execution in $\mathcal{G}_2$ that is constructed based on a given execution in $\mathcal{G}_3$ with the index projection. Observe that Constraints 1 and 2 are maintained in this projection.

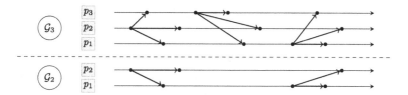

Fig. 1. Given execution in $\mathcal{G}_3$, construct an execution in $\mathcal{G}_2$ by index projection.

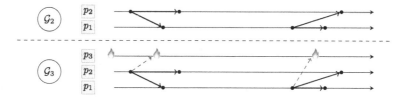

Fig. 2. Construct an execution in $\mathcal{G}_3$ based on a given execution in $\mathcal{G}_2$.

– Let π_2 be an execution in $\mathcal{G}_2$. We construct an execution π_N in $\mathcal{G}_N$ based on π_2 such that all processes $3..N$ crash from the beginning, and π_2 is an index projection of π_N [34]. For example, Fig. 2 demonstrates an execution in $\mathcal{G}_3$ that is constructed based on an given execution in $\mathcal{G}_2$. If Constraints 1 and 2 hold on π_2, these constraints also hold on π_N.

4   Encoding the Chandra and Toueg Failure Detector

In this section, we first discuss why it is sufficient to verify the failure detector by checking a system with only one sender and one receiver by applying the cutoffs presented in Sect. 3. Next, we introduce two approaches to encoding the message buffer, and an abstraction of in-transit messages that are older than Δ time-units. Finally, we present how to encode the relative speed of processes with counters over natural numbers. These techniques allow us to tune our models to the strength of the verification tools: FAST, Ivy, and model checkers for TLA$^+$.

4.1   The System with One Sender and One Receiver

We discussed our cutoff results in Sect. 3. These results allow us to verify the Chandra and Toueg failure detector under partial synchrony by checking only instances with two processes. In the following, we discuss the model with two processes, and formalize the properties with two-process indexes. By process symmetry, it is sufficient to verify Strong Accuracy, Eventually Strong Accuracy, and Strong Completeness by checking the following properties.

$$\mathbf{G}((\mathit{Correct}(1) \wedge \mathit{Correct}(2)) \Rightarrow \neg\mathit{Suspected}(2,1)) \tag{1}$$

$$\mathbf{FG}((\mathit{Correct}(1) \wedge \mathit{Correct}(2)) \Rightarrow \neg\mathit{Suspected}(2,1)) \tag{2}$$

$$\mathbf{FG}((\neg\mathit{Correct}(1) \wedge \mathit{Correct}(2)) \Rightarrow \mathit{Suspected}(2,1)) \tag{3}$$

We can take a further step towards facilitating verification of the failure detector. First, every process typically has a local variable to store messages that it needs to send to itself, instead of using a real communication channel. Hence, we can assume that there is no delay for those messages, and that each correct process never suspects itself. Second, local variables in Algorithm 1 are arrays whose elements correspond one-to-one with a remote process, e.g., $timeout[2, 1]$ and $suspected[2, 1]$. Third, communication between processes is point-to-point. When this is not the case, one can use cryptography to establish one-to-one communication. Hence, reasoning about Properties 1–3 requires no information about messages from process 1 to itself, local variables of process 1, and messages from process 2.

Due to the above characteristics, it is sufficient to consider process 1 as a sender, and process 2 as a receiver. In detail, the sender follows Task 1 in Algorithm 1, but does nothing in Task 2 and Task 3. The sender does not need the initialization step, and local variables $suspected$ and $timeout$. In contrast, the receiver has local variables corresponding to the sender, and follows only the initialization step, and Task 2, and Task 3 in Algorithm 1. The receiver can increase its waiting time in Task 1, but does not send any message.

4.2   Encoding the Message Buffer

Algorithm 1 assumes unbounded message buffers between processes that produce an infinite state space. Moreover, a sent message might be in-transit for a long time before it is delivered. We first introduce two approaches to encode the message buffer based on a logical predicate, and a counter over natural numbers. The first approach works for TLA^+ and Ivy, but not for counter automata (FAST). The latter is supported by all mentioned tools, but it is less efficient as it requires more transitions. Then, we present an abstraction of in-transit messages that are older than Δ time-units. This technique reduces the state space, and allows us to tune our models to the strength of the verification tools.

Encoding the Message Buffer with a Predicate. In Algorithm 1, only "alive" messages are sent, and the message delivery depends only on the age of in-transit messages. Moreover, the computation of the receiver does not depend on the contents of its received messages. Hence, we can encode a message buffer by using a logical predicate existsMsgOfAge(x). For every $k \geq 0$, predicate existsMsgOfAge(k) refers to whether there exists an in-transit message that is k time-units old. The number 0 refers to the age of a fresh message in the buffer.

It is convenient to encode the message buffer's behaviors in this approach. For instance, Formulas 4 and 5 show constraints on the message buffer when a new message is sent:

$$\text{existsMsgOfAge}'(0) = \top \tag{4}$$

$$\forall x \in \mathbb{N} . \, x > 0 \Rightarrow \text{existsMsgOfAge}'(x) = \text{existsMsgOfAge}(x) \tag{5}$$

Age indexes   0 1 2 3 4 $\cdots$ Age indexes   0 1 2 3 4 $\cdots$

Messages in **buf** [|⊠|⊠| |]   [| |⊠|⊠|]

Fig. 3. The message buffer after increasing message ages in case of **buf** = 6

where existsMsgOfAge$'$ refers to the value of existsMsgOfAge in the next state. Formula 4 implies that a fresh message has been added to the message buffer. Formula 5 ensures that other in-transit messages are unchanged.

Another example is the relation between existsMsgOfAge and existsMsgOfAge$'$ after the message delivery. This relation is formalized with Formulas 6–9. Formula 6 requires that there exists an in-transit message in existsMsgOfAge that can be delivered. Formula 7 ensures that no old messages are in transit after the delivery. Formula 8 guarantees that no message is created out of thin air. Formula 9 implies that at least one message is delivered.

$$\exists x \in \mathbb{N} \,.\, \mathsf{existsMsgOfAge}(x) \tag{6}$$

$$\forall x \in \mathbb{N} \,.\, x \geq \Delta \Rightarrow \neg\mathsf{existsMsgOfAge}'(x) \tag{7}$$

$$\forall x \in \mathbb{N} \,.\, \mathsf{existsMsgOfAge}'(x) \Rightarrow \mathsf{existsMsgOfAge}(x) \tag{8}$$

$$\exists x \in \mathbb{N} \,.\, \mathsf{existsMsgOfAge}'(x) \neq \mathsf{existsMsgOfAge}(x) \tag{9}$$

This encoding works for TLA^+ and Ivy, but not for FAST, because the input language of FAST does not support functions.

Encoding the Message Buffer with a Counter. In the following, we present an encoding technique for the buffer that can be applied in all tools TLA^+, Ivy, and FAST. This approach encodes the message buffer with a counter **buf** over natural numbers. The kth bit refers to whether there exists an in-transit message with k time-units old.

In this approach, message behaviors are formalized with operations in Presburger arithmetic. For example, assume $\Delta > 0$, we write **buf**$'$ = **buf** + 1 to add a fresh message in the buffer. Notice that the increase of **buf** by 1 turns on the 0th bit, and keeps the other bits unchanged.

To encode the increase of the age of every in-transit message by 1, we simply write **buf**$'$ = **buf** $\times$ 2. Assume that we use the least significant bit (LSB) first encoding, and the left-most bit is the 0th bit. By multiplying **buf** by 2, we have updated **buf**$'$ by shifting to the right every bit in **buf** by 1. For example, Fig. 3 demonstrates the message buffer after the increase of message ages in case of **buf** = 6. We have **buf**$'$ = **buf** $\times$ 2 = 12. It is easy to see that the 1st and 2nd bits in **buf** are on, and the 2nd and 3rd bits in **buf**$'$ are on.

Recall that Presburger arithmetic does not allow one to divide by a variable. Therefore, to guarantee the constraint in Formula 8, we need to enumerate all constraints on possible values of **buf** and **buf**$'$ after the message delivery. For example, assume **buf** = 3, and $\Delta = 1$. After the message delivery, **buf**$'$ is either 0 or 1. If **buf** = 2 and $\Delta = 1$, **buf**$'$ must be 0 after the message delivery.

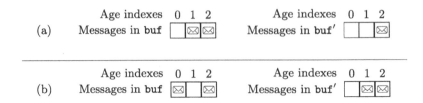

Fig. 4. The increase of message ages with the abstraction of old messages. In the case (a), we have $\Delta = 2$, buf $= 6$, and buf$' = 4$. In the case (b), we have $\Delta = 2$, buf $= 5$, and buf$' = 6$.

Importantly, the number of transitions for the message delivery depends on the value of Δ.

To avoid the enumeration of all possible cases, Formula 8 can be rewritten with bit-vector arithmetic. However, bit-vector arithmetic are currently not supported in all verification tools TLA$^+$, FAST, and Ivy.

The advantage of this encoding is that when bound Δ is fixed, every constraint in the system behaviors can be rewritten in Presburger arithmetic. Thus, we can use FAST, which accepts constraints in Presburger arithmetic. To specify cases with arbitrary Δ, the user can use TLA$^+$ or Ivy.

Abstraction of Old Messages. Algorithm 1 assumes underlying unbounded message buffers between processes. Moreover, a sent message might be in transit for a long time before it is delivered. To reduce the state space, we develop an abstraction of in-transit messages that are older than Δ time-units; we call such messages "old". This abstraction makes the message buffer between the sender and the receiver bounded. In detail, the message buffer has a size of Δ. Importantly, we can apply this abstraction to two above encoding techniques for the message buffer.

In partial synchrony, if process p executes Receive at some time point from the Global Stabilization Time, *every* old message sent to p will be delivered immediately. Moreover, the computation of a process in Algorithm 1 does not depend on the content of received messages. Hence, instead of tracking all old messages, our abstraction keeps only one old message that is Δ time-units old, does not increase its age, and throws away other old messages.

In the following, we discuss how to integrate this abstraction into the encoding techniques of the message buffer. We demonstrate our ideas by showing the pseudo-code of the increase of message ages. It is straightforward to adopt this abstraction to the message delivery, and to the sending of a new message.

Figure 4(a) presents the increase of message ages with this abstraction in a case of $\Delta = 2$, and buf $= 6$. Unlike Fig. 3, there exists no in-transit message that is 3 time-units old in Fig. 4(a). Moreover, the message buffer in Fig. 4(a) has a size of 3. In addition, buf$'$ has only one in-transit message that is 2 time-units old. We have buf$' = 4$ in this case. Figure 4(b) demonstrates another case of $\Delta = 2$, buf $= 5$, and buf$' = 6$.

```
1: if buf < 2^Δ then buf' ← buf × 2
2: else
3:     if buf ≥ 2^Δ + 2^{Δ-1} then buf' ← buf × 2 - 2^{Δ+1}
4:     else buf' ← buf × 2 - 2^{Δ+1} + 2^Δ
```

Fig. 5. Encoding the increase of message ages with a counter buf, and the abstraction of old messages.

Formally, Fig. 5 presents the pseudo-code of the increase of message ages that is encoded with a counter buf, and the abstraction of old messages. There are three cases. In the first case (Line 1), there exist no old messages in buf, and we simply set $\mathsf{buf}' = \mathsf{buf} \times 2$. In other cases (Lines 3 and 4), buf contains an old message. Figure 4(a) demonstrates the second case (Line 3). We subtract $2^{\Delta+1}$ to remove an old message with $\Delta + 1$ time-units old from the buffer. Figure 4(b) demonstrates the third case (Line 4). In the third case, we also need to remove an old message with $\Delta + 1$ time-units old from the buffer. Moreover, we need to put an old message with Δ time-units old to the buffer by adding 2^Δ.

Now we discuss how to integrate the abstraction of old messages in the encoding of the message buffer with a predicate. Formulas 10–13 present the relation between existsMsgOfAge and existsMsgOfAge′ when message ages are increased by 1, and this abstraction is applied. Formula 10 ensures that no fresh message will be added to existsMsgOfAge′. Formula 11 ensures that the age of every message that is until $(\Delta - 2)$ time-units old will be increased by 1. Formulas 12–13 are introduced by this abstraction. Formula 12 implies that if there exists an old message or a message with $(\Delta - 1)$ time-units old in existsMsgOfAge, there will be an old message that is Δ time-units old in existsMsgOfAge′. Formula 13 ensures that there exists no message that is older than Δ time-units old.

$$\neg\mathsf{existsMsgOfAge}'(0) \tag{10}$$

$$\forall x \in \mathbb{N} . (0 \leq x \leq \Delta - 2)$$
$$\Rightarrow \mathsf{existsMsgOfAge}'(x+1) = \mathsf{existsMsgOfAge}(x) \tag{11}$$

$$\mathsf{existsMsgOfAge}'(\Delta) = \mathsf{existsMsgOfAge}(\Delta) \vee \mathsf{existsMsgOfAge}(\Delta - 1) \tag{12}$$

$$\forall x \in \mathbb{N} . x > \Delta \Rightarrow \mathsf{existsMsgOfAge}'(x) = \bot \tag{13}$$

4.3   Encoding the Relative Speed of Processes

Recall that we focus on the case of unknown bounds Δ and Φ. In this case, every correct process must take at least one step in every contiguous time interval containing Φ time-units [15].

To maintain this constraint on executions generated by the verification tools, we introduced two additional control variables sTimer and rTimer for the sender and the receiver, respectively. These variables work as timers to keep track of how long a process has not taken a step, and when a process can take a step. Since these timers play similar roles, we here focus on rTimer. In our encoding, only the environment can update rTimer. To schedule the receiver, the environment non-deterministically executes one of two actions: (i) resets rTimer to 0, and (ii) if rTimer $< \Phi$, increases rTimer by 1. Moreover, the receiver must take a step whenever rTimer $= 0$.

5   Reduce Liveness Properties to Safety Properties

To verify the liveness properties Eventually Strong Accuracy and Strong Completeness with Ivy, we first need to reduce them to safety properties. Intuitively, these liveness properties are bounded; therefore, they become safety ones. This section demonstrates how to reduce Eventually Strong Accuracy to a safety one.

By cutoffs discussed in Sect. 3, it is sufficient to verify Eventually Strong Accuracy on the Chandra and Toueg failure detector by checking the following property on instances with 2 processes.

$$\mathbf{FG}((Correct(1) \land Correct(2)) \Rightarrow \neg Suspected(2,1)) \qquad (14)$$

In the failure detector [11], the receiver suspects the sender only if its waiting time reaches the timeout (see Line 6 in Algorithm 1). To reduce Formula 14 to a safety property, we found a specific guard g for timeout such that if timeout $\geq g$ and the sender is correct, then waitingtime $< g$. Hence, it is sufficient to verify Formula 14 by checking the following property.

$$\mathbf{G}\big(\text{timeout} \geq g \Rightarrow ((Correct(1) \land Correct(2)) \Rightarrow \neg Suspected(2,1))\big)$$

6   Experiments for Small Δ and Φ

In this section, we describe our experiments with TLA$^+$ and FAST. We ran the following experiments on a virtual machine with Core i7-6600U CPU and 8GB DDR4. Our specifications can be found at [33].

6.1   Model Checkers For TLA$^+$: TLC and APALACHE

In our work, we use TLA$^+$ [22] to specify the failure detector with both encoding techniques for the message buffer, and the abstraction in Sect. 4. Then, we use the model checkers TLC [36] and APALACHE [18] to verify instances with fixed bounds Δ and Φ, and the GST $T_0 = 1$. This approach helps us to search constraints in inductive invariants in case of fixed parameters. The main reason is that counterexamples and inductive invariants in case of fixed parameters, e.g., $\Delta \leq 1$ and $\Phi \leq 1$, are simpler than in case of arbitrary parameters. Hence, if a counterexample is found, we can quickly analyze it, and change constraints in an inductive invariant candidate. After obtaining inductive invariants in small cases, we can generalize them for cases of arbitrary bounds, and check with theorem provers, e.g., Ivy (Sect. 7).

TLA$^+$ offers a rich syntax for sets, functions, tuples, records, sequences, and control structures [22]. Hence, it is straightforward to apply the encoding

```
1:  SSnd  ≜  ∧ ePC = "SSnd"
2:            ∧ IF (sTimer = 0 ∧ sPC = "SSnd")
3:               THEN buf' = buf + 1
4:               ELSE UNCHANGED buf
5:            ∧ ePC' = "RNoSnd"
6:            ∧ UNCHANGED ⟨sTimer, rTimer...⟩
```

Fig. 6. Sending a new message in TLA$^+$ in case of $\Delta > 0$

techniques and the abstraction presented in Sect. 4 in TLA$^+$. For example, Fig. 6 represents a TLA$^+$ action *SSnd* for sending a new message in case of $\Delta > 0$. Variables *ePC* and *sPC* are program counters for the environment and the sender, respectively. Line 1 is a precondition, and refers to that the environment is in subround Send. Lines 2–3 say that if the sender is active in subround Send, the counter *buf'* is increased by 1. Otherwise, two counters *buf* and *buf'* are the same (Line 4). Line 5 implies that the environment is still in the subround Send, but it is now the receiver's turn. Line 6 guarantees that other variables are unchanged in this action. (The details are left for the full report [32].)

Now we present the experiments with TLC and APALACHE. We used these tools to verify (i) the safety property Strong Accuracy, and (ii) an inductive invariant for Strong Accuracy, and (iii) an inductive invariant for a safety property reduced from the liveness property Strong Completeness in case of fixed bounds, and GST = 1 (initial stabilization). The structure of the inductive invariants verified here are very close to one in case of arbitrary bounds Δ and Φ.

Table 1 shows the results in verification of Strong Accuracy in case of the initial stabilization, and fixed bounds Δ and Φ. Table 1 shows the experiments with the three tools TLC, APALACHE, and FAST. The column "#states" shows the number of distinct states explored by TLC. The column "#depth" shows the maximum execution length reached by TLC and APALACHE. The column "buf" shows how to encode the message buffer. The column "LOC" shows the number of lines in the specification of the system behaviors (without comments). The symbol "-" (minus) refers to that the experiments are intentionally missing since FAST does not support the encoding of the message buffer with a predicate. The abbreviation "pred" refers to the encoding of the message buffer with a predicate. The abbreviation "cntr" refers to the encoding of the message buffer with a counter. The abbreviation "TO" means a timeout of 6 h. In these experiments, we initially set timeout = $6 \times \Phi + \Delta$, and Strong Accuracy is satisfied. The experiments show that TLC finishes its tasks faster than the others, and APALACHE prefers the encoding of the message buffer with a predicate.

Table 2 summarizes the results in verification of Strong Accuracy with the tools TLC, APALACHE, and FAST in case of the initial stabilization, and small bounds Δ and Φ, and initially timeout = $\Delta + 1$. Since timeout is initialized with a too small value, there exists a case in which sent messages are delivered after the timeout expires. The tools reported an error execution where Strong

Table 1. Showing strong accuracy for fixed parameters.

#	Δ	Φ	buf	TLC				APALACHE		FAST	
				Time	#states	Depth	LOC	Time	Depth	Time	LOC
1	2	4	Pred	3 s	10.2 K	176	190	8 m	176	-	-
2			cntr	3 s	10.2 K	176	266	9 m	176	16 m	387
3	4	4	pred	3 s	16.6 K	183	190	12 m	183	-	-
4			cntr	3 s	16.6 K	183	487	35 m	183	TO	2103
5	4	*5	pred	3 s	44.7 K	267	190	TO	222	-	-
6			cntr	3 s	44.7 K	267	487	TO	223	TO	2103

Table 2. Violating strong accuracy for fixed parameters.

#	Δ	Φ	buf	TLC			APALACHE		FAST
				Time	#states	Depth	Time	Depth	Time
1	2	4	pred	1 s	840	43	11 s	42	-
2			cntr	1 s	945	43	12 s	42	10 m
3	4	4	pred	2 s	1.3 K	48	15 s	42	-
4			cntr	2 s	2.4 K	56	16 s	42	TO
5	20	20	pred	TO	22.1 K	77	1 h 15 m	168	-

Accuracy is violated. In these experiments, APALACHE is the winner. The abbreviation "TO" means a timeout of 6 h. The meaning of other columns and abbreviations is the same as in Table 1.

Table 3 shows the results in verification of inductive invariants for Strong Accuracy and Strong Completeness with TLC and APALACHE in case of the initial stabilization, and slightly larger bounds Δ and Φ. The message buffer was encoded with a predicate in these experiments. In these experiments, inductive invariants hold, and APALACHE is faster than TLC in verifying them.

As one sees from the tables, APALACHE is fast at proving inductive invariants, and at finding a counterexample when a desired safety property is violated. TLC is a better option in cases where a safety property is satisfied.

In order to prove correctness of the failure detector in cases where parameters Δ and Φ are arbitrary, the user can use the interactive theorem prover TLA$^+$

Table 3. Proving inductive invariants with TLC and APALACHE.

#	Δ	Φ	Property	TLC		APALACHE
				Time	#states	Time
1	4	40	Strong accuracy	33 m	347.3 M	12 s
2	4	10	Strong completeness	44 m	13.4 M	17 s

```
1: transition SSnd_Active := {
2:     from := incMsgAge;
3:     to := ssnd;
4:     guard := sTimer = 0;
5:     action := buf' = buf + 1; };
```

Fig. 7. Sending a new message in FAST in case of $\Delta > 0$

Proof System (TLAPS) [12]. A shortcoming of TLAPS is that it does not provide a counterexample when an inductive invariant candidate is violated. Moreover, proving the failure detector with TLAPS requires more human effort than with Ivy. Therefore, we provide Ivy proofs in Sect. 7.

6.2 FAST

A shortcoming of the model checkers TLC and APALACHE is that parameters Δ and Φ must be fixed before running these tools. FAST is a tool designed to reason about safety properties of counter systems, i.e. automata extended with unbounded integer variables [6]. If Δ is fixed, and the message buffer is encoded with a counter, the failure detector becomes a counter system. We specified the failure detector in FAST, and made experiments with different parameter values to understand the limit of FAST: (i) the initial stabilization, and small bounds Δ and Φ, and (ii) the initial stabilization, fixed Δ, but unknown Φ.

Figure 7 represents a FAST transition for sending a new message in case of $\Delta > 0$. Line 2 describes the (symbolic) source state of the transition, and region incMsgAge is a set of configurations in the failure detector that is reachable from a transition for increasing message ages. Line 3 mentions the (symbolic) destination state of the transition, and region sSnd is a set of configurations in the failure detector that is reachable from a transition named "SSnd_Active" for sending a new message. Line 4 represents the guard of this transition. Line 5 is an action. Every unprimed variable that is not written in Line 5 is unchanged.

The input language of FAST is based on Presburger arithmetics for both system and properties specification. Hence, we cannot apply the encoding of the message buffer with a predicate in FAST.

Tables 1 and 2 described in the previous subsection summarize the experiments with FAST, and other tools where all parameters are fixed. Moreover, we ran FAST to verify Strong Accuracy in case of the initial stabilization, $\Delta \leq 4$, and arbitrary Φ. FAST is a semi-decision procedure; therefore, it does not terminate on some inputs. Unfortunately, FAST could not prove Strong Accuracy in case of arbitrary Φ, and crashed after 30 min.

7   Ivy Proofs for Parametric Δ and Φ

While TLC, APALACHE, and FAST can automatically verify some instances of the failure detector with fixed parameters, these tools cannot handle cases with unknown bounds Δ and Φ. To overcome this problem, we specify and prove correctness of the failure detector with the interactive theorem prover Ivy [28]. In the following, we first discuss the encoding of the failure detector, and then presents the experiments with Ivy.

The encoding of the message buffer with a counter requires that bound Δ is fixed. We here focus on cases where bound Δ is unknown. Hence, we encode the message buffer with a predicate in our Ivy specifications.

In Ivy, we declare `relation` existsMsgOfAge(X: num). Type `num` is interpreted as integers. Since Ivy does not support primed variables, we need an additional relation tmpExistsMsgOfAge(X : num). Intuitively, we first compute and store the value of existsMsgOfAge in the next state in tmpExistsMsgOfAge, then copy the value of tmpExistsMsgOfAge back to existsMsgOfAge. We do not consider the requirement of tmpExistsMsgOfAge as a shortcoming of Ivy since it is still straightforward to transform the ideas in Sect. 4 to Ivy.

Figure 2 represents how to add a fresh message in the message buffer in Ivy. Line 1 means that tmpExistsMsgOfAge is assigned an arbitrary value. Line 2 guarantees the appearance of a fresh message. Line 3 ensures that every in-transit message in existsMsgOfAge is preserved in tmpExistsMsgOfAge. Line 4 copies the value of tmpExistsMsgOfAge back to existsMsgOfAge.

Algorithm 2. Adding a fresh message in Ivy

1: tmpExistsMsgOfAge(X) := *;
2: **assume** tmpExistsMsgOfAge(0);
3: **assume forall** X: num . $0 < X \rightarrow$ existsMsgOfAge(X) = tmpExistsMsgOfAge(X);
4: existsMsgOfAge(X) := tmpExistsMsgOfAge(X);

Importantly, our specifications are not in decidable theories supported by Ivy. In Formula 11, the interpreted function " $+$ " (addition) is applied to a universally quantified variable x.

The standard way to check whether a safety property *Prop* holds in an Ivy specification is to find an inductive invariant *IndInv* with *Prop*, and to (interactively) prove that *Indinv* holds in the specification. To verify the liveness properties Eventually Strong Accuracy, and Strong Completeness, we reduced them into safety properties by applying a reduction technique in Sect. 5, and found inductive invariants containing the resulted safety properties. These inductive invariants are the generalization of the inductive invariants in case of fixed parameters that were found in the previous experiments.

Table 4 shows the experiments on verification of the failure detector with Ivy in case of unknown Δ and Φ. The symbol $\star$ refers to that the initial value of time-out is arbitrary. The column "#line$_I$" shows the number of lines of an inductive invariant, and the column "#strengthening steps" shows the number of lines of strengthening steps that we provided for Ivy. The meaning of other columns is the same as in Table 1. While our specifications are not in the decidable theories

Table 4. Proving inductive invariants with Ivy for arbitrary Δ and Φ.

#	Property	timeout$_{init}$	Time	LOC	#line$_I$	#strengthening steps
1	Strong accuracy	$= 6 \times \Phi + \Delta$	4 s	183	30	0
2	Eventually strong accuracy	$= \star$	4 s	186	35	0
3	Strong completeness	$= 6 \times \Phi + \Delta$	8 s	203	111	0
4		$\geq 6 \times \Phi + \Delta$	22 s	207	124	15
5		$= \star$	44 s	207	129	0

supported in Ivy, our experiments show that Ivy needs no user-given strengthening steps to prove most of our inductive invariants. Hence, it took us about 4 weeks to learn Ivy from scratch, and to prove these inductive invariants.

The most important thing to prove a property satisfied in an Ivy specification is to find an inductive invariant. Our inductive invariants use non-linear integers, quantifiers, and uninterpreted functions. (The inductive invariants in Table 4 are given in the full report [32].)

8 Conclusion

We have presented verification of both safety and liveness of the Chandra and Toeug failure detector by using the verification tools: model checkers for TLA$^+$ (TLC and APALACHE), counter automata (FAST), and the theorem prover Ivy. To do that, we first prove the cutoff results that can apply to the failure detector under partial synchrony. Next, we develop the encoding techniques to efficiently specify the failure detector, and to tune our models to the strength of the mentioned tools. We verified safety in case of fixed parameters by running the tools TLC, APALACHE, and FAST. To cope with cases of arbitrary bounds Δ and Φ, we reduced liveness properties to safety properties, and proved inductive invariants with desired properties in Ivy. While our specifications are not in the decidable theories supported in Ivy, our experiments show that Ivy needs no additional user assistance to prove most of our inductive invariants.

Modeling the failure detector in TLA$^+$ helps us understand and find inductive invariants in case of fixed parameters. Their structure is simpler but similar to the structure of parameterized inductive invariants. We found that the TLA$^+$ Toolbox [21] has convenient features, e.g., Profiler and Trace Exploration. A strong point of Ivy is in producing a counterexample quickly when a property is violated, even if all parameters are arbitrary. In contrast, FAST reports no counterexample in any case. Hence, debugging in FAST is very challenging.

While our specification describes executions of the Chandra and Toueg failure detector, we conjecture that many time constraints on network behaviors, correct processes, and failures in our inductive invariants can be reused to prove other algorithms under partial synchrony. We also conjecture that correctness of other partially synchronous algorithms may be proven by following the presented methodology. For future work, we would like to extend the above results for cases where GST is arbitrary. It is also interesting to investigate how to express discrete partial synchrony in timed automata [3], e.g., UPPAAL [25].

Acknowledgments. Supported by Interchain Foundation (Switzerland) and the Austrian Science Fund (FWF) via the Doctoral College LogiCS W1255.

References

1. Aguilera, M.K., Delporte-Gallet, C., Fauconnier, H., Toueg, S.: On implementing omega in systems with weak reliability and synchrony assumptions. Distrib. Comput. **21**(4), 285–314 (2008)
2. Aguilera, M.K., Delporte-Gallet, C., Fauconnier, H., Toueg, S.: Consensus with Byzantine failures and little system synchrony. In: International Conference on Dependable Systems and Networks (DSN), pp. 147–155. IEEE (2006)
3. Alur, R., Dill, D.L.: A theory of timed automata. Theor. Comput. Sci. **126**(2), 183–235 (1994)
4. André, É., Fribourg, L., Kühne, U., Soulat, R.: IMITATOR 2.5: a tool for analyzing robustness in scheduling problems. In: Giannakopoulou, D., Méry, D. (eds.) FM 2012. LNCS, vol. 7436, pp. 33–36. Springer, Heidelberg (2012). https://doi.org/10.1007/978-3-642-32759-9_6
5. Atif, M., Mousavi, M.R., Osaiweran, A.: Formal verification of unreliable failure detectors in partially synchronous systems. In: Proceedings of the 27th ACM Symposium on Applied Computing (SAC), pp. 478–485 (2012). https://doi.org/10.1145/2245276.2245369
6. Bardin, S., Leroux, J., Point, G.: FAST extended release. In: Ball, T., Jones, R. (eds.) CAV 2006. LNCS, vol. 4144, pp. 63–66. Springer, Heidelberg (2006). https://doi.org/10.1007/11817963_9
7. Bloem, R., et al.: Decidability of Computing Theory. Morgan & Claypool Publishers (2015). https://doi.org/10.2200/S00658ED1V01Y201508DCT013
8. Bravo, M., Chockler, G., Gotsman, A.: Making Byzantine consensus live. In: DISC. Schloss Dagstuhl-Leibniz-Zentrum für Informatik (2020)
9. Buchman, E., Kwon, J., Milosevic, Z.: The latest gossip on BFT consensus. arXiv preprint arXiv:1807.04938 (2018)
10. Bunte, O.: The mCRL2 toolset for analysing concurrent systems. In: Vojnar, T., Zhang, L. (eds.) TACAS 2019. LNCS, vol. 11428, pp. 21–39. Springer, Cham (2019). https://doi.org/10.1007/978-3-030-17465-1_2
11. Chandra, T.D., Toueg, S.: Unreliable failure detectors for reliable distributed systems. J. ACM **43**(2), 225–267 (1996)
12. Chaudhuri, K., Doligez, D., Lamport, L., Merz, S.: The TLA⁺ proof system: building a heterogeneous verification platform. In: Cavalcanti, A., Deharbe, D., Gaudel, M.-C., Woodcock, J. (eds.) ICTAC 2010. LNCS, vol. 6255, p. 44. Springer, Heidelberg (2010). https://doi.org/10.1007/978-3-642-14808-8_3
13. Cousineau, D., Doligez, D., Lamport, L., Merz, S., Ricketts, D., Vanzetto, H.: TLA⁺ proofs. In: Giannakopoulou, D., Méry, D. (eds.) FM 2012. LNCS, vol. 7436, pp. 147–154. Springer, Heidelberg (2012). https://doi.org/10.1007/978-3-642-32759-9_14
14. Drăgoi, C., Widder, J., Zufferey, D.: Programming at the edge of synchrony. In: Proceedings of the ACM on Programming Languages 4 (OOPSLA), pp. 1–30 (2020)

15. Dwork, C., Lynch, N., Stockmeyer, L.: Consensus in the presence of partial synchrony. J. ACM **35**(2), 288–323 (1988)
16. Emerson, E.A., Namjoshi, K.S.: Reasoning about rings. In: POPL, pp. 85–94 (1995)
17. Galois, I.: Ivy proofs of tendermint. https://github.com/tendermint/spec/tree/master/ivy-proofs. Accessed December 2020
18. Konnov, I., Kukovec, J., Tran, T.H.: TLA$^+$ model checking made symbolic. In: Proceedings of the ACM on Programming Languages 3 (OOPSLA), pp. 1–30 (2019)
19. Konnov, I., Lazić, M., Veith, H., Widder, J.: Para^2: parameterized path reduction, acceleration, and SMT for reachability in threshold-guarded distributed algorithms. Formal Methods Syst. Design **51**(2), 270–307 (2017)
20. Konnov, I., Lazić, M., Veith, H., Widder, J.: A short counterexample property for safety and liveness verification of fault-tolerant distributed algorithms. In: POPL, pp. 719–734 (2017)
21. Kuppe, M.A., Lamport, L., Ricketts, D.: The TLA$^+$ toolbox. arXiv preprint arXiv:1912.10633 (2019)
22. Lamport, L.: Specifying Systems: The TLA$^+$ Language and Tools for Hardware and Software Engineers. Addison-Wesley, Boston (2002)
23. Lamport, L.: Using TLC to check inductive invariance (2018)
24. Larrea, M., Arevalo, S., Fernndez, A.: Efficient algorithms to implement unreliable failure detectors in partially synchronous systems. In: Jayanti, P. (ed.) DISC 1999. LNCS, vol. 1693, pp. 34–49. Springer, Heidelberg (1999). https://doi.org/10.1007/3-540-48169-9_3
25. Larsen, K.G., Pettersson, P., Yi, W.: UPPAAL in a nutshell. Int. J. Softw. Tools Technol. Transfer **1**(1–2), 134–152 (1997)
26. Lime, D., Roux, O.H., Seidner, C., Traonouez, L.-M.: Romeo: a parametric model-checker for petri nets with stopwatches. In: Kowalewski, S., Philippou, A. (eds.) TACAS 2009. LNCS, vol. 5505, pp. 54–57. Springer, Heidelberg (2009). https://doi.org/10.1007/978-3-642-00768-2_6
27. Lynch, N.A., Tuttle, M.R.: An Introduction to Input/Output Automata. Laboratory for Computer Science, Massachusetts Institute of Technology (1988)
28. McMillan, K.L.: Ivy. https://microsoft.github.io/ivy/. Accessed December 2020
29. McMillan, K.L., Padon, O.: Ivy: a multi-modal verification tool for distributed algorithms. In: Lahiri, S.K., Wang, C. (eds.) CAV 2020. LNCS, vol. 12225, pp. 190–202. Springer, Cham (2020). https://doi.org/10.1007/978-3-030-53291-8_12
30. Roscoe, A.W.: Understanding Concurrent Systems. Springer, Cham (2010)
31. Stoilkovska, I., Konnov, I., Widder, J., Zuleger, F.: Verifying safety of synchronous fault-tolerant algorithms by bounded model checking. In: Vojnar, T., Zhang, L. (eds.) TACAS 2019. LNCS, vol. 11428, pp. 357–374. Springer, Cham (2019). https://doi.org/10.1007/978-3-030-17465-1_20
32. Tran, T.H., Konnov, I., Widder, J.: FORTE2021-FD. https://github.com/banhday/forte2021-fd. Accessed April 2021
33. Tran, T.H., Konnov, I., Widder, J.: Specifications of the Chandra and Toueg failure detector in TLA$^+$, and Ivy. https://zenodo.org/record/4687714#.YHcBeBKxVH4. Accessed April 2021
34. Tran, T.-H., Konnov, I., Widder, J.: Cutoffs for symmetric point-to-point distributed algorithms. In: Georgiou, C., Majumdar, R. (eds.) NETYS 2020. LNCS, vol. 12129, pp. 329–346. Springer, Cham (2021). https://doi.org/10.1007/978-3-030-67087-0_21

35. Yin, M., Malkhi, D., Reiter, M.K., Gueta, G.G., Abraham, I.: Hotstuff: BFT consensus with linearity and responsiveness. In: PODC, pp. 347–356 (2019)
36. Yu, Y., Manolios, P., Lamport, L.: Model checking TLA$^+$ specifications. In: Pierre, L., Kropf, T. (eds.) CHARME 1999. LNCS, vol. 1703, pp. 54–66. Springer, Heidelberg (1999). https://doi.org/10.1007/3-540-48153-2_6

π with Leftovers: A Mechanisation in Agda

Uma Zalakain$^{(\boxtimes)}$ ⓘ and Ornela Dardha ⓘ

University of Glasgow, Glasgow, Scotland
u.zalakain.1@research.gla.ac.uk, ornela.dardha@glasgow.ac.uk

Abstract. Linear type systems need to keep track of how programs use their resources. The standard approach is to use *context splits* specifying how resources are (disjointly) split across subterms. In this approach, context splits redundantly echo information which is already present within subterms. An alternative approach is to use *leftover typing* [2,23], where in addition to the usual (input) usage context, typing judgments have also an output usage context: the leftovers. In this approach, the leftovers of one typing derivation are fed as input to the next, threading through linear resources while avoiding context splits. We use leftover typing to define a type system for a resource-aware π-calculus [26,27], a process algebra used to model concurrent systems. Our type system is parametrised over a set of *usage algebras* [20,34] that are general enough to encompass *shared types* (free to reuse and discard), *graded types* (use exactly n number of times) and *linear types* (use exactly once). Linear types are important in the π-calculus: they ensure privacy and safety of communication and avoid race conditions, while graded and shared types allow for more flexible programming. We provide a framing theorem for our type system, generalise the weakening and strengthening theorems to include linear types, and prove subject reduction. Our formalisation is fully mechanised in about 1850 lines of Agda [37].

Keywords: Pi-calculus · Linear types · Leftover typing · Concurrency · Mechanisation · Agda

1 Introduction

The π-calculus [26,27] is a computational model for communication and concurrency that boils concurrent processing down to the sending and receiving of data over communication channels. Notably, it features channel mobility: channels themselves are first class values and can be sent and received. Kobayashi et al. [22] introduced a typed version of the π-calculus with linear channel types,

This work is supported by the EU HORIZON 2020 MSCA RISE project 778233 "Behavioural Application Program Interfaces" (BehAPI).

© IFIP International Federation for Information Processing 2021
Published by Springer Nature Switzerland AG 2021
K. Peters and T. A. C. Willemse (Eds.): FORTE 2021, LNCS 12719, pp. 157–174, 2021.
https://doi.org/10.1007/978-3-030-78089-0_9

where channels must be used *exactly* once. Linearity in the π-calculus guarantees privacy and safety of communication and avoids race conditions.

More broadly, linearity allows for resource-aware programming and more *efficient* implementations [35], and it inspired unique types (as in Clean [4]), and ownership types (as in Rust [24]). A linear type system must keep track of what resources are used in which parts of the program, and guarantee that they are *neither duplicated nor discarded.* To do so, the standard approach is to use context splits: typing rules for terms with multiple subterms add an extra side condition specifying what resources to allocate to each of the subterms. The typing derivations for the subterms must then use the entirety of their allocated resources. A key observation here is that each subterm already *knows* about the resources it needs. *Context splits contain usage information that is already present in the subterms.* Moreover, the subterms cannot be typed until the context splits have been defined. On top of that, using binary context splits means that typing rules with n subterms require $n-1$ context splits, which considerably clutters the type system.

An alternative approach is *leftover typing*, a technique used to formulate intuitionistic linear logic [23] and to mechanise the linear λ-calculus [2]. Leftover typing changes the shape of the typing judgments and includes a second *leftover* output context that contains the resources that were left unused by the term. As a result, typing rules *thread* the resources *through* subterms without needing context splits: each subterm uses the resources it needs, and leaves the rest for its siblings. The first subterm in this chain of resources immediately knows what resources it has available.

In this paper, we use leftover typing to define for the first time a resource-aware type system for the π-calculus, and we fully mechanise our work in Agda [37]. All previous work on mechanisation of linear process calculi uses context splits instead [8,15,16,18,33]. We will further highlight the benefits of leftover typing as opposed to context splits in contributions and the rest of the paper.

Below we present two alternative typing rules for parallel composition in the linear π-calculus: the one on the left uses context splits, while the one on the right does not, and uses leftover typing instead:

$$\frac{\Gamma := \Delta \otimes \Xi \quad \Delta \vdash P \quad \Xi \vdash Q}{\Gamma \vdash P \,\|\, Q} \qquad\qquad \frac{\Gamma \vdash P \triangleright \Delta \quad \Delta \vdash Q \triangleright \Xi}{\Gamma \vdash P \,\|\, Q \triangleright \Xi}$$

Contributions and Structure of the Paper

1. **Leftover typing for resource-aware π-calculus.** Our type system uses leftover typing to model the resource-aware π-calculus (Sect. 4.3) and satisfies subject reduction (Theorem 5). In addition to making context splits unnecessary, leftover typing allows for a *framing* theorem (Theorem 1) to be stated and is naturally associative, making type safety properties considerably easier to reason about (Sect. 5). Thanks to leftover typing, we can now state *weakening* (Theorem 2) and *strengthening* (Theorem 3) for the whole framework,

not just the shared fragment. This give a uniform and complete presentation of all the meta-theory for the resource-aware π-calculus.

2. **Shared, graded and linear unified π-calculus.** We generalise resource counting to a set of usage algebras that can be mixed within the same type system. We do not instantiate our type system to only work with linear resources, instead we present an algebra-agnostic type system, and admit a mix of user-defined *resource aware* algebras [20,34] (Sect. 4.1). Any *partial commutative monoid* that is *decidable, deterministic, cancellative* and has a *minimal element* is a valid such algebra. Multiple algebras can be mixed in the type system—usage contexts keep information about what algebra to use for each type (Sect. 4.2). In particular, this allows for type systems combining linear (use exactly once), graded (exact number of n times) and shared (free to reuse and discard) types under the same framework.

3. **Full mechanisation in Agda.** The formalisation of the π-calculus with leftover typing, from the syntax to the semantics and the type system, has been fully mechanised in Agda in about 1850 lines of code, and is publicly available at [37]. We have fully mechanised all meta-theory and the details of a proof of subject reduction can be found in our extended paper [36] and repository [37].

We use type level de Bruijn indices [11,14] to define a syntax of π-calculus processes that is *well scoped by construction*: every free variable is accounted for in the type of the process that uses it (Sect. 2). We then provide an operational semantics for the π-calculus, prior to any typing (Sect. 3). This operational semantics is defined as a reduction relation on processes. The reduction relation tracks at the type level the channel on which communication occurs. This information is later used to state the subject reduction theorem. The reduction relation is defined modulo *structural congruence*—a relation defined on processes that acts as a quotient type to remove unnecessary syntactic minutiae introduced by the syntax of the π-calculus. We then define an interface for resource-aware algebras (Sect. 4.1) and use it to parametrise a type system based on leftover typing (Sect. 4.3). Finally, we present the meta theoretical properties of our type system in Sect. 5.

Notation. Data type definitions ($\mathbb{N}$) use double inference lines and index-free synonyms (NAT) as rule names for ease of reference. Constructors (0 and $1+$) are used as inference rule names. We maintain a close correspondence between the definitions presented in this paper and our mechanised definitions in Agda: inference rules become type constructors, premises become argument types and conclusions return types. Universe levels and universe polymorphism are omitted for brevity—all our types are of type SET. Implicit arguments are mentioned in type definitions but omitted by constructors.

$$\frac{}{\mathbb{N} : \text{SET}} \text{NAT} \qquad \frac{}{0 : \mathbb{N}} \qquad \frac{n : \mathbb{N}}{1{+}n : \mathbb{N}}$$

We use colours to further distinguish the different entities in this paper. TYPES are blue and uppercased, with indices as subscripts, constructors are orange, functions are teal, variables are black, and some constructor names are overloaded—and disambiguated by context.

2 Syntax

In order to mechanise the π-calculus syntax in Agda, we need to deal with bound names in continuation processes. Names are cumbersome to mechanise: they are not inherently well scoped, one has to deal with alpha-conversion, and inserting new variables into a context entails proving that their names differ from all other names in context. To overcome these challenges, we use de Bruijn indices [11], where a natural number n (aka *index*) is used to refer to the variable introduced n binders ago. That is, binders no longer introduce names; terms at different *depths* use different indices to refer to the same binding.

While de Bruijn indices are useful for mechanisation, they are not as readable as names. To overcome this difficulty and demonstrate the correspondence between a π-calculus that uses names and one that uses de Bruijn indices, we provide *conversion functions* in both directions and prove that they are inverses of each other up to α-conversion. Further details can be found in our extended paper [36] and repository [37].

Definition 1 (VAR and PROCESS). A variable reference occurring under n binders can refer to n distinct variables. We introduce the indexed family of types [14] VAR_n: for all naturals n, the type VAR_n has n distinct elements. We index processes according to their *depth*: for all naturals n, a process of type $\mathrm{PROCESS}_n$ contains free variables that can refer to n distinct elements. Every time we go under a binder, we increase the index of the continuation process, allowing the variable references within to refer to one more thing.

$$\frac{n : \mathbb{N}}{\mathrm{VAR}_n : \mathrm{SET}}\ \mathrm{VAR} \qquad \frac{n : \mathbb{N}}{0 : \mathrm{VAR}_{1+n}} \qquad \frac{x : \mathrm{VAR}_n}{1+x : \mathrm{VAR}_{1+n}}$$

$$\frac{n : \mathbb{N}}{\mathrm{PROCESS}_n : \mathrm{SET}}\ \mathrm{PROCESS}$$

$$\begin{aligned}
\mathrm{PROCESS}_n ::=\ &0 &&\text{(inaction)}\\
\mid\ &\nu\,\mathrm{PROCESS}_{1+n} &&\text{(restriction)}\\
\mid\ &\mathrm{PROCESS}_n \parallel \mathrm{PROCESS}_n &&\text{(parallel)}\\
\mid\ &\mathrm{VAR}_n\,(\,)\,\mathrm{PROCESS}_{1+n} &&\text{(input)}\\
\mid\ &\mathrm{VAR}_n\,\langle\,\mathrm{VAR}_n\,\rangle\,\mathrm{PROCESS}_n &&\text{(output)}
\end{aligned}$$

Process 0 denotes the terminated process, where no further communications can occur; process $\nu\,P$ creates a new channel and binds it at index 0 in the continuation process P; process $P \parallel Q$ composes P and Q in parallel; process

$x\,(\ )\,P$ receives data along channel x and makes that data available at index 0 in the continuation process P; process $x\,\langle\, y\,\rangle\, P$ sends variable y over channel x and continues as process P.

Example 1 (The courier system).
We present a courier system that consists of three *roles*: a sender, who wants to send a package; a receiver, who receives the package sent by the sender; and a courier, who carries the package from the sender to the receiver.

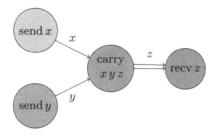

Our courier system is defined by four π-calculus processes composed in parallel instantiating the above three roles: we have two sender processes, send x and send y, sending data over channels x and y, respectively; one receiver process, recv z, which receives over channel z the data sent from each of the senders – hence receives twice; and a courier process carry $x\,y\,z$, which synchronises communication among the senders and the receiver. The courier process first receives data from the two senders along its input channels x and y, and then sends the two received bits of data to the receiver along its output channel z.

The sender and receiver *roles* are defined below, parametrised by the channels on which they operate. The sender creates a new channel to be sent as data, and sends it over channel c, and then terminates. Processes send x and send y are an instantiation of send c. The receiver receives data *twice* on a channel c and then terminates. The receiver process recv z is an instantiation of recv c.

$$\text{send } c = \nu\,(1{+}c\,\langle\,0\,\rangle\,\mathbb{0}) \qquad\qquad \text{recv } c = c\,(\ )\,(1{+}c)\,(\ )\,\mathbb{0}$$

The courier role is defined below as carry $x\ y\ z$. It sequentially receives on the two input channels x and y, instantiated as $in0$ and $in1$, and then outputs the two pieces of received data on the output channel z, instantiated as out. Finally, we create three communication channels and compose all four processes together: the first channel is shared between the one sender and the courier, the second between the other sender and the courier, and the third between the receiver and the courier. The result is the courier system defined below.

$$\text{carry } in0\ in1\ out\ = in0\,(\ )\,(1{+}\,in1)\,(\ )\,(1{+}1{+}\,\textbf{\textit{out}})\,\langle\,1{+}0\,\rangle\,(1{+}1{+}\,\textbf{\textit{out}})\,\langle\,0\,\rangle\,\mathbb{0}$$

$$\text{system} = \nu\,(\text{send } 0\,\|\,\nu\,(\text{send } 0\,\|\,\nu\,(\text{recv } 0\,\|\,\text{carry } (1{+}1{+}0)\,(1{+}0)\,0)))$$

We continue this running example in Sect. 4.3, where we provide typing derivations for the above processes and use a mix of linear, graded and shared typing to type the courier system.

3   Operational Semantics

Thanks to our well-scoped grammar in Sect. 2, we now define the semantics of our language on the totality of the syntax.

Definition 2 (UNUSED).   We consider a variable i to be unused in P (UNUSED$_i$ P) if none of the inputs nor the outputs refer to it. UNUSED$_i$ P is defined as a recursive predicate on P, incrementing i every time we go under a binder, and using $i \not\equiv x$ (which unfolds to the negation of propositional equality on VAR, i.e. $i \equiv x \rightarrow \bot$) to compare variables.

Definition 3 (STRUCTCONG).   We define the base cases of a structural congruence relation $\cong$ as follows:

$$\frac{}{P \cong Q : \text{SET}} \text{STRUCTCONG} \qquad \frac{}{\text{comp-assoc} : P \parallel (Q \parallel R) \cong (P \parallel Q) \parallel R}$$

$$\frac{}{\text{comp-sym} : P \parallel Q \cong Q \parallel P} \qquad \frac{}{\text{comp-id} : P \parallel 0_n \cong P}$$

$$\frac{}{\text{scope-end} : \nu\, 0_{1+n} \cong 0_n} \qquad \frac{uQ : \text{UNUSED}_0\ Q}{\text{scope-ext} : \nu\,(P \parallel Q) \cong (\nu\, P) \parallel \text{lower}_0\ Q\ uQ}$$

$$\frac{}{\text{scope-comm} : \nu\,\nu\, P \cong \nu\,\nu\, \text{exchange}_0\ P}$$

The first three rules (comp$-*$) state associativity, symmetry, and 0 as being the neutral element of parallel composition, respectively. The last three rules (scope$-*$) state garbage collection, scope extrusion and commutativity of restrictions, respectively. In scope-ext the side condition UNUSED$_i$ Q makes sure that i is unused in Q (see Definition 2). The function lower$_i$ Q uQ traverses Q decrementing every index greater than i. In scope-comm the function exchange$_i$ P traverses P (of type PROCESS$_{1+1+n}$) and swaps variable references i and $1+i$. In all the above, i is incremented every time we go under a binder.

Definition 4 (EQUALS).   We lift the relation STRUCTCONG $\cong$ and close it under equivalence and congruence in $\simeq$. This relation is structurally congruent under a context $\mathcal{C}[\cdot]$ [31] and is reflexive, symmetric and transitive.

Definition 5 (REDUCES).   The operational semantics of the π-calculus is defined as a reduction relation $\longrightarrow_c$ indexed by the channel c on which communication occurs. We keep track of channel c so we can state subject reduction (Theorem 5).

$$\frac{n : \mathbb{N}}{\text{CHANNEL}_n : \text{SET}} \; \text{CHANNEL} \qquad \qquad \overline{\text{internal} : \text{CHANNEL}_n}$$

$$\frac{i : \text{VAR}_n}{\text{external } i : \text{CHANNEL}_n} \qquad \frac{c : \text{CHANNEL}_n \quad P\, Q : \text{PROCESS}_n}{P \longrightarrow_c Q : \text{SET}} \; \text{REDUCES}$$

$$\frac{i\, j : \text{VAR}_n \quad P : \text{PROCESS}_{1+n} \quad Q : \text{PROCESS}_n}{\text{comm} : i\,(\;)\, P \parallel i\,\langle\, j\,\rangle\, Q \longrightarrow_{\text{external } i} \text{lower}_0\,(P\,[\,0 \mapsto 1+j\,])\; u P' \parallel Q}$$

$$\frac{red : P \longrightarrow_c P'}{\text{par } red : P \parallel Q \longrightarrow_c P' \parallel Q} \qquad \frac{red : P \longrightarrow_c Q}{\text{res } red : \nu\, P \longrightarrow_{\text{dec } c} \nu\, Q}$$

$$\frac{eq_1 : P \simeq P' \quad red : P' \longrightarrow_c Q' \quad eq_2 : Q' \simeq Q}{\text{struct } eq\, red : P \longrightarrow_c Q}$$

We distinguish between channels that are created inside the process (internal), and channels that are created outside (external i), where i is the index of the channel variable. In rule comm, parallel processes reduce when they communicate over a common channel with index i. As a result of that communication, the continuation of the input process P has all the references to its most immediate variable substituted with references to $1+j$, the variable sent by the output process $i\,\langle\, j\,\rangle\, Q$. After this substitution, $P\,[\,0 \mapsto 1+j\,]$ is *lowered*—all variable references are decreased by one (and we derive the proof $\text{UNUSED}_0\,(P\,[\,0 \mapsto 1+j\,])$). Reduction is closed under parallel composition (rule par), restriction (rule res) and structural congruence (rule struct)—notably, not under input nor output, as doing so would not preserve the sequencing of actions [31]. Rule res uses dec to decrement the index of channel c as we wrap processes P and Q inside a binder. It is defined as expected below:

$$\begin{aligned}
\text{dec internal} \quad &= \text{internal} \\
\text{dec (external } 0) \quad &= \text{internal} \\
\text{dec (external } (1+n)) \quad &= \text{external } n
\end{aligned}$$

4   Resource-Aware Type System

In Sect. 4.1 we characterise a usage algebra for our type system. It defines how resources are *split* in parallel composition and *consumed* in input and output. We define typing and usage contexts in Sect. 4.2. We provide a type system for a resource-aware π-calculus in Sect. 4.3.

4.1   Multiplicities and Capabilities

In the linear π-calculus each channel has an input and an output *capability*, and each capability has a given *multiplicity* of 0 (exhausted) or 1 (available).

We generalise over this notion by defining an algebra for multiplicities [20,34] that is satisfied by linear, graded and shared types alike. We then use pairs of multiplicities as usage annotations for a channel's input and output capabilities.

Definition 6 (ALGEBRA). *A usage algebra is a ternary relation $x := y \cdot z$ that is partial (as not any two multiplicities can be combined), deterministic and cancellative (to aid equational reasoning) and associative and commutative (following directly from subject congruence for parallel composition). In addition, we ask that the leftovers can be computed so that we can automatically update the usage context every time input and output occurs—this is purely for usability. It has a neutral element $\cdot 0$ that is absorbed on either side, and that is also minimal (so that new resources cannot arbitrarily spring into life). It has an element $\cdot 1$ that is used to count inputs and outputs.* Below we define such an algebra as a record $\mathrm{ALGEBRA}_C$ on a carrier C. (We use $\forall$ for universal quantification. The dependent product $\exists$ uses the value of its first argument in the type of its second. The type DEC P is a witness of either P or $P \to \bot$, where $\bot$ is the empty type with no constructors.)

$$
\begin{array}{lll}
\_ := \_ \cdot \_ & : & C \to C \to C \to \mathrm{SET} \\
\cdot\text{-unique} & : \forall x x' y z & \to x' := y \cdot z \to x := y \cdot z \to x' \equiv x \\
\cdot\text{-unique}^{\mathrm{l}} & : \forall x y y' z & \to x := y' \cdot z \to x := y \cdot z \to y' \equiv y \\
\cdot\text{-assoc} & : \forall x y z u v & \to x := y \cdot z \to y := u \cdot v \to \exists w \ (x := u \cdot w \times w := v \cdot z) \\
\cdot\text{-comm} & : \forall x y z & \to x := y \cdot z \to x := z \cdot y \\
\cdot\text{-compute}^{\mathrm{r}} & : \forall x y & \to \mathrm{DEC} \ (\exists z \ (x := y \cdot z)) \\
\cdot 0 & : & C \\
\cdot\text{-id}^{\mathrm{l}} & : \forall x & \to x := \cdot 0 \cdot x \\
\cdot\text{-min}^{\mathrm{l}} & : \forall y z & \to \cdot 0 := y \cdot z \to y \equiv \cdot 0 \\
\cdot 1 & : & C
\end{array}
$$

We sketch the implementation of linear, graded and shared types as instances of our usage algebra below. Their use in typing derivations is illustrated in Example 3.

		carrier	operation
linear		0 : Lin 1 : Lin	$0 := 0 \cdot 0$ $1 := 1 \cdot 0$ $1 := 0 \cdot 1$
graded		0 : Gra $1+$: Gra $\to$ Gra	$\forall x\, y\, z$ $\to x \equiv y + z$ $\to x := y \cdot z$
shared		ω : Sha	$\omega := \omega \cdot \omega$

4.2 Typing Contexts

We use indexed sets of usage algebras to allow several usage algebras to coexist in our type system with leftovers (Sect. 4.3).

Definition 7 (ALGEBRAS). An *indexed set of usage algebras* is a type IDX of indices that is nonempty ($\exists$IDX) together with an interpretation USAGE of indices into types, and an interpretation ALGEBRAS of indices into usage algebras of the corresponding type.

$$
\begin{aligned}
\text{IDX} &\quad : \text{SET} \\
\exists \text{IDX} &\quad : \text{IDX} \\
\text{USAGE} &\quad : \text{IDX} \to \text{SET} \\
\text{ALGEBRAS} &\quad : (idx : \text{IDX}) \to \text{ALGEBRA}_{\text{USAGE}_{idx}}
\end{aligned}
$$

We keep typing contexts (PRECTX) and usage contexts (CTX) separate. The former are preserved throughout typing derivations; the latter are transformed as a result of input, output, and context splits.

Definition 8 (TYPE and PRECTX: types and typing contexts). A *type* is either a unit type ($\mathbb{1}$), or a channel type ($\mathsf{C}[\,t\,;\,x\,]$).

$$
\frac{}{\text{TYPE} : \text{SET}} \; \text{TYPE} \qquad \frac{}{\mathbb{1} : \text{TYPE}} \qquad \frac{t : \text{TYPE} \quad \begin{array}{c} idx : \text{IDX} \\ x : \text{USAGE}^2_{idx} \end{array}}{\mathsf{C}[\,t\,;\,x\,] : \text{TYPE}}
$$

The unit type $\mathbb{1}$ serves as a base case for types. The type $\mathsf{C}[\,t\,;\,x\,]$ of a channel determines what type t of data and what usage annotations x are sent over that channel—we use the notation C^2 to stand for a $\mathsf{C} \times \mathsf{C}$ pair of input and output multiplicities, respectively. This channel notation aligns with $[t]\;\mathbf{chan}_{(i^y, o^z)}$, where y, z are the input and output multiplicities, respectively [21]. Henceforth, we use $\ell_\emptyset$ to denote the multiplicity pair $-0, -0$, ℓ_i for the pair $-1, -0$, ℓ_o for $-0, -1$, and $\ell_\#$ for $-1, -1$. This notation was originally used in the linear π-calculus [22,31]. A *typing context* PRECTX_n is a length-indexed list of types that is either empty ($[]$) or the result of appending a type $t : \text{TYPE}$ to an existing context (γ, t).

Definition 9 (IDXS and CTX: contexts of indices and usage contexts). A context of indices IDXS_n is a length-indexed list that is either empty ($[]$) or the result of appending an index $i : \text{IDX}$ to an existing context $(idxs.i)$. A *usage context* is a context CTX_{idxs} indexed by a context of indices $idxs : \text{IDXS}_n$ that is either empty ($[]$) or the result or appending a usage annotation pair $u : \text{USAGE}^2_{idx}$ with index $idx : \text{IDX}$ to an existing context (Γ, u).

4.3 Typing with Leftovers

We present a resource-aware type system for the π-calculus based on *leftover typing* [2], a technique that, in addition to the usual typing context PRECTX_n and (input) usage context CTX_{idxs}, adds an extra *(output) usage context* CTX_{idxs} to the typing rules. This output context contains the *leftovers* (the unused multiplicities) of the process being typed. These leftovers can then be used as input to another typing derivation.

Leftover typing inverts the information flow of usage annotations so that it is the typing derivations of subprocesses which determine how resources are allocated. As a result, context split proofs are no longer necessary. Leftover typing also allows *framing* to be stated, and *weakening* and *strengthening* to cover linear types too.

Our type system is composed of two typing judgments: one for variable references (Definition 10) and one for processes (Definition 11). Both judgments are indexed by a typing context γ, an input usage context Γ, and an output usage context Δ (the leftovers). The **typing judgement for variables** $\gamma ; \Gamma \ni_i t ; y \triangleright \Delta$ asserts that "index i in typing context γ is of type t, and subtracting y at position i from input usage context Γ results in leftovers Δ". The **typing judgement for processes** $\gamma ; \Gamma \vdash P \triangleright \Delta$ asserts that "process P is well typed under typing context γ, usage input context Γ and leftovers Δ".

Definition 10 (VARREF: typing variable references). The VARREF typing relation for variable references is presented below.

$$\frac{\begin{array}{c} t : \text{TYPE} \\ idx : \text{IDX} \qquad idxs : \text{IDXS}_n \\ \gamma : \text{PRECTX}_n \quad i : \text{VAR}_n \quad y : \text{USAGE}_{idx}^2 \quad \Gamma \, \Delta : \text{CTX}_{idxs} \end{array}}{\gamma ; \Gamma \ni_i t ; y \triangleright \Delta : \text{SET}} \text{VARREF}$$

$$\frac{x := y \cdot^2 z}{0 : \gamma , t ; \Gamma , x \ni_0 t ; y \triangleright \Gamma , z} \qquad \frac{v : \gamma ; \Gamma \ni_i t ; x \triangleright \Delta}{1 + v : \gamma , t' ; \Gamma , x' \ni_{1+i} t ; x \triangleright \Delta , x'}$$

We lift the operation $x := y \cdot z$ and its algebraic properties to an operation $(x_l , x_r) := (y_l , y_r) \cdot^2 (z_l , z_r)$ on pairs of multiplicities. The base case 0 splits the usage annotation x of type USAGE_{idx}^2 into y and z (the leftovers). Note that the remaining context Γ is preserved unused as a leftover. This splitting $x := y \cdot^2 z$ is as per the usage algebra provided by the developer for the index idx. In our Agda implementation, $x := y \cdot^2 z$ is actually a trivially satisfiable implicit argument if $x := y \cdot^2 z$ is inhabited and an unsatisfiable argument otherwise. The inductive case $1+$ appends the type t' to the typing context, and the usage annotation x' to both the input and output usage contexts.

Example 2 (Variable reference). egVar defines a variable reference $1+0$ with type $C[\, 1 ; \ell_i \,]$ and usage ℓ_i. We must show that this variable is well typed in an environment with a typing context $\gamma = [\,] , C[\, 1 : \ell_i \,] , 1$ and a usage context $\Gamma = [\,] , \ell_\# , \ell_\#$. The VARREF constructors are completely determined by the variable index $1+ 0$ in the type. The constructor $1+$ steps under the outermost variable in the context, preserving its usage annotation $\ell_\#$ from input to output. The constructor 0 asserts that the next variable is of type $C[\, 1 : \ell_i \,]$, and that the usage annotation $\ell_\#$ can be split such that $\ell_\# := \ell_i \cdot \ell_o$ — using ·-compute^r to automatically fulfill the proof obligation.

$$\text{egVar} : ([\,] , C[\, 1 ; \ell_i \,] , 1) ; ([\,] , \ell_\# , \ell_\#) \ni_{1+0} C[\, 1 : \ell_i \,] ; \ell_i \triangleright ([\,] , \ell_o , \ell_\#)$$
$$\text{egVar} = 1+0$$

Definition 11 (TYPES: typing processes). The TYPES typing relation for the resource-aware π-calculus processes is presented below. For convenience, we reuse the constructor names introduced for the syntax in Sect. 2.

$$\frac{\gamma : \mathrm{PRECTX}_n \qquad P : \mathrm{PROCESS}_n \qquad \begin{array}{c} idxs : \mathrm{IDXS}_n \\ \Gamma\,\Delta : \mathrm{CTX}_{idxs} \end{array}}{\gamma\,;\Gamma \vdash P \triangleright \Delta : \mathrm{SET}} \text{ TYPES}$$

$$\frac{}{0 : \gamma\,;\Gamma \vdash 0 \triangleright \Gamma} \qquad \frac{t : \mathrm{TYPE} \qquad x : \mathrm{USAGE}^2_{idx} \qquad y : \mathrm{USAGE}_{idx'} \\ cont : \gamma\,,\mathrm{C}[\,t\,;x\,]\,;\Gamma\,,(y\,,y) \vdash P \triangleright \Delta\,,\ell_\emptyset}{\nu\ t\ x\ y\ cont : \gamma\,;\Gamma \vdash \nu\,P \triangleright \Delta}$$

$$\frac{chan : \gamma \quad ;\Gamma \quad \ni_i \mathrm{C}[\,t\,;x\,]\,;\ell_\mathrm{i} \triangleright \Xi \\ cont : \gamma\,,t\,;\Xi\,,x \vdash P \qquad\qquad \triangleright \Theta\,,\ell_\emptyset}{chan\,(\ )\,cont : \gamma\,;\Gamma \vdash i\,(\ )\,P \triangleright \Theta}$$

$$\frac{chan : \gamma\,;\Gamma \ni_i \mathrm{C}[\,t\,;x\,]\,;\ell_\mathrm{o} \triangleright \Delta \\ loc \ : \gamma\,;\Delta \ni_j t \qquad\quad ;x \triangleright \Xi \\ cont \ : \gamma\,;\Xi \vdash \ P \qquad\qquad \triangleright \Theta}{chan\,\langle loc\rangle\,cont : \gamma\,;\Gamma \vdash i\,\langle j\rangle\,P \triangleright \Theta}$$

$$\frac{l : \gamma\,;\Gamma \vdash P \ \triangleright \Delta \\ r : \gamma\,;\Delta \vdash Q \triangleright \Xi}{l\,\|\,r : \gamma\,;\Gamma \vdash P\,\|\,Q \triangleright \Xi}$$

The inaction process in rule 0 does not change usage annotations. The scope restriction in rule ν expects three arguments: the type t of data being transmitted; the usage annotation x of what is being transmitted; and the multiplicity y given to the channel itself. This multiplicity y is used for both input and output, so that they are balanced. The continuation process P is provided with the new channel with usage annotation $y\,,y$, which it must completely exhaust. The input process in rule $(\ )$ requires a channel $chan$ at index i with usage ℓ_i available, such that data with type t and usage x can be sent over it. Note that the index i is determined by the syntax of the typed process. We use the leftovers Ξ to type the continuation process, which is also provided with the received element—of type t and multiplicity x—at index 0. The received element x must be completely exhausted by the continuation process. Similarly to input, the output process in rule $\langle\ \rangle$ requires a channel $chan$ at index i with usage ℓ_o available, such that data with type t and usage x can be sent over it. We use the leftover context Δ to type the transmitted data, which needs an element loc at index j with type t and usage x, as per the type of the channel $chan$. The leftovers Ξ are used to type the continuation process. Note that both indices i and j are determined by the syntax of the typed process. Parallel composition in rule $\|$ uses the leftovers of the left-hand process to type the right-hand process. Indeed, Theorem 4 shows that an alternative rule where the resources are first threaded through Q is admissible too.

Example 3 (Typing derivation (Continued)). We provide the typing derivation for the courier system defined in Example 1. For the sake of simplicity, we instantiate these processes with concrete variable references before typing them.

The receiver defined by the recv process receives data along the channel with index 0, which needs to be of type $C[t \colon u]$ for some t and u. After receiving twice, the process ends: we must not be left with any unused multiplicities, thus $u = \ell_\emptyset$. We will use graded types to keep track of the exact number of times communication happens. Whatever the input multiplicity of the channel, we will consume 2 of it and leave the remaining as leftovers. The sender defined by the send process sends data along the channel with index 0, which needs to be of type $C[t \colon u]$ for some t and u. We instantiate t (the type of data that the sender sends) to the trivial channel $C[\mathbb{1} \colon \omega]$. As per the type of the process recv, $u = \ell_\emptyset$. We will transmit once, thus use $1{+}0$ output multiplicity, and leave the rest as leftovers. Agda can uniquely determine the arguments required by the ν constructor.

$$\text{recvwt} \; : \; \gamma, C[t \colon \ell_\emptyset] \, ; \Gamma, (1{+}1{+}l \, , r) \vdash \text{recv } 0 \triangleright \Gamma, (l \, , r)$$
$$\text{recvwt} \; = \; 0\,(\;)\,(1{+}0)\,(\;)\,0$$
$$\text{sendwt} \; : \; \gamma, C[C[\mathbb{1} \colon \omega] \colon \ell_\emptyset] \, ; \Gamma, (l \, , 1{+}r) \vdash \text{send } 0 \triangleright \Gamma, (l \, , r)$$
$$\text{sendwt} \; = \; \nu \; \_\_ \cdot\!\cdot 0\,(1{+}0\,\langle\,0\,\rangle\,0)$$

Dually, the courier defined by the carry process expects input multiplicities for the channels shared with send and output multiplicities for the channel shared with recv. We can now compose these processes in parallel and type the courier system.

$$
\begin{aligned}
\text{carrywt} \quad &: \gamma, C[t \colon \ell_\emptyset], C[t \colon \ell_\emptyset], C[t \colon \ell_\emptyset] \\
&; \Gamma, (1{+}lx \, , rx), (1{+}ly \, , ry), (lz \, , 1{+}1{+}rz) \\
&\vdash \text{carry } (1{+}1{+}0)\,(1{+}0)\,0 \\
&\triangleright \Gamma, (lx \, , rx), (ly \, , ry), (lz \, , rz) \\
\text{carrywt} \quad &= (1{+}1{+}0)\,(\;)\,(1{+}1{+}0)\,(\;)\,(1{+}1{+}0)\,\langle\,1{+}0\,\rangle\,(1{+}1{+}0)\,\langle\,0\,\rangle\,0 \\
\text{systemwt} \quad &: [\,]\,; [\,] \vdash \text{system} \triangleright [\,] \\
\text{systemwt} \quad &= \nu \; \_\_\_ (\text{sendwt} \,\|\, \nu \; \_\_\_ (\text{sendwt} \,\|\, \nu \; \_\_\_ (\text{recvwt} \,\|\, \text{carrywt})))
\end{aligned}
$$

5   Meta-Theory

We have mechanised subject reduction for our π-calculus with leftovers in 850 lines of Agda code. The meta-theory of resource-aware type systems often needs to reason on typing derivations modulo associativity in the allocation of resources. For type systems using context splitting side conditions, this means applying associativity lemmas to recompute context splits; for type systems using leftover typing it does not. As an example, the proof that comp-asssoc preserves typing proceeds by deconstructing the input derivation into $P \| (Q \| R)$ and reassembling it as $(P \| Q) \| R$ without the need of any extra reasoning.

All the reasoning carried out in our type safety proofs is based on the algebraic properties introduced in Sect. 4.1 – the exception to this is ·compute^r, only there for the user's convenience. We lift the operation $x := y \cdot^2 z$ and its algebraic properties to an operation $\Gamma := \Delta \otimes \Xi$ on usage contexts that have the same underlying context of indices. The algebraic properties of the algebras allow us to see a typing derivation $\gamma \,;\, \Gamma \vdash P \triangleright \Delta$ as a unique *arrow* from Γ to Δ, and to freely compose and reason with arrows with the same typing context and a matching output and input usage contexts.

Leftover typing also allows us to state a *framing* theorem showing that adding or subtracting arbitrary usage annotations to the input and output usage contexts preserves typing – one can understand a typing derivation independently from its unused resources. With framing one can show that comp-comm preserves typing: in $P \parallel Q$ the typing of P and Q is independent of one another.

Theorem 1 (Framing). Let $\gamma \,;\, \Gamma_l \vdash P \triangleright \Xi_l$. Let Δ be such that $\Gamma_l := \Delta \otimes \Xi_l$. Then for any Γ_r and Ξ_r where $\Gamma_r := \Delta \otimes \Xi_r$ it holds that $\gamma \,;\, \Gamma_r \vdash P \triangleright \Xi_r$.

Leftover typing allows *weakening* and *strengthening* to acquire a more general form where linear variables can freely be added or removed from context too – as long as they are added and removed to and from both the input and output contexts.

Theorem 2 (Weakening). Let ins_i insert an element into a context at position i. Let P be well typed in $\gamma \,;\, \Gamma \vdash P \triangleright \Xi$. Then, lifting every variable greater than or equal to i in P is well typed in $\text{ins}_i\, t\, \gamma \,;\, \text{ins}_i\, x\, \Gamma \vdash \text{lift}_i\, P \triangleright \text{ins}_i\, x\, \Xi$.

Theorem 3 (Strengthening). Let del_i delete the element at position i from a context. Let P be well typed in $\gamma \,;\, \Gamma \vdash P \triangleright \Xi$. Let i be a variable not in P, such that $uP \,:\, \text{UNUSED}_i\, P$. Then lowering every variable greater than i in P is well typed in $\text{del}_i\, \gamma \,;\, \text{del}_i\, \Gamma \vdash \text{lower}_i\, P\, uP \triangleright \text{del}_i\, \Xi$.

Subject congruence states that structural congruence (Definition 4) preserves the well-typedness of a process.

Theorem 4 (Subject Congruence). Let P and Q be processes. If $P \simeq Q$ and $\gamma \,;\, \Gamma \vdash P \triangleright \Xi$, then $\gamma \,;\, \Gamma \vdash Q \triangleright \Xi$.

Finally, subject reduction states that reducing on a channel c (Definition 5) preserves the well-typedness of a process—after consuming $\ell_\#$ from c if c is an external channel. Below we use $\Gamma \ni_i x \triangleright \Delta$ to stand for $\gamma \,;\, \Gamma \ni_i t \,;\, x \triangleright \Delta$ for some γ and t.

Theorem 5 (Subject Reduction). Let $\gamma \,;\, \Gamma \vdash P \triangleright \Xi$ and $P \longrightarrow_c Q$. If c is internal, then $\gamma \,;\, \Gamma \vdash Q \triangleright \Xi$. If c is external i and $\Gamma \ni_i \ell_\# \triangleright \Delta$, then $\gamma \,;\, \Delta \vdash Q \triangleright \Xi$.

We refer to our extended paper [36] and repository [37] for a more detailed account of the mechanised proofs.

6   Conclusions, Related and Future Work

Extrinsic Encodings. Extrinsic encodings define a syntax (often well-scoped) and a runtime semantics prior to any type system. This allows one to talk about ill-typed terms, and defers the proof of subject reduction to a later stage. To the best of our knowledge, leftover typing makes its appearance in 1994, when Ian Mackie first uses it to formulate intuitionistic linear logic [23]. Allais [2] uses leftover typing to mechanise in Agda a bidirectional type system for the linear λ-calculus. He proves type preservation and provides a decision procedure for type checking and type inference. In this paper, we follow Allais [2] and apply leftover typing to the π-calculus for the first time. We generalise the usage algebra, leading to linear, graded and shared type systems. Drawing from quantitative type theory (by McBride and Atkey [3,25]), in our work we too are able to talk about fully consumed resources — e.g., we can transmit $\ell_\emptyset$ multiplicities of a fully exhausted channel. Recent years have seen an increase in the efforts to mechanise resource-aware process algebras, but one of the earliest works is the mechanisation of the linear π-calculus in Isabelle/HOL by Gay [15]. Gay encodes the π-calculus with linear and shared types using de Bruijn indices, a reduction relation and a type system posterior to the syntax. However, in his work typing rules demand user-provided context splits, and variables with consumed usage annotations are erased from context. We remove the demand for context splits, preserve the ability to talk about consumed resources, and adopt a more general usage algebra. Orchard et al. introduce Granule [28], a fully-fledged functional language with graded modal types, linear types, indexed types and polymorphism. Modalities include exact usages, security levels and intervals; resource algebras are pre-ordered semirings with partial addition. The authors provide bidirectional typing rules, and show the type safety of their semantics. The work by Goto et al. [18] is, to the best of our knowledge, the first formalisation of session types which comes along with a mechanised proof of type safety in Coq. The authors extend session types with polymorphism and pattern matching. They use a locally-nameless encoding for variable references, a syntax prior to types, and an LTS semantics that encodes session-typed processes into the π-calculus. Their type system uses reordering of contexts and extrinsic context splits, which are not needed in our work.

Intrinsic Encodings. Intrinsic encodings merge syntax and type system. As a result, one can only ever talk about well-typed terms, and the reduction relation by construction carries a proof of subject reduction. Significantly, by merging the syntax and static semantics of the object language one can fully use the expressive power of the host language. Thiemann formalises in Agda the MicroSession (minimal GV [16]) calculus with support for recursion and subtyping [33]. As Gay does in [15], context splits are given extrinsically, and exhausted resources are removed from typing contexts altogether. The runtime semantics are given as an intrinsically typed CEK machine with a global context of session-typed channels. In their recent paper, Ciccone and Padovani mechanise a dependently-typed linear π-calculus in Agda [8]. Their intrinsic encoding allows them to leverage Agda's dependent types to provide a dependently-typed interpretation of messages—to avoid linearity violations the interpretation of channel types is erased. Message input is modeled as a dependent function in Agda, and as a result message predicates, branching, and variable-length conversations can be encoded. In contrast to our work, their algebra is on the multiplicities 0, 1, ω, and top-down context splitting proofs must be provided. In another recent work, Rouvoet et al. provide an intrinsic type system for a λ-calculus with session types [30]. They use proof relevant separation logic and a notion of a supply and demand *market* to make context splits transparent to the user. Their separation logic is based on a partial commutative monoid that need not be deterministic nor cancellative. Their typing rules preserve the balance between supply and demand, and are extremely elegant. They distill their typing rules even further by modelling the supply and demand market as a state monad.

Other Work. Castro et al. [6] provide tooling for locally-nameless representations of process calculi in Coq, where de Bruijn indices are less popular than in Agda or Idris. They use their tool to help automate proofs of subject reduction for a type system with session types. Orchard and Yoshida [29] embed a small effecftul imperative language into the session-typed π-calculus, showing that session types are expressive enough to encode effect systems. Based on contextual type theory, LINCX [17] extends the linear logical framework LLF [7] by internalising the notion of bindings and contexts. The result is a meta-theory in which HOAS encodings with both linear and dependent types can be described. The developer obtains for free an equational theory of substitution and decidable typechecking without having to encode context splits within the object language. Further work on mechanisation of the π-calculus [1,5,12,13,19], focuses on non-linear variations, differently from our range of linear, graded and shared types.

Conclusions and Future Work. We provide a well-scoped syntax and a semantics for the π-calculus, extrinsically define a type system on top of the syntax capable of handling linear, graded and shared types under the same unified framework and show subject reduction. We avoid extrinsic context splits by defining a type system based on leftover typing [2]. As a result, theorems like framing, weakening and strengthening can now be stated also for the linear π-calculus. Our work is fully mechanised in around 1850 lines of code in Agda [37].

As future work we intend to expand our framework to include infinite behaviour by adding process replication, which is challenging, as to prove subject congruence one needs to uniquely determine the resources consumed by a process—e.g., by adding type annotations to the syntax. Orthogonally, we aim to investigate making our typing rules bidirectional which would allow us to provide a decision procedure for type checking processes in a given set of algebras. Finally, we will use our π-calculus with leftovers as an underlying framework on top of which we can implement session types, via their encodings into linear types [9,10,32] and other advanced type theories.

Acknowledgments. We want to thank Erika Kreuter, Wen Kokke, James Wood, Guillaume Allais, Bob Atkey, and Conor McBride for their valuable suggestions.

References

1. Affeldt, R., Kobayashi, N.: A Coq Library for Verification of Concurrent Programs. Electron. Notes Theor. Comput. Sci. **199**, 17–32 (2008). https://doi.org/10.1016/j.entcs.2007.11.010
2. Allais, G.: Typing with leftovers - a mechanization of intuitionistic multiplicative-additive linear logic. In: Types for Proofs and Programs, TYPES. LIPIcs, vol. 104, pp. 1:1–1:22. Schloss Dagstuhl - Leibniz-Zentrum für Informatik (2017). https://doi.org/10.4230/LIPIcs.TYPES.2017.1
3. Atkey, R.: Syntax and semantics of quantitative type theory. In: Logic in Computer Science, LICS, pp. 56–65. ACM (2018). https://doi.org/10.1145/3209108.3209189
4. Barendsen, E., Smetsers, S.: Uniqueness typing for functional languages with graph rewriting semantics. Math. Struct. Comput. Sci. **6**(6), 579–612 (1996)
5. Bengtson, J.: The pi-calculus in nominal logic, vol. 2012 (2012). https://www.isa-afp.org/entries/Pi_Calculus.shtml
6. Castro, D., Ferreira, F., Yoshida, N.: EMTST: engineering the meta-theory of session types. TACAS 2020. LNCS, vol. 12079, pp. 278–285. Springer, Cham (2020). https://doi.org/10.1007/978-3-030-45237-7_17
7. Cervesato, I., Pfenning, F.: A linear logical framework. In: Logic in Computer Science, LICS, pp. 264–275. IEEE Computer Society (1996). https://doi.org/10.1109/LICS.1996.561339
8. Ciccone, L., Padovani, L.: A dependently typed linear π-calculus in Agda. In: PPDP 2020: 22nd International Symposium on Principles and Practice of Declarative Programming, pp. 8:1–8:14. ACM (2020). https://doi.org/10.1145/3414080.3414109
9. Dardha, O.: Recursive session types revisited. In: Carbone, M. (ed.) Workshop on Behavioural Types, BEAT. EPTCS, vol. 162, pp. 27–34 (2014). https://doi.org/10.4204/EPTCS.162.4
10. Dardha, O., Giachino, E., Sangiorgi, D.: Session types revisited. Inf. Comput. **256**, 253–286 (2017). https://doi.org/10.1016/j.ic.2017.06.002. Extended version of [10]
11. de Bruijn, N.G.: Lambda calculus notation with nameless dummies, a tool for automatic formula manipulation, with application to the Church-Rosser theorem. In: Indagationes Mathematicae (Proceedings), vol. 75, pp. 381–392. Elsevier (1972)
12. Deransart, P., Smaus, J.: Subject reduction of logic programs as proof-theoretic property, vol. 2002 (2002). http://danae.uni-muenster.de/lehre/kuchen/JFLP/articles/2002/S02-01/JFLP-A02-02.pdf
13. Despeyroux, J.: A higher-order specification of the π-Calculus. In: van Leeuwen, J., Watanabe, O., Hagiya, M., Mosses, P.D., Ito, T. (eds.) TCS 2000. LNCS, vol. 1872, pp. 425–439. Springer, Heidelberg (2000). https://doi.org/10.1007/3-540-44929-9_30
14. Dybjer, P.: Inductive families. Formal Asp. Comput. **6**(4), 440–465 (1994). https://doi.org/10.1007/BF01211308
15. Gay, S.J.: A framework for the formalisation of pi calculus type systems in Isabelle/HOL. In: Boulton, R.J., Jackson, P.B. (eds.) TPHOLs 2001. LNCS, vol. 2152, pp. 217–232. Springer, Heidelberg (2001). https://doi.org/10.1007/3-540-44755-5_16

16. Gay, S.J., Vasconcelos, V.T.: Linear type theory for asynchronous session types. J. Funct. Program. **20**(1), 19–50 (2010). https://doi.org/10.1017/S0956796809990268

17. Georges, A.L., Murawska, A., Otis, S., Pientka, B.: LINCX: a linear logical framework with first-class contexts. In: Yang, H. (ed.) ESOP 2017. LNCS, vol. 10201, pp. 530–555. Springer, Heidelberg (2017). https://doi.org/10.1007/978-3-662-54434-1_20

18. Goto, M.A., Jagadeesan, R., Jeffrey, A., Pitcher, C., Riely, J.: An extensible approach to session polymorphism. Math. Struct. Comput. Sci. **26**(3), 465–509 (2016). https://doi.org/10.1017/S0960129514000231

19. Honsell, F., Miculan, M., Scagnetto, I.: pi-calculus in (Co)inductive-type theory. Theor. Comput. Sci. **253**(2), 239–285 (2001). https://doi.org/10.1016/S0304-3975(00)00095-5

20. Jung, R., et al.: Iris: monoids and invariants as an orthogonal basis for concurrent reasoning. In: Rajamani, S.K., Walker, D. (eds.) Symposium on Principles of Programming Languages, POPL 2015, pp. 637–650. ACM (2015). https://doi.org/10.1145/2676726.2676980

21. Kobayashi, N.: Type systems for concurrent programs (2007). http://www.kb.ecei.tohoku.ac.jp/~koba/papers/tutorial-type-extended.pdf

22. Kobayashi, N., Pierce, B.C., Turner, D.N.: Linearity and the Pi-Calculus. In: Symposium on Principles of Programming Languages, POPL, pp. 358–371. ACM Press (1996). https://doi.org/10.1145/237721.237804

23. Mackie, I.: Lilac: a functional programming language based on linear logic. J. Funct. Program. **4**(4), 395–433 (1994). https://doi.org/10.1017/S0956796800001131

24. Matsakis, N.D., II, F.S.K.: The rust language. In: High Integrity Language Technology, HILT, pp. 103–104. ACM (2014). https://doi.org/10.1145/2663171.2663188

25. McBride, C.: I got plenty o' Nuttin'. In: Lindley, S., McBride, C., Trinder, P., Sannella, D. (eds.) A List of Successes That Can Change the World. LNCS, vol. 9600, pp. 207–233. Springer, Cham (2016). https://doi.org/10.1007/978-3-319-30936-1_12

26. Milner, R.: Communicating and Mobile Systems - The Pi-calculus. Cambridge University Press, Cambridge (1999)

27. Milner, R., Parrow, J., Walker, D.: A calculus of mobile processes, Parts I and II. Inf. Comput. **100**(1), 1–40 (1992). https://doi.org/10.1016/0890-5401(92)90008-4

28. Orchard, D., Liepelt, V., III, H.E.: Quantitative program reasoning with graded modal types. Proc. ACM Program. Lang. **3**(ICFP), 110:1–110:30 (2019). https://doi.org/10.1145/3341714

29. Orchard, D.A., Yoshida, N.: Using session types as an effect system. In: Gay, S., Alglave, J. (eds.) Programming Language Approaches to Concurrency- and Communication-cEntric Software, PLACES 2015. EPTCS, vol. 203, pp. 1–13 (2015). https://doi.org/10.4204/EPTCS.203.1

30. Rouvoet, A., Poulsen, C.B., Krebbers, R., Visser, E.: Intrinsically-typed definitional interpreters for linear, session-typed languages. In: Certified Programs and Proofs, CPP, pp. 284–298. ACM (2020). https://doi.org/10.1145/3372885.3373818

31. Sangiorgi, D., Walker, D.: The Pi-Calculus - A Theory of Mobile Processes. Cambridge University Press, Cambridge (2001)

32. Scalas, A., Dardha, O., Hu, R., Yoshida, N.: A linear decomposition of multiparty sessions for safe distributed programming. In: European Conference on Object-Oriented Programming, ECOOP. LIPIcs, vol. 74, pp. 24:1–24:31. Schloss Dagstuhl - Leibniz-Zentrum für Informatik (2017). https://doi.org/10.4230/LIPIcs.ECOOP.2017.24

33. Thiemann, P.: Intrinsically-typed mechanized semantics for session types, pp. 19:1–19:15 (2019). https://doi.org/10.1145/3354166.3354184
34. Turon, A.J., Thamsborg, J., Ahmed, A., Birkedal, L., Dreyer, D.: Logical relations for fine-grained concurrency. In: Giacobazzi, R., Cousot, R. (eds.) Symposium on Principles of Programming Languages, POPL 2013, pp. 343–356. ACM (2013). https://doi.org/10.1145/2429069.2429111
35. Wadler, P.: Linear types can change the world! In: Programming Concepts and Methods, p. 561. North-Holland (1990)
36. Zalakain, U., Dardha, O.: π with leftovers: a mechanisation in Agda. CoRR abs/2005.05902 (2020). https://arxiv.org/abs/2005.05902
37. Zalakain, U., Dardha, O.: Typing the linear π-Calculus – formalisation in Agda (2021). https://github.com/umazalakain/typing-linear-pi

Short and Journal-First Papers

Supervisory Synthesis of Configurable Behavioural Contracts with Modalities

Davide Basile[1]([⊠]) [iD], Maurice H. ter Beek[1] [iD], Pierpaolo Degano[2] [iD],
Axel Legay[3] [iD], Gian-Luigi Ferrari[2] [iD], Stefania Gnesi[1] [iD],
and Felicita Di Giandomenico[1] [iD]

[1] ISTI–CNR, Pisa, Italy
davide.basile@isti.cnr.it
[2] University of Pisa, Pisa, Italy
[3] UCLouvain, Louvain-la-Neuve, Belgium

Abstract. Service contracts characterise the desired behavioural compliance of a composition of services, typically defined by the fulfilment of all service requests through service offers. Contract automata are a formalism for specifying behavioural service contracts. Based on the notion of synthesis of the most permissive controller from Supervisory Control Theory, a safe orchestration of contract automata can be computed that refines a composition into a compliant one. This short paper summarises the contributions published in [8], where we endow contract automata with two orthogonal layers of variability: (i) at the structural level, constraints over service requests and offers define different configurations of a contract automaton, depending on which requests and offers are selected or discarded; and (ii) at the behavioural level, service requests of different levels of criticality can be declared, which induces the novel notion of semi-controllability. The synthesis of orchestrations is thus extended to respect both the structural and the behavioural variability constraints. Finally, we show how to efficiently compute the orchestration of all configurations from only a subset of these configurations. A recently redesigned and refactored tool supports the developed theory.

Extended Abstract

A contract automaton [4] represents a single service (a *principal*) or a multi-party composition of services [2,11]. Each principal's goal is to reach an accepting state by matching its service request actions with corresponding service offer actions of other principals. An orchestration is synthesised from the principals to only allow finite executions in agreement, i.e., each request action a is fulfilled by an offer action $\overline{a}$. Technically, such an orchestration is synthesised as the *most permissive controller* (mpc) known from Supervisory Control Theory (SCT) [15,23].

Automata $\mathcal{A}_1$ and $\mathcal{A}_2$ in Fig. 1(left) interact on a service action a. Their composition $\mathcal{A}_1 \otimes \mathcal{A}_2$ in Fig. 1(right) models two possible ways to fulfill service request a from $\mathcal{A}_1$ by matching it with a service offer $\overline{a}$ of $\mathcal{A}_2$, represented as

© IFIP International Federation for Information Processing 2021
Published by Springer Nature Switzerland AG 2021
K. Peters and T. A. C. Willemse (Eds.): FORTE 2021, LNCS 12719, pp. 177–181, 2021.
https://doi.org/10.1007/978-3-030-78089-0_10

$(a, \overline{a})$. Assume that a must be matched with $\overline{a}$ to obtain agreement, and that for some reason the state ϟ is to be avoided in favour of state ✓. In most automata-based formalisms, including the contract automata of [4,7], this is typically not allowed by the notion of *uncontrollability*, and thus the resulting mpc is empty.

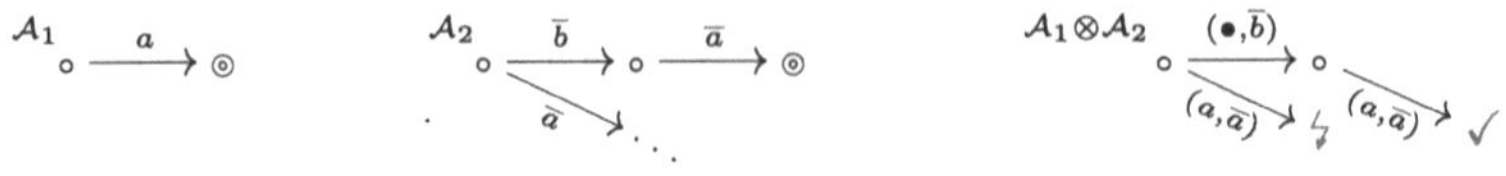

Fig. 1. Two automata $\mathcal{A}_1$ and $\mathcal{A}_2$ and a possible composition $\mathcal{A}_1 \otimes \mathcal{A}_2$

In [8], we introduce a way to express that a must *eventually* be matched, rather than *always*, by defining contract automata in which it is possible to orchestrate the composition of $\mathcal{A}_1$ and $\mathcal{A}_2$ such that the result is similar to the composition $\mathcal{A}_1 \otimes \mathcal{A}_2$ depicted in Fig. 1 but *without state* ϟ, i.e., a is only matched with $\overline{a}$ *after* the occurrence of an unmatched service offer $\overline{b}$ of $\mathcal{A}_2$, i.e., $(\bullet, \overline{b})$.

Technically, in [8] we extend contract automata with action modalities to distinguish *permitted* from *necessary* service requests (borrowed from [7]). Permitted and necessary request actions differ in that the latter *must* be fulfilled, while the former *may* also be omitted. As in [7], we assume service offer actions to be always permitted because a service contract may *always* withdraw its offers that are not needed to reach an agreement. Furthermore, we endow contract automata with two orthogonal *variability* mechanisms.

The first variability mechanism concerns constraints operating on the entire service contract, i.e., at the *structural* level, to define different configurations. This is important because services are typically reused in configurations that vary over time and need to be adapted to changing environments. Such configurations are characterised by which service actions are *mandatory* and which *forbidden*. The *valid* configurations are those respecting all structural constraints. We follow the well-established paradigm of Software Product Line Engineering (SPLE), which aims at efficiently managing a family of highly configurable systems to allow for mass customisation [1,22]. To compactly represent a *product line*, i.e., the set of *valid* product configurations, we use a so-called *feature constraint*, a propositional formula φ whose atoms are features [10,16,19] and we identify features as service actions (offers as well as requests). Usually, in SPLE, each feature is either selected or discarded to configure a product, i.e., all variability is resolved and the interpretation of the atoms of φ is *total*. Instead, we consider as valid those products (called *sub-families* in SPLE terms) that are defined by a *partial* assignment satisfying φ. This enables to synthesise the orchestration of an entire product line by considering a few valid products only (those such that their union contains all possible behaviour of the product line's orchestration), rather than computing *all* the valid ones (and retaining unnecessary complexity due to duplicated behaviour). This is one of the main results of [8].

The second variability mechanism is defined inside service contracts, i.e., at the *behavioural* level, to declare necessary request actions to be either *urgent* or *lazy*. These modalities drive the orchestrator to fulfill *all the occurrences* of an urgent action, which is the classical notion of uncontrollability from SCT, while it is required to fulfill *at least one occurrence* of lazy actions, which is the novel notion of *semi-controllability* useful for orchestration synthesis. The simplistic example above has no urgent action; the only necessary one is the lazy request a. Intuitively, the matching of a lazy request may be delayed whereas this is not the case for urgent requests. Obviously, a must not be forbidden, either directly or because it is not part of any valid configuration.

To effectively use the variability mechanisms, we refine the classical synthesis algorithm from SCT [23]. We compute the orchestrations of a single *valid* configuration, i.e., including all mandatory and none of the forbidden actions, besides fulfilling all the necessary and the maximal number of permitted requests (i.e., if the orchestration were to fulfill another permitted request, then one of the other requirements would no longer be fulfilled).

Summarising, the main contributions of [8] are as follows:

1. A novel formalism for behavioural service contracts, called *Featured Modal Contract Automata* (FMCA), which offers support for both structural and behavioural variability not available before in the literature.
2. The new notion of *semi-controllability* (related to lazy actions), which refines both the notion of controllability (related to permitted actions) and that of uncontrollability (related to urgent actions) as used in classical synthesis algorithms from SCT. This new notion is fundamental to handle different service requests in the orchestration synthesis for FMCA.
3. A revised algorithm for synthesising an orchestration of services for a single valid product configuration. Each FMCA $\mathcal{A}$ is a pair made of an automaton and a feature constraint φ, which is related to the automaton in the following way. The labels on the arcs of the automaton identify the actions for requests and offers, a subset of which corresponds to all features in φ. The FMCA $\mathcal{A}$ is said to *respect* a product p whenever all features declared *mandatory* (*forbidden*, respectively) by p correspond to actions that are reachable (unreachable, respectively) from the initial state of $\mathcal{A}$.
4. An algorithm to compute the orchestration of an entire product line by joining the orchestrations of a small selected subset of valid product configurations, *without* computing the orchestration for each of its valid product configurations. Since the number of valid product configurations is known to be exponential in the number of features [13], only using few of them greatly improves performance and guarantees scalability of the novel framework of contract automata presented in [8]. The algorithm is thus more efficient than the standard ones available in the literature (e.g., cf. [12]).
5. The open-source prototypical Contract Automata Tool [5] extended to include FMCA is briefly surveyed and evaluated (cf. [6] for more details of the FMCA tool). It exploits FeatureIDE [21], an open-source framework for feature-oriented software development based on Eclipse, offering a variety of feature model editing and management tools.

The research on the formalism and its associated tool has evolved since [8]. As reported in [3], the tool has recently been redesigned according to the principles of model-based systems engineering [18,24] and of writing clean and readable code [14,20], and it has been refactored using lambda expressions and Java Streams as available in Java 8 [17,25], exploiting parallelism. Also, the abstract parametric synthesis algorithm from [9] has been implemented. The current version is available at https://github.com/davidebasile/ContractAutomataTool and previous implementations are still available in other branches of the repository.

References

1. Apel, S., Batory, D.S., Kästner, C., Saake, G.: Feature-Oriented Software Product Lines: Concepts and Implementation. Springer, Heidelberg (2013). https://doi.org/10.1007/978-3-642-37521-7
2. Bartoletti, M., Cimoli, T., Zunino, R.: Compliance in behavioural contracts: a brief survey. In: Bodei, C., Ferrari, G.-L., Priami, C. (eds.) Programming Languages with Applications to Biology and Security. LNCS, vol. 9465, pp. 103–121. Springer, Cham (2015). https://doi.org/10.1007/978-3-319-25527-9_9
3. Basile, D., ter Beek, M.H.: A clean and efficient implementation of choreography synthesis for behavioural contracts. In: Damiani, F., Dardha, O. (eds.) COORDINATION 2021. LNCS, vol. 12717 (2021). https://doi.org/10.1007/978-3-030-78142-2_14
4. Basile, D., Degano, P., Ferrari, G.L.: Automata for specifying and orchestrating service contracts. Log. Meth. Comput. Sci. **12**(4), 1–51 (2016). https://doi.org/10.2168/LMCS-12(4:6)2016
5. Basile, D., Degano, P., Ferrari, G.-L., Tuosto, E.: Playing with our CAT and communication-centric applications. In: Albert, E., Lanese, I. (eds.) FORTE 2016. LNCS, vol. 9688, pp. 62–73. Springer, Cham (2016). https://doi.org/10.1007/978-3-319-39570-8_5
6. Basile, D., Di Giandomenico, F., Gnesi, S.: FMCAT: supporting dynamic service-based product lines. In: SPLC, pp. 3–8. ACM (2017). https://doi.org/10.1145/3109729.3109760
7. Basile, D., Di Giandomenico, F., Gnesi, S., Degano, P., Ferrari, G.L.: Specifying variability in service contracts. In: VaMoS, pp. 20–27. ACM (2017). https://doi.org/10.1145/3023956.3023965
8. Basile, D., ter Beek, M.H., Degano, P., Legay, A., Ferrari, G.L., Gnesi, S., Di Giandomenico, F.: Controller synthesis of service contracts with variability. Sci. Comput. Program. **187** (2020). https://doi.org/10.1016/j.scico.2019.102344
9. Basile, D., ter Beek, M.H., Pugliese, R.: Synthesis of orchestrations and choreographies: bridging the gap between supervisory control and coordination of services. Log. Methods Comput. Sci. **16**(2) (2020). https://doi.org/10.23638/LMCS-16(2:9)2020
10. Batory, D.S.: Feature models, grammars, and propositional formulas. In: Obbink, H., Pohl, K. (eds.) SPLC 2005. LNCS, vol. 3714, pp. 7–20. Springer, Heidelberg (2005). https://doi.org/10.1007/11554844_3
11. ter Beek, M.H., Bucchiarone, A., Gnesi, S.: Web service composition approaches: from industrial standards to formal methods. In: ICIW. IEEE (2007). https://doi.org/10.1109/ICIW.2007.71
12. ter Beek, M.H., Reniers, M.A., de Vink, E.P.: Supervisory controller synthesis for product lines using CIF 3. In: Margaria, T., Steffen, B. (eds.) ISoLA 2016. LNCS, vol. 9952, pp. 856–873. Springer, Cham (2016). https://doi.org/10.1007/978-3-319-47166-2_59

13. Benavides, D., Segura, S., Ruiz-Cortés, A.: Automated analysis of feature models 20 years later: a literature review. Inf. Syst. **35**(6), 615–636 (2010). https://doi.org/10.1016/j.is.2010.01.001

14. Boswell, D., Foucher, T.: The Art of Readable Code. O'Reilly, Sebastopol (2011)

15. Caillaud, B., Darondeau, P., Lavagno, L., Xie, X. (eds.): Synthesis and Control of Discrete Event Systems. Springer, Dordtrecht (2002). https://doi.org/10.1007/978-1-4757-6656-1

16. Czarnecki, K., Wąsowski, A.: Feature diagrams and logics: there and back again. In: SPLC, pp. 23–34. IEEE (2007). https://doi.org/10.1109/SPLINE.2007.24

17. Goetz, B., Peierls, T., Bloch, J., Bowbeer, J., Holmes, D., Lea, D.: Java Concurrency in Practice. Addison-Wesley, Upper Saddle River (2006)

18. Henderson, K., Salado, A.: Value and benefits of model-based systems engineering (MBSE): evidence from the literature. Syst. Eng. **24**(1), 51–66 (2021). https://doi.org/10.1002/sys.21566

19. Mannion, M.: Using first-order logic for product line model validation. In: Chastek, G.J. (ed.) SPLC 2002. LNCS, vol. 2379, pp. 176–187. Springer, Heidelberg (2002). https://doi.org/10.1007/3-540-45652-X_11

20. Martin, R.C.: Clean Code. Prentice Hall, Upper Saddle River (2008)

21. Meinicke, J., Thüm, T., Schröter, R., Benduhn, F., Leich, T., Saake, G.: Mastering Software Variability with FeatureIDE. Springer, Cham (2017). https://doi.org/10.1007/978-3-319-61443-4

22. Pohl, K., Böckle, G., van der Linden, F.J.: Software Product Line Engineering: Foundations, Principles, and Techniques. Springer, Heidelberg (2005). https://doi.org/10.1007/3-540-28901-1

23. Ramadge, P.J., Wonham, W.M.: Supervisory control of a class of discrete event processes. SIAM J. Control Optim. **25**(1), 206–230 (1987). https://doi.org/10.1137/0325013

24. Tockey, S.: How to Engineer Software: A Model-Based Approach. Wiley, Chichester (2019)

25. Warburton, R.: Java 8 Lambdas: Pragmatic Functional Programming. O'Reilly, New York (2014)

Off-the-Shelf Automated Analysis of Liveness Properties for Just Paths
(Extended Abstract)

Mark Bouwman, Bas Luttik, and Tim Willemse[✉]

Eindhoven University of Technology, Eindhoven, The Netherlands
{m.s.bouwman,s.p.luttik,t.a.c.willemse}@tue.nl

Abstract. Recent work by van Glabbeek and coauthors suggests that the liveness property for Peterson's mutual exclusion algorithm, which states that any process wanting to enter the critical section will eventually enter it, cannot be analysed in CCS and related formalisms. In our article, we explore the formal underpinning of this suggestion and its ramifications. In particular, we show that the liveness property for Peterson's algorithm *can* be established convincingly with the mCRL2 toolset, which has a conventional ACP-style process-algebra based specification formalism.

1 Introduction

A process-algebraic specification of a distributed algorithm or system typically includes unrealistic finite or infinite computations in which progress in some component halts. Their mere presence often sits in the way of a proof that the algorithm or system satisfies a set of desirable liveness properties. The go-to solution is to exclude these unrealistic computations from consideration by imposing additional assumptions such as *progress* and *fairness* (see [7] for a comprehensive overview).

For the analysis of so-called *fair schedulers*—of which Peterson's mutual exclusion algorithm is an example—one should, however, use such fairness assumptions cautiously, as fair schedulers themselves are intended to realise the very aspect of fairness in a system. In [6], Van Glabbeek and Höfner propose *justness* as a criterion that is just strong enough to exclude unrealistic computation of fair schedulers:

"Once a transition is enabled that stems from a set of parallel components, one (or more) of these components eventually partake in a transition" [7].

The semantics of process calculi are usually defined by associating with every expression a labelled transition system. Thus, the semantic mapping is forgetful with respect to the notion of component, which is crucial for the definition of which computations are just. To facilitate reasoning about liveness properties

© IFIP International Federation for Information Processing 2021
Published by Springer Nature Switzerland AG 2021
K. Peters and T. A. C. Willemse (Eds.): FORTE 2021, LNCS 12719, pp. 182–187, 2021.
https://doi.org/10.1007/978-3-030-78089-0_11

of processes specified in process calculi while taking justness into account, two solutions are proposed in the literature: In [4], the definition of *just path* uses that the states of the labelled transition system associated with an expression of the calculus (by its structural operational semantics) are themselves expressions of the calculus; hence, these expressions reflect component information. In [5], the operational semantics is revised so that it yields a labelled transition system enriched with a concurrency relation that reflects component information; the notion of *just path* is then formulated referring to the concurrency relation.

The disadvantage of both the aforementioned definitions of just path is that they preclude the use of state-of-the-art verification technology for process calculi that has been developed over the past two decades relying firmly on the forgetful semantic mapping to labelled transition systems. Our aim, in [1], is to present a method by which the mCRL2 toolset [2] can be used to verify that all just paths associated with an mCRL2 specification satisfy a liveness property, and to establish its correctness. Our method does not require an adaptation of the mCRL2 toolset itself. Instead, it relies on a disciplined use of labels in an mCRL2 process specification, by which from the label it can be inferred exactly which components contribute to the execution of the transition with that label. Then a justness assumption can be built into the μ-calculus formula expressing the liveness property. In the remainder of this article we first summarise the main ideas regarding the disciplined use of labels in the mCRL2 process specification, and then we discuss how to formalise liveness for all just paths in the μ-calculus.

2 Label-Based Justness for mCRL2

In mCRL2 a process is specified as a parallel composition of sequential components. The language has a flexible mechanism, inherited from ACP, to define interaction between those components. One can, e.g., specify that if one component can execute a transition labelled with a and another component can execute a transition labelled with b, then the two components may synchronise, resulting in a transition labelled with c. This is achieved by including the specification of a so-called *communication function*, expressing that labels a and b communicate to c. (In contrast, the language CCS has a fixed communication function that only allows transitions labelled with complementary labels (a and $\bar{a}$) to synchronise and the resulting transition is always labelled with the special label τ.) The flexibility to specify a communication function is crucial to our definition of justness, because it allows one to encode in the label c which components are contributing to the interaction it represents.

Our formalisation of the notion of *just path* in the context of an mCRL2 specification, which is inspired by the formalisation in [5], assumes that the following two mappings are defined:

1. *npc* associates with every label a set of *necessary participants*, i.e., components that participate in the interaction; and
2. *afc* associates with every label a set of *affected components*, i.e., components that are thought to be affected by the interaction.

The pair (npc, afc) is called a *component assignment* for the specification.

The distinction between necessary participant and affected component is important when, for example, modelling shared variables in process calculi. Since process calculi adhere strictly to the message passing paradigm, shared variables should be modelled as separate components. As pointed out in [4], to get a realistic notion of justness (e.g., for a model of Peterson's algorithm), the activity of reading the value of the variable should, however, not be treated as affecting the component representing the shared variable. To facilitate this, we partition the set of labels used in the mCRL2 specification into a set $\mathcal{S}$ of *signals* and a set $\mathcal{A}$ of *actions*. Transitions labelled by signals are special in that they should not change state; this is a property that needs to be established for the mCRL2 specification at hand. Moreover, the communication function should respect signals in the sense that it should not yield a signal, unless it was applied exclusively to signals.

For a correct definition of just path, it is important that the component assignment truly reflects the component structure; such a component assignment we shall call *consistent*. It is not possible to associate a consistent component assignment with every mCRL2 specification, but in [1] we give fairly liberal sufficient conditions that ensure that a consistent component assignment does indeed exist. Roughly, these conditions require that the sets of labels occurring in components are disjoint, that the component assignment assigns each such label to the correct component, and that the communication function is consistent with the component assignment. It is worth reiterating that the flexible communication mechanism of mCRL2 is crucial for the latter. We refer to [1] for the formalities. It is also argued in [1], that Peterson's algorithm for mutual exclusion can be modelled by an mCRL2 specification for which there exists a consistent component assignment.

Finally, as argued in [5], justness is used to specify which paths represent a *complete computation* of the system. The distinction between *blocking* and *non-blocking* actions is relevant for determining when a computation is complete. Blocking actions are not entirely under the control of the specified system; their execution may depend on interaction with the environment. A non-blocking action is assumed to be completely under the control of the system. Complete computations may therefore only end in a state in which only blocking actions are enabled.

We now proceed to define the notion of *just path*. First, to define the notion of path we refer to the labelled transition system associated with an mCRL2 specification. A *path* in this transition system is a finite or infinite alternating sequence $s_0 a_1 s_1 a_2 s_2 \cdots$ of states and actions, starting with a state and if it is finite also ending with a state, such that $s_i \xrightarrow{a_{i+1}} s_{i+1}$ for all relevant i. Furthermore, we say that an action a is *enabled* in a state s if there exists s' such that $s \xrightarrow{a} s'$. The component assignment for the mCRL2 specification induces a *concurrency relation* on labels: we define that $\lambda_1 \smile \lambda_2$ if, and only if, $npc(\lambda_1) \cap afc(\lambda_2) = \emptyset$. This concurrency relation is used to define the notion of just path.

Definition 1. *Let* $\mathcal{B} \subseteq \mathcal{A}$ *be a set of* blocking actions. *A path* π *is* $\mathcal{B}$-just *if for every action* $a \notin \mathcal{B}$ *that is enabled in some state* s *on* π, *an action* a' *occurs in the suffix of* π *starting at* s *such that* $a \not\smile^\bullet a'$.

Indeed, the above formalisation of a just path captures the essence of the informal definition. For every action transition involving some set of parallel components that is enabled at some point on the path, ultimately some other transition is executed by a set of parallel components interfering with those enabling the first transition.

3   Off-the-Shelf Verification of Liveness

Due to the nature of justness, which 'dynamically' checks for enabledness of actions along a path, and their future elimination, it is not obvious whether one can express liveness properties restricted to just paths only, in a suitable modal logic. In spite of this dynamic nature, we show that the modal μ-calculus (supported by mCRL2) can be used to express typical liveness properties under justness.

As an illustration, Table 1 displays a template formula asserting (the violation of) the property **a-b**-liveness, stating that on all just paths, an action **a** is inevitably followed by action **b**. This template formula can be instantiated by a user wishing to carry out a liveness verification of an algorithm: it only requires information concerning which labels are designated as signals. As a result, mCRL2 can also be used to verify liveness properties of algorithms such as Peterson's.[1] As a bonus, counterexamples [3] can be provided in case of liveness violations.

Notice that the template formula asserts the *existence* of a just path violating the liveness property, which is conceptually simpler than the dual problem of asserting that the liveness property holds true on all just paths. A just path constitutes a violation to our liveness property exactly when (1) this path has a prefix leading to a state, reached by an **a**-labelled transition, and (2) along the just suffix of this path action **b** never takes place.

Formula violate simply characterises the set of states satisfying property (1), i.e., those states admitting paths in which an **a**-action enters a state admitting a path satisfying property (2). The states that meet the latter property are represented by formula invariant, characterising exactly those states that allow for a **b**-free just path. The justness of that path is captured by the fact that, along that path, each enabled, non-blocking action λ is eliminated along that path. The latter is expressed by elim(λ), asserting that there is a finite **b**-free path consisting of actions that do not interfere with λ, and an action λ', which *does* interfere with λ, leads to a state again satisfying invariant. It is a property of the transition system associated with mCRL2 specifications with a consistent component assignment that on this finite path towards the action interfering

---

[1] The mCRL2 sources can be found in the *academic* example directory of the mCRL2 repository, see https://github.com/mCRL2org/mCRL2, revision b45856d9a.

Table 1. Template formula that characterises the set of states that admit a just path violating a-b-liveness. The user provides the sets $\mathcal{A}$ and $\overline{\mathcal{B}} = \mathcal{A} \setminus \mathcal{B}$, the relation $\rightsquigarrow^{\bullet}$ and the pair of actions **a** and **b** to instantiate/generate the formula for checking a transition system associated with an mCRL2 specification with a consistent component assignment. Note that the modality $\langle\lambda\rangle\phi$ asserts that there is a λ-labelled transition leading to a state satisfying ϕ. The fixed points indicate that one is interested in the least (μ) or largest (ν) set of states satisfying the formula.

$$\mathsf{violate} = \quad \mu W. \; (\langle \mathbf{a} \rangle \mathsf{invariant} \vee \bigvee_{\lambda \in \mathcal{A}} \langle \lambda \rangle W \,)$$

$$\mathsf{invariant} = \nu Y. \; \bigwedge_{\lambda \in \overline{\mathcal{B}}} (\langle \lambda \rangle \top \Rightarrow \mathsf{elim}(\lambda))$$

$$\mathsf{elim}(\lambda) = \quad \mu Q. \; (\bigvee_{\lambda' \in \#\lambda \setminus \{\mathbf{b}\}} \langle \lambda' \rangle Y \vee \bigvee_{\lambda' \in \mathcal{A} \setminus (\#\lambda \cup \{\mathbf{b}\})} \langle \lambda' \rangle Q \,)$$

$$\text{where } \#\lambda = \{ \lambda' \mid \lambda \not\rightsquigarrow^{\bullet} \lambda' \}$$

with λ, all other enabled non-blocking actions remain enabled so long as no action interfering with them is executed.

Theorem 1. *For an mCRL2 specification with a consistent component assignment it holds that all just paths starting in state s satisfy* **a-b**-*liveness iff* $s \notin [\![\mathsf{violate}]\!]$.

References

1. Bouwman, M., Luttik, B., Willemse, T.: Off-the-shelf automated analysis of liveness properties for just paths. Acta Informatica **57**, 551–590 (2020). https://doi.org/10.1007/s00236-020-00371-w
2. Bunte, O., et al.: The mCRL2 toolset for analysing concurrent systems. In: Vojnar, T., Zhang, L. (eds.) TACAS 2019. LNCS, vol. 11428, pp. 21–39. Springer, Cham (2019). https://doi.org/10.1007/978-3-030-17465-1_2
3. Cranen, S., Luttik, B., Willemse, T.A.C.: Evidence for fixpoint logic. In: CSL. Volume 41 of LIPIcs, pp. 78–93. Schloß Dagstuhl - Leibniz-Zentrum für Informatik (2015)
4. Dyseryn, V., van Glabbeek, R.J., Höfner, P.: Analysing mutual exclusion using process algebra with signals. In: Peters, K., Tini, S. (eds.) Proceedings of EXPRESS/SOS 2017. Volume 255 of EPTCS, pp. 18–34 (2017)
5. Glabbeek, R.: Justness. In: Bojańczyk, M., Simpson, A. (eds.) FoSSaCS 2019. LNCS, vol. 11425, pp. 505–522. Springer, Cham (2019). https://doi.org/10.1007/978-3-030-17127-8_29

6. van Glabbeek, R.J., Höfner, P.: CCS: it's not fair! - fair schedulers cannot be implemented in CCS-like languages even under progress and certain fairness assumptions. Acta Informatica **52**(2–3), 175–205 (2015). https://doi.org/10.1007/s00236-015-0221-6
7. van Glabbeek, R.J., Höfner, P.: Progress, justness, and fairness. ACM Comput. Surv. **52**(4), 69:1–69:38 (2019)

Towards a Spatial Model Checker on GPU

Laura Bussi[1], Vincenzo Ciancia[2(✉)], and Fabio Gadducci[1]

[1] Dipartimento di Informatica, Università di Pisa, Pisa, Italy
laura.bussi@phd.unipi.it,fabio.gadducci@unipi.it
[2] Istituto di Scienza e Tecnologie dell'Informazione, CNR, Pisa, Italy
vincenzo.ciancia@isti.cnr.it

Abstract. The tool `VoxLogicA` merges the state-of-the-art library of computational imaging algorithms `ITK` with the combination of declarative specification and optimised execution provided by spatial logic model checking. The analysis of an existing benchmark for segmentation of brain tumours via a simple logical specification reached very high accuracy. We introduce a new, GPU-based version of `VoxLogicA` and present preliminary results on its implementation, scalability, and applications.

Keywords: Spatial logics · Model checking · GPU computation

1 Introduction and Background

Spatial and Spatio-temporal model checking have gained an increasing interest in recent years in various application domains, including collective adaptive [11,12] and networked systems [5], runtime monitoring [4,15,17], modelling of cyber-physical systems [20] and medical imaging [3,13]. Introduced in [7], `VoxLogicA` (*Voxel-based Logical Analyser*)[1] caters for a declarative approach to (medical) image segmentation, supported by spatial model checking. A spatial logic is defined, tailored to high-level imaging features, such as regions, contact, texture, proximity, distance. Spatial operators are mostly derived from the *Spatial Logic of Closure Spaces* (SLCS, see Fig. 1). Models of the spatial logic are (pixels of) images, with atomic propositions given by imaging features (e.g. colour, intensity), and spatial structure obtained via adjacency of pixels. SLCS features a modal operator *near*, denoting adjacency of pixels, and a reachability operator $\rho \, \phi_1[\phi_2]$, holding at pixel x whenever there is a path from x to a pixel y satisfying ϕ_1, with all intermediate points, except the extremes, satisfying ϕ_2.

Research partially supported by the MIUR Project PRIN 2017FTXR7S "IT- MaT-TerS" and by POR FESR Toscana 2014–2020 As. 1 - Az. 1.1.5 – S.A. A1 N. 7165 project STINGRAY. The authors are thankful to: Raffaele Perego, Franco Maria Nardini and the HPC-Lab at ISTI-CNR for a powerful GPU used in early development; Gina Belmonte, Diego Latella, and Mieke Massink, for fruitful discussions. The authors are listed in alphabetical order, having equally contributed to this work.

[1] `VoxLogicA`: see https://github.com/vincenzoml/VoxLogicA.

K. Peters and T. A. C. Willemse (Eds.): FORTE 2021, LNCS 12719, pp. 188–196, 2021.
https://doi.org/10.1007/978-3-030-78089-0_12

$$\phi ::= p \mid \neg \phi \mid \phi_1 \wedge \phi_2 \mid \mathcal{N}\phi \mid \rho \; \phi_1[\phi_2]$$

Fig. 1. SLCS syntax. Atomic propositions p correspond to image properties (e.g. *intensity*, colour); boolean operators act pixel-wise; the *near* operator $\mathcal{N}$ denotes pixel *adjacency* (using *8-adjacency*: the pixels having a vertex in common with a given one).

The main case study of [7] is brain tumour segmentation for radiotherapy, using the BraTS 2017 public dataset of medical images [2]. An high-level specification for glioblastoma segmentation was proposed and tested using VoxLogicA, resulting in a procedure that competes in accuracy with state-of-the-art techniques. In [6], also an accurate specification for nevus segmentation was presented. This paper introduces a novel development in the direction of taking advantage of Graphical Processing Units: high-performance, massively parallel computational devices. GPU computing differs from the *multi-core* paradigm of modern CPUs in many respects: the execution model is *Single Instruction Multiple Data*; the number of computation cores is high; the memory model is highly localised and synchronisation among parallel threads is very expensive. Each GPU core performs the same operation on different coordinates (a single pixel, in our case). The dimension of the problem (e.g. the size of an image) is provided to the GPU when the program (*kernel*) is launched, yet the number of threads does not scale with the problem size, being bounded by the number of computing units in the GPU. Currently, such a number is in the order of thousands, whereas the problem size may include millions of tasks. The problems that benefit the most of such architecture are the inherently massively parallel ones. In that case, the main issue is to minimise read/write operations from and to the GPU memory, and to turn a problem into a highly parallel implementation.

A substantial redesign is thus required to port existing algorithms to GPUs. VoxLogicA-GPU implements the core logical primitives of VoxLogicA on GPU, sharing motivation with a recent trend on implementing formal methods on GPU [8,16,18,21,22]. This paper aims to describe the tool architecture, including asynchronous execution of logical primitives on GPU and garbage collection, and to demonstrate a consistent efficiency improvement. In doing so, we had to overcome two major issues: implementing connected component labelling on GPUs and minimising the number of (computationally expensive) CPU $\leftrightarrow$ GPU memory transfers. Our current results are very encouraging, obtaining a (task-dependent) speed-up of one or two orders of magnitude.

2   Functional Description and Implementation

VoxLogicA-GPU[2] is a *global, explicit state* model checker, aiming at high efficiency and maximum portability. It is implemented in FSharp, using the NET

---

[2] VoxLogicA-GPU is Free and Open Source software. Its source code is currently available at https://github.com/vincenzoml/VoxLogicA/tree/experimental-gpu.

Core infrastructure, and the *General-Purpose GPU computing* library OpenCL[3]. The choice of OpenCL is motivated by portability to different GPU brands. VoxLogicA-GPU is a command line tool, accepting as input a text file describing the analysis, and a number of input images. The text file contains a set of logic formulas and parametrised, non-recursive macro abbreviations. As in [7], the tool expands macros, identifies the *ground* formulas (that is, without variables), and constructs a directed acyclic graph of *tasks* and *dependencies*. Such a graph is equivalent to the syntax tree, but it enjoys *maximal sharing*: no sub-formula is ever computed twice. In the CPU version, the tasks run in parallel on the available CPU cores, yielding a speed-up proportional to the degree of parallelism of the task graph and to the number of cores. In the GPU version, the tasks are currently executed asynchronously with respect to the main CPU execution thread, but sequentially: so-called *out-of-order execution* is left for future work.

The focus of this first release of VoxLogicA-GPU is on the *design* of a free and open source GPU-based infrastructure, with proven scalability. Thus, development has been narrowed to a core implementation that is powerful enough to reach the stated objectives, although not as feature-complete as VoxLogicA. In particular, the implemented primitives are those of SLCS plus basic arithmetics, and computation is restricted to 2D and integer-valued images. Implementation-wise, VoxLogicA-GPU is a command-line tool. It takes only one parameter, a text file containing the specification to be executed, i.e., a sequence of commands. Five commands are currently implemented: let, load, save, print, import. The model checking algorithm of VoxLogicA-GPU is shared with VoxLogicA. After parsing, parametric macros are expanded, while at the same time (to avoid explosion of the syntax tree) the aforementioned task graph is computed. A major issue is that each task allocates a memory area proportional to the size of the input image to store its results, thus *garbage collection* is required. The current strategy is a simple reference counting, as the number of reverse dependencies of each task (i.e. the tasks taking the given one as argument) is known before execution, and no task is created at run time. This problem is more relevant to the GPU implementation: as a GPU memory is usually smaller than a CPU one, and GPU buffers are explicitly allocated by the programmer, large formulas can easily lead to *Out of Memory* errors at run time. If a reference counter turns to 0, no more tasks take the given one as an input, and the pointer referencing the buffer can be disposed. As no pointer longer refers that GPU memory area, this can be reused. A task is an operator of the language or an output instruction. The semantics of the former is delegated to the GPU implementation of the VoxLogicA API, defining the core type Value, which is instantiated as a shorthand for a type called GPUImage. Such type represents a computation, asynchronously running on a GPU, whose purpose is to fill an image buffer. GPUImage contains a pointer to a buffer stored in the GPU, its imaging features, and an *OpenCL event* (an handle to the asynchronous computation). The latter is used to wait for termination before transferring the results

---

[3] **FSharp:** see https://fsharp.org. **NET Core:** see https://dotnet.microsoft.com. **OpenCL:** see https://www.khronos.org/opencl. **ITK:** see https://itk.org.

to the CPU and to make task dependencies explicit to the GPU for proper sequencing. Since commands, parameters, and results must be transferred from the CPU to the GPU and back, keeping pointers to GPU buffers minimises this overhead, allowing for the reuse of partial results. Thus, data is transferred only at the beginning of the computation and when retrieving results to be saved to disk. The model checker is responsible for decreasing reference counts after each task terminates, and for scheduling garbage collection when a reference counter reaches 0. Each operator is implemented in a small module running on CPU, whose only purpose is to prepare memory buffers and launch one or more *kernels* (i.e. functions running on GPU). As in `VoxLogicA`, the reachability operator $\rho \phi_1[\phi_2]$ is implemented using connected components labelling.

2.1   Connected Components Labelling in `VoxLogicA-GPU`

We designed a simple algorithm for connected component labelling, biased towards implementation simplicity, although efficient enough for our proto-type. Similarly to the classic result in [19], the algorithm exploits the *pointer jumping* technique[4]: see Algorithm 1 for the pseudo-code of the kernels (ter-mination checking is omitted) and Fig. 2 for an example. After initialisation, *mainIteration* is iterated. By pointer jumping, it converges in logarithmic time with respect to the number of pixels N, but it may fail to correctly label con-nected components with corners in specific directions (see Fig. 2, Iteration 13). Then *reconnect* is called, checking if there are two adjacent pixels with different labels, and changing one of them (deterministically chosen) so that the two labels now coincide The way *reconnect* changes the image ensures that *mainIteration* will restart and will be enabled to converge again. The termination condition is reached when *reconnect* does not change the image, which requires a global check on its input and output. For checking termination we adopted a *reduce*-type operation[5]: it takes $log(N)$ iterations, since it divides the image size at each iteration until a single-pixel image containing a boolean flag is obtained. If the termination condition is false, the algorithm restarts from *mainIteration*[6]. In most cases, *reconnect* is called a very small number of times before convergence, and the total number of iterations is in the order of $log(N)$ (see [10] for details).

---

[4] *Pointer jumping* or *path doubling* is a design technique for parallel algorithms that operate on pointer structures, such as linked lists and directed graphs. It allows an algorithm to follow paths with a time complexity that is logarithmic with respect to the length of the longest path. It does this by "jumping" to the end of the path computed by neighbors. See https://en.wikipedia.org/wiki/Pointer_jumping.

[5] See e.g. https://en.wikipedia.org/wiki/MapReduce.

[6] Since checking termination takes $log(N)$ iterations, instead of waiting for *mainIteration* to converge, *reconnect* is called each k itcrations ($k = 8$ in the current implementation, which experimentally proved to be a reasonable compromise).

Algorithm 1: Pseudocode for connected components labelling

---

1 **initialization**(start: image of bool, output: image of int × int)
2   // parallel for on GPU
3   **for** $(i, j) \in Coords$ **do**
4   **if** $start(i,j)$ **then**
5       $output(i, j) = (i, j)$ // null otherwise
6 **mainIteration**(start: image of bool, input, output: image of int × int)
7   // parallel for on GPU
8   **for** $(i, j) \in Coords$ **do**
9   **if** $start(i,j)$ **then**
10       $(i', j') = input(i, j)$ // pointer jumping
11       $output(i, j) = maxNeighbour(input, i', j')$
12 **reconnect**(start: image of bool, input, output: image of int × int)
13   // parallel for on GPU
14   **for** $(i, j) \in Coords$ **do**
15   **if** $start(i,j)$ **then**
16       $(i', j') = input(i, j)$
17       $(a, b) = maxNeighbour(input, i, j)$
18       $(c, d) = input(i', j')$
19   **if** $(a, b) > (c, d)$ **then**
20         $output(i', j') = (a, b)$ // Requires atomic write

---

3   Preliminary Evaluation

This section illustrates the scalability results obtained in our preliminary tests[7]. Experiments have been executed on a machine equipped with an `Intel Core i9-9900K` and a `NVIDIA RTX 3080` GPU. This is indicative of the attainable speed-up as both CPU and GPU are current, high-end (workstation-oriented) devices. It is important to remark that CPU and GPU execution times are subject to high variability. Indeed, a highly parallel test may run about 8 times faster on CPU with 16 cores (a current high-end desktop workstation) than a machine with 2 cores (a current travelling laptop), as witnessed by the law on theoretical speed-up given by parallel machines [14]. Since the range of current CPUs is highly variable, so are the execution times in our tests. This fact also explains the different speedup in our tests comparing CPU and GPU on sequential and parallel tasks (see Fig. 3 and Fig. 4). In the parallel test, all the 16 cores of the chosen CPU are exploited, thus the CPU is more efficient.

    We built two kinds of large formulas for stressing the tool: sequential (i.e. of shape $f(g(\dots(x))))$) and "parallel" ones, where the operators are composed in order to maximise parallelism. More precisely, formulas are written in order to have many independent sub-formulas (i.e., having shape $f(g(\dots, \dots), h(\dots, \dots)))$). In the CPU implementation, such sub-formulas can be computed in parallel, up to the number of available cores. Note again that

---

[7] All the tests we present, and the script to run them, are available in the source code repository https://github.com/vincenzoml/VoxLogicA/tree/experimental-gpu.

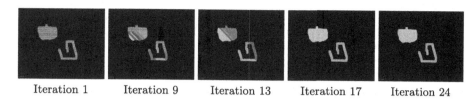

| Iteration 1 | Iteration 9 | Iteration 13 | Iteration 17 | Iteration 24 |

Fig. 2. CC-labelling of a 2048 × 2048 pixels image in 24 iterations. Different colours represent different labels. Reconnect is called every 8 main iterations. Iteration 13: the main iterations converged; the image does not change until iteration 16 (reconnect). Iteration 17: label propagation after reconnect. Iteration 24: termination.

No. of Tasks	CPU	GPU			GPU-GC	
	Time	Time	Speed-up		Time	Speed-up
11	410ms	190ms	2.15		200ms	2.05
35	1470ms	190ms	7.73		230ms	6.39
67	1800ms	190ms	9.47		230ms	7.82
195	8200ms	200ms	41.00		320ms	25.62
259	10900ms	210ms	51.90		360ms	30.27
1027	43600ms	350ms	124.57		980ms	44.48
4099	174600ms	Out of memory	-		4100ms	42.58
8195	479000ms	Out of memory	-		12000ms	39.91

Fig. 3. Execution times for the sequential test.

maximising CPU usage entails a smaller speedup for the GPU. Figure 3 and 4 report execution times for each type of test. Each row reports the number of tasks to execute (i.e., the number of nodes in the directed acyclic graph described in Sect. 2), and the obtained speed-up for the two GPU algorithms. In all cases, VoxLogicA-GPU achieves a relevant speed-up. The CPU version performs better on very small formulas, due to the overhead needed to set up GPU computation. The version with garbage collection is much slower than the version without. This is due to garbage collection being run in the current implementation as soon as reference counts reach 0, and recall that memory deallocation and reallocation is particularly expensive on GPUs. Obvious improvements are expected by scheduling garbage collection to be run only when a memory usage threshold is reached. However, we plan to design a garbage collector which is more specific to the execution patterns of a model checker. We also carried out a preliminary assessment of the brain tumour segmentation case study of [7]. Given the current restrictions of VoxLogicA-GPU to 2D images and the core logical primitives (see Sect. 2), it is only possible to use a simplified dataset and specification, obtaining too small tasks for interesting measurements. We omit the full results (see [10]), but we note that a mild speed-up was obtained: this is interesting, as the CPU version uses a state-of-the-art imaging library designed for high efficiency.

No. of Tasks	CPU	GPU			GPU-GC	
	Time	Time	Speed-up		Time	Speed-up
10	70ms	180ms	0.38		200ms	0.35
26	260ms	180ms	1.44		200ms	1.30
43	310ms	180ms	1.72		210ms	1.47
61	500ms	190ms	2.63		210ms	2.38
73	510ms	190ms	2.68		220ms	2.31
174	860ms	200ms	4.30		270ms	3.18
323	1600ms	220ms	7.27		340ms	4.70
472	2400ms	290ms	8.27		430ms	5.58
621	3000ms	360ms	8.33		510ms	5.88
1813	8800ms	Out of memory	-		1200ms	7.33
3005	14600ms	Out of memory	-		2000ms	7.30

Fig. 4. Execution times for the parallel test.

4   Conclusions and Future Work

Our preliminary evaluation of spatial model checking on GPU is encouraging: large formulas benefit most, with significant speedups. Connected components labelling will be a focus for future work: indeed, the topic is very active, and our simple, proof-of-concept algorithm might well be replaced by state-of-the-art procedures (see e.g. the recent [1]). The currently attained speed-up can be used, for instance, for interactive calibration of parameters or for automated parameter optimisation, e.g. using gradient descent algorithms. However, given the peak performance of recent GPUs, our results are just the tip of the iceberg of what can be achieved. Future work will concentrate on fully exploiting more powerful GPUs, using out-of-order execution to permit the execution of more independent tasks at the same time, and taking into account GPU-specific architectural features (memory banking, number of channels, etc.). Making VoxLogicA-GPU feature-complete with respect to VoxLogicA is also a goal. In this respect, we remark that although in this work we decided to go through the "GPU-only" route, future developments will also consider a *hybrid* execution mode with some operations executed on the CPU, so that existing primitives in VoxLogicA can be run in parallel with those that have a GPU implementation. Usability of VoxLogicA-GPU would be greatly enhanced by a user interface. However, understanding modal logical formulas is generally considered a difficult task, and cognitive/human aspects may become predominant with respect to technological concerns. Formal methods could be used to mitigate such concerns (see e.g. [9]).

References

1. Allegretti, S., Bolelli, F., Grana, C.: Optimized block-based algorithms to label connected components on GPUs. IEEE Trans. Parallel Distrib. Syst. **31**(2), 423–438 (2020)
2. Bakas, S., et al.: Advancing the cancer genome atlas glioma MRI collections with expert segmentation labels and radiomic features. Sci. Data **4**, 1–13 (2017)
3. Banci Buonamici, F., Belmonte, G., Ciancia, V., Latella, D., Massink, M.: Spatial logics and model checking for medical imaging. Softw. Tools Technol. Transf. **22**(2), 195–217 (2020). https://doi.org/10.1007/s10009-019-00511-9
4. Bartocci, E., Bortolussi, L., Loreti, M., Nenzi, L.: Monitoring mobile and spatially distributed cyber-physical systems. In: Talpin, J., Derler, P., Schneider, K. (eds.) MEMOCODE 2017, pp. 146–155. ACM (2017)
5. Bartocci, E., Gol, E., Haghighi, I., Belta, C.: A formal methods approach to pattern recognition and synthesis in reaction diffusion networks. IEEE Trans. Control Netw. Syst. **5**(1), 308–320 (2016)
6. Belmonte, G., Broccia, G., Ciancia, V., Latella, D., Massink, M.: Feasibility of spatial model checking for nevus segmentation. In: Bliudze, S., Semini, L. (eds.) FORMALISE@ICSE 2021 (2021, to appear)
7. Belmonte, G., Ciancia, V., Latella, D., Massink, M.: VoxLogicA: a spatial model checker for declarative image analysis. In: Vojnar, T., Zhang, L. (eds.) TACAS 2019. LNCS, vol. 11427, pp. 281–298. Springer, Cham (2019). https://doi.org/10.1007/978-3-030-17462-0_16
8. Berkovich, S., Bonakdarpour, B., Fischmeister, S.: GPU-based runtime verification. In: IPDPS 2013, pp. 1025–1036. IEEE Computer Society (2013)
9. Broccia, G., Milazzo, P., Ölveczky, P.C.: Formal modeling and analysis of safety-critical human multitasking. Innovations Syst. Softw. Eng. **15**(3–4), 169–190 (2019). https://doi.org/10.1007/s11334-019-00333-7
10. Bussi, L., Ciancia, V., Gadducci, F.: A spatial model checker in GPU (extended version). CoRR abs/2010.07284 (2020)
11. Ciancia, V., Latella, D., Massink, M., Paškauskas, R., Vandin, A.: A tool-chain for statistical spatio-temporal model checking of bike sharing systems. In: Margaria, T., Steffen, B. (eds.) ISoLA 2016. LNCS, vol. 9952, pp. 657–673. Springer, Cham (2016). https://doi.org/10.1007/978-3-319-47166-2_46
12. Ciancia, V., Gilmore, S., Grilletti, G., Latella, D., Loreti, M., Massink, M.: Spatio-temporal model checking of vehicular movement in public transport systems. Softw. Tools Technol. Transf. **20**(3), 289–311 (2018). https://doi.org/10.1007/s10009-018-0483-8
13. Grosu, R., Smolka, S., Corradini, F., Wasilewska, A., Entcheva, E., Bartocci, E.: Learning and detecting emergent behavior in networks of cardiac myocytes. Commun. ACM **52**(3), 97–105 (2009)
14. Gustafson, J.L.: Reevaluating Amdahl's law. Commun. ACM **31**(5), 532–533 (1988)
15. Ma, M., Bartocci, E., Lifland, E., Stankovic, J., Feng, L.: SaSTl: spatial aggregation signal temporal logic for runtime monitoring in smart cities. In: ICCPS 2020, pp. 51–62. IEEE (2020)
16. Neele, T., Wijs, A., Bošnački, D., van de Pol, J.: Partial-order reduction for GPU model checking. In: Artho, C., Legay, A., Peled, D. (eds.) ATVA 2016. LNCS, vol. 9938, pp. 357–374. Springer, Cham (2016). https://doi.org/10.1007/978-3-319-46520-3_23

17. Nenzi, L., Bortolussi, L., Ciancia, V., Loreti, M., Massink, M.: Qualitative and quantitative monitoring of spatio-temporal properties with SSTL. Log. Methods Comput. Sci. **14**(4), 2:1–2:38 (2018)
18. Osama, M., Wijs, A.: Parallel SAT simplification on GPU architectures. In: Vojnar, T., Zhang, L. (eds.) TACAS 2019. LNCS, vol. 11427, pp. 21–40. Springer, Cham (2019). https://doi.org/10.1007/978-3-030-17462-0_2
19. Shiloach, Y., Vishkin, U.: An O(logn) parallel connectivity algorithm. J. Algorithms **3**(1), 57–67 (1982)
20. Tsigkanos, C., Kehrer, T., Ghezzi, C.: Modeling and verification of evolving cyber-physical spaces. In: Bodden, E., Schäfer, W., van Deursen, A., Zisman, A. (eds.) ESEC/FSE 2017, pp. 38–48. ACM (2017)
21. Wijs, A., Bošnački, D.: Many-core on-the-fly model checking of safety properties using GPUs. Softw. Tools Technol. Transf. **18**(2), 169–185 (2016). https://doi.org/10.1007/s10009-015-0379-9
22. Wijs, A., Neele, T., Bošnački, D.: GPUexplore 2.0: unleashing GPU explicit-state model checking. In: Fitzgerald, J., Heitmeyer, C., Gnesi, S., Philippou, A. (eds.) FM 2016. LNCS, vol. 9995, pp. 694–701. Springer, Cham (2016). https://doi.org/10.1007/978-3-319-48989-6_42

Formal Verification of HotStuff

Leander Jehl[(✉)]

University of Stavanger, Stavanger, Norway
leander.jehl@uis.no

Abstract. HotStuff is a recent algorithm for repeated distributed consensus used in permissioned blockchains. We present a simplified version of the HotStuff algorithm and verify its safety using both Ivy and the TLA Proof Systems tools.

We show that HotStuff deviates from the traditional view-instance model used in other consensus algorithms and instead follows a novel tree model to solve this fundamental problem. We argue that the tree model results in more complex verification tasks than the traditional view-instance model. Our verification efforts provide initial evidence towards this claim.

1 Introduction

The advent of blockchain technology has significantly increased interest in Byzantine Fault Tolerant (BFT) systems. BFT systems tolerate arbitrary misbehavior of a fraction of participating nodes. Therefore, these systems build a key component for recent permissioned [2,23] and federated [17] blockchain systems. However, BFT algorithms are notoriously difficult to design or implement and thus have been the subject of numerous efforts in formal methods [1,13,14,22]

In this paper we verify safety properties of the HotStuff [23] algorithm using both the TLA Proof System (TLAPS) [5] and the recent Ivy tool [18]. The HotStuff algorithm is used in the Diem, formerly libra blockchain, and was, to the best of our knowledge, not previously formally verified. Drawing inspiration from blockchain technology, the HotStuff algorithm presents a novel paradigm in consensus algorithms, replacing views and instances with a tree model. We show that this new paradigm complicates the verification of safety properties compared to the more traditional view-instance model. We find that the need to ensure the tree structure and a reachability or ancestor predicate for the tree model in inductive invariants complicates verification. The different nature of the formal frameworks used allows us to address this issue in different ways. Additionally, our efforts led us to discover a simplified version of the HotStuff protocol, which restricts the use of reachability to the safety properties and removes it from the algorithm.

In addition to contributing to the formal verification and better understanding of the HotStuff algorithm and its novel paradigm, this work also presents a case study and comparison of the two verification systems, TLAPS and Ivy. Both

© IFIP International Federation for Information Processing 2021
Published by Springer Nature Switzerland AG 2021
K. Peters and T. A. C. Willemse (Eds.): FORTE 2021, LNCS 12719, pp. 197–204, 2021.
https://doi.org/10.1007/978-3-030-78089-0_13

of these aim to make formal verification available to practitioners and engineers and at least the TLA+ model checker is actively used in industry [19]. Thus, our comparison of the recent Ivy tool with more mature TLAPS forms an important evaluation of that system. We are not aware of a previous use of the Ivy tool that did not involve the original authors.

2   View-Instance and Tree Model for Repeated Consensus

This section presents the different paradigms used to implement repeated consensus. Repeated consensus is a fundamental problem in distributed computing. The problem requires processes to maintain an append-only log in the presence of faults. Repeated consensus allows to implement arbitrary objects in a fault tolerant manner, by maintaining a log of deterministic operations, applied to the object. This state machine approach [21] is widely deployed in the cloud, e.g. in Zookeeper [10], but also builds the basis for recent blockchain systems.

A common model for algorithms solving repeated consensus (e.g. [3,16]) is, what we call the *view-instance model*. As shown in Fig. 1, this model uses a matrix of possibly infinitely many slots (or fields), each indexed by two integers, instance and view. In each slot a value can be proposed by a leader and committed by the processes. We say that the slot is committed. Since fault tolerant consensus is impossible to solve in an asynchronous system [6], some slots may remain uncommitted, or even without a proposed value, e.g. due to leader failure or network partitions. Indeed, in some algorithms, such as FlexiblePaxos [9] or PBFT [3], slots have a preassigned leader that may fail. In such cases, values may be committed in additional slots, using higher views. The instances represent the entries in the distributed log. If a value is committed in one slot, it is adopted for that instance. Algorithms thus need to ensure, that values committed in different slots belonging to one instance are the same. This property forms the main safety condition for algorithm using the view-instance model.

The view-instance model has advantages both for algorithm design and performance. For algorithm design, the slots belonging to one instance can be viewed as solving a single instance of the consensus problem. Most prominently the Paxos algorithm has been presented in this way [16]. Similarly, the safety condition stated above can be applied to a single instance. Thus, in the formal verification of these algorithms, a common approach is to first validate the algorithm in a single instance [14,15] and subsequently extend the model to cover multiple instances [4,20]. Additionally, the view-instance model allows optimizations that apply operations across instances. For example a new view may be started simultaneously in all instances [16].

Tree Model: HotStuff [23] introduces a new way to implement repeated consensus, that organizes slots in a rooted tree, as shown in Fig. 2. We call this the *tree model*. If a slot is committed, the slot and its ancestors are used as the current prefix of the log. Thus, the depths of slots in the tree-model corresponds to the instance in the view-instance model. If a slot cannot be committed, the processes can try to commit one of its descendants instead. Alternatively, a new slot can

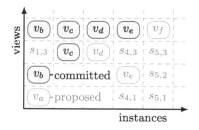
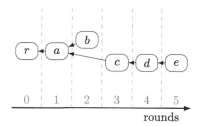

Fig. 1. View-instance matrix with proposed and committed slots.

Fig. 2. Slots with parent relation in the tree model.

be proposed at the same depth. To distinguish slots at the same depth, the tree model uses rounds, similar to views in the view-instance model. E.g. if a slot was not proposed at a certain depth, a new slot can be proposed at the same depth, but with a larger round. Different from views, rounds do not restart for every depth, but instead the round of a slot is larger than the round of its parent. The core safety property in the tree model is that any two committed slots are on the same branch of the tree, or equivalent for two committed slots, one is the ancestor of the other.

The advantage of the tree model is that it is not necessary to ensure that different slots commit the same value. Thus, this model is suited to allow simple leader change procedures, since it is not necessary to ensure that the new leader proposes the same value as an old leader. Indeed, leader change with linear communication complexity is the main contribution of the HotStuff protocol. An additional advantage is that not every slot has to be actually committed, as long as a descendant of the slot is eventually committed. Similar to the view-instance model, the tree model allows specific optimizations. HotStuff spreads the process of committing a slot over the slot's descendants.

The verification of an algorithm in the tree model poses some novel challenges. First, the existing approach to first prove safety of a single instance and then extend this proof to multiple instance does not apply. Second, in the tree model, the safety property, and in some cases also the algorithm contain a reachability predicate. Reachability on finite graphs, i.e. the transitive closure of the neighbor relation, cannot be expressed in first order logic [12].

3   Simplified HotStuff Algorithm

We now present our simplified version of the HotStuff algorithm and explain how it differs from HotStuff [23]. As common in BFT algorithms, HotStuff assumes $3f+1$ processes, of which at most f may be faulty. Faulty processes may stop but also violate the protocol. However, faulty nodes cannot subvert cryptographic primitives. Especially, a digital signature scheme is used to authenticate messages, preventing impersonation of correct (non-faulty) processes.

The algorithm contains the following two operations. The leader of a round may propose new slots. If the followers accept the proposed slot they sign it. The leader collects these signatures into a certificate, containing $2f + 1$ signatures. We say that a slot is *certified*, if there exists a certificate for that slot. The root slot has a certificate at startup. Slots are propagated using reliable broadcast [8], ensuring that either all or none of the correct nodes receive it.

To correctly *propose a slot*, it needs to include the signature of the leader, and the round in which it is proposed. Additionally, the slot must specify its parent and contain a certificate for that parent slot. Thus, any slot that has a child is certified. Finally, the slot's round must be bigger than its parent's round.

Two rules govern whether a correct process *signs a slot*. Rule 1 allows processes to only sign in increasing rounds. E.g. after signing in round 3, a correct process would no longer sign a slot in round 2. For Rule 2, processes maintain a locked slot l and only sign a new slot s_{new}, if the round of the parent of s_{new} is greater or equal to the round of l.

$$\text{parent}(s_{new}).round \overset{?}{\geq} l.round \qquad \text{(Rule 2)}$$

The lock l of every process is initially set to the root of the tree. On signing a new slot s_{new} a process checks whether the grandparent of s_{new} has a higher round than the process's current lock l. If that is the case, l is updated to said grandparent.

A slot s is *committed*, if it has a grandchild s_{gc}, which is certified, and no rounds are omitted between s and s_{gc}: $s.round + 2 = s_{gc}.round$. In practice, on receiving a new slot, processes check whether its great-grand parent is committed. For example, in Fig. 2, slot d does not commit slot a, because round 2 was omitted between them. However, if slot e is certified, slot c will be committed.

The two rules stated guarantee safety. To propose a slot that can be signed by all correct processes, a new leader collects the last certificate (with highest round) from $2f + 1$ processes, selects the one with the highest round among them, and uses the certified slot as parent in a new proposal.

3.1   Original HotStuff

Here we explain how the original HotStuff algorithm differs from our simplified version presented above. We refer the reader to [23] for an in depth specification of the original algorithm. Original HotStuff does not require the parent of a new slot to be certified. Instead of a certificate for its parent, the slot includes a certificate for one of its ancestors. Thus, HotStuff has a more complex tree structure with each slot specifying both a link to its parent and a link to a certified ancestor. It is these ancestor links that correspond to parent links in our simplified version.

To be able to commit a slot in original HotStuff, parent and certified links must point to the same slot, resulting in the same condition as in our simplified version. Thus, during normal operation and when original HotStuff can commit slots, it is identical to our simplified version.

The removal of uncertified parent links significantly simplifies Rule 2. The original rule is as follows, where $parent_c(s)$ is the certified ancestor of s.

$$\text{ancestor}(l, s_{new}) \vee parent_c(s_{new}).round > l.round \qquad \text{(ORule 2)}$$

4 Verification

In the following we report on our effort to verify safety of simplified HotStuff in both TLAPS and Ivy. Models and proofs are available online [11]. We mainly focus on our experience with these tools, their ease of use, as perceived by us and how we modelled the ancestor relationship.

Ivy is a recent tool for the verification of distributed algorithms [18]. The default tactic in Ivy, used to verify safety properties, is a proof of an inductive invariant using an SMT solver. To avoid the SMT solver diverging, Ivy requires the specification to be written in uninterpreted first order logic. This prohibits the use of interpreted theories, e.g. integers. The required rewriting is quite straightforward. For example, instead of using integers, we require that rounds are totally ordered and use an intersection property on sets of processes, instead of process counts [20].

Additionally, Ivy requires that the verification condition must lie in a decidable logical fragment called FAU [7]. This requires the elimination of certain functions and quantifier combinations in the model. On submitting a model, Ivy checks if the given model lies within FAU and if not, specifies which functions or formulas violate conditions. Given a verification condition in FAU, Ivy either proves the invariant, or produces a counterexample to inductivity. Counterexamples can be ruled out through additional invariants.

Violations through functions can be removed by replacing functions with relations. For example, we had to rewrite the function relating a proposed slot to its parent. Violations through quantifier combinations are more difficult to remove. Following Padon et al. [20], resolving this issue requires to introduce new relations into the model. Understanding which relations we could add required a good understanding of our model and the quantifier restrictions it implies. For example consider the following condition expressing that for a given slot z and process n, there exists a slot s in a higher round, which n has signed and which is not a descendant of z:

$$\exists s \in \text{Slot} : s.round > z.round \wedge \neg\text{ancestor}(z, s) \wedge \text{signed}(n, s) \qquad (1)$$

If z or n are free variables or under universal quantification, this expression does violate FAU in our model. Realizing that we could existentially quantify over rounds, we introduced two predicates, $\text{signedIn}(N, R)$ and $\text{signedAncIn}(N, R, S)$

and replaced Expression (1) with the following, that specifies that n has signed a slot in some round r but has not signed an ancestor of z in round r:

$$\exists r \in \text{Rounds} : r > z.round \land \text{signedIn}(n, r) \land \neg\text{signedAncIn}(n, r, z) \qquad (2)$$

To model the ancestor relation we included this relation as predicate in our model and update it whenever a new slot is added to the tree. This is similar to the signedIn and signedAncIn relations added to express invariants in FAU. The drawback with this approach is that it requires many invariants to be added. We were able to prove that our ancestor relation can be implemented by checking parent links of individual slots inside a while loop.

Our finished model contains 36 auxiliary invariants, 18 or which are only concerned with the tree structure and ancestor relation. While developing the proof, we struggled with long running times in Ivy. In some cases the tool timed out, without a proof or counterexample. After a decomposition and some final simplification, however, Ivy verifies our proof in a few seconds.

The complexity of our model is significantly larger than what we found in consensus algorithms using the instance-view model. For example, to verify both Paxos or Multipaxos, the authors of the Ivy tool required only 4 auxiliary invariants. While some complexity may be due to our inexperience, we believe this also shows an increased complexity of the tree model.

One of the main advantages with Ivy was that it was easy to modularize or refactor our model. After several such refactorings, the proof was still running, or at least easy to reestablish.

HotStuff in TLA+. Based on our experience in Ivy we specified simplified Hot-Stuff in TLA+ and proved safety using TLAPS. We also verified the safety condition for small instances of our model using the TLC model checker. In using TLAPS, we benefited from an active and helpful user group.

In TLA+ we could use both integers and functions, where our Ivy model uses totally ordered sets and relations. While the availability of integers made little difference, the ability to use functions significantly simplified the formulation of the model and the inductive invariant.

TLA+ and TLAPS allowed us to define the ancestor relation inductively. However, we were unable to apply lemmas about the ancestor relation to primed variables. This however was necessary to use the ancestor relation in the inductive invariant. Instead, we discovered Rule 2, which allowed us to formulate both the model and the inductive invariant without the notion of ancestry. The ancestor relation thus only appears in the safety property, which we prove follows from the inductive invariant.

Again we found that the verification was quite complex, due to the complex safety condition in the tree model, but also due to the need to ensure a correct tree structure in the inductive invariant. In total our proof amounts to approximately 1000 lines, plus additional 200 lines to prove properties about the ancestor relation. For comparison, previous work proved safety of Paxos [15] and Multi-Paxos [4] in around 500 lines.

5   Conclusion

We have presented the tree model for repeated consensus and a simplified version of the HotStuff algorithm. The advantages of the model, especially similarity to techniques from permissionless blockchains encourages further investigation.

Our verification efforts using both Ivy and TLAPS highlight both advantages and disadvantages of these tool and suggest that the tree model may result in more complex verification tasks, than the traditional view-instance model.

References

1. Dillig, I., Tasiran, S. (eds.): CAV 2019. LNCS, vol. 11561. Springer, Cham (2019). https://doi.org/10.1007/978-3-030-25540-4
2. Buchman, E.: Tendermint: Byzantine fault tolerance in the age of blockchains. Ph.D. thesis (2016)
3. Castro, M., Liskov, B.: Practical byzantine fault tolerance and proactive recovery. ACM Trans. Comput. Syst. **20**(4), 398–461 (2002)
4. Chand, S., Liu, Y.A., Stoller, S.D.: Formal verification of multi-paxos for distributed consensus. In: Fitzgerald, J., Heitmeyer, C., Gnesi, S., Philippou, A. (eds.) FM 2016. LNCS, vol. 9995, pp. 119–136. Springer, Cham (2016). https://doi.org/10.1007/978-3-319-48989-6_8
5. Chaudhuri, K., Doligez, D., Lamport, L., Merz, S.: Verifying safety properties with the TLA$^+$ proof system. In: Giesl, J., Hähnle, R. (eds.) IJCAR 2010. LNCS (LNAI), vol. 6173, pp. 142–148. Springer, Heidelberg (2010). https://doi.org/10.1007/978-3-642-14203-1_12
6. Fischer, M.J., Lynch, N.A., Paterson, M.S.: Impossibility of distributed consensus with one faulty process. J. ACM (JACM) **32**(2), 374–382 (1985)
7. Ge, Y., de Moura, L.: Complete instantiation for quantified formulas in satisfiabiliby modulo theories. In: Bouajjani, A., Maler, O. (eds.) CAV 2009. LNCS, vol. 5643, pp. 306–320. Springer, Heidelberg (2009). https://doi.org/10.1007/978-3-642-02658-4_25
8. Hadzilacos, V., Toueg, S.: Fault-Tolerant Broadcasts and Related Problems, pp. 97–145. ACM Press/Addison-Wesley Publishing Co., New York (1993)
9. Howard, H., Malkhi, D., Spiegelman, A.: Flexible paxos: Quorum intersection revisited. In: 20th International Conference on Principles of Distributed Systems (OPODIS 2016). Schloss Dagstuhl-Leibniz-Zentrum fuer Informatik (2017)
10. Hunt, P., Konar, M., Junqueira, F.P., Reed, B.: Zookeeper: Wait-free coordination for internet-scale systems. In: USENIX Annual Technical Conference, vol. 8 (2010)
11. Jehl, L.: Verifying simplified hotstuff (2021). https://doi.org/10.5281/zenodo.4711071
12. Kolaitis, P.G.: On the expressive power of logics on finite models. In: Finite Model Theory and Its Applications, pp. 27–123. Springer, Heidelberg (2007). https://doi.org/10.1007/3-540-68804-8_2
13. Konnov, I., Veith, H., Widder, J.: SMT and POR beat counter abstraction: parameterized model checking of threshold-based distributed algorithms. In: Kroening, D., Păsăreanu, C.S. (eds.) CAV 2015. LNCS, vol. 9206, pp. 85–102. Springer, Cham (2015). https://doi.org/10.1007/978-3-319-21690-4_6
14. Peleg, D. (ed.): DISC 2011. LNCS, vol. 6950. Springer, Heidelberg (2011). https://doi.org/10.1007/978-3-642-24100-0

15. Lamport, L., Merz, S., Doligez, D.: Paxos.tla (2014). https://github.com/tlaplus/tlapm/blob/master/examples/paxos/Paxos.tla
16. Lamport, L., et al.: Paxos made simple. ACM SIGACT News **32**(4), 18–25 (2001)
17. Lokhava, M., et al.: Fast and secure global payments with stellar. In: Proceedings of the 27th ACM Symposium on Operating Systems Principles, SOSP (2019)
18. McMillan, K.L., Padon, O.: Ivy: a multi-modal verification tool for distributed algorithms. In: Lahiri, S.K., Wang, C. (eds.) CAV 2020. LNCS, vol. 12225, pp. 190–202. Springer, Cham (2020). https://doi.org/10.1007/978-3-030-53291-8_12
19. Newcombe, C., Rath, T., Zhang, F., Munteanu, B., Brooker, M., Deardeuff, M.: How amazon web services uses formal methods. Commun. ACM **58**(4), 66–73 (2015)
20. Padon, O., Losa, G., Sagiv, M., Shoham, S.: Paxos made EPR: decidable reasoning about distributed protocols. In: Proceedings of the ACM on Programming Languages 1(OOPSLA) (2017)
21. Schneider, F.B.: Implementing fault-tolerant services using the state machine approach: a tutorial. ACM Comput. Surv. **22**(4), 299–319 (1990)
22. Vukotic, I., Rahli, V., Esteves-Veríssimo, P.: Asphalion: trustworthy shielding against byzantine faults. Proc. ACM Program. Lang. **3**(OOPSLA), 1–3 (2019)
23. Yin, M., Malkhi, D., Reiter, M.K., Gueta, G.G., Abraham, I.: Hotstuff: BFT consensus with linearity and responsiveness. In: Proceedings of the 2019 ACM Symposium on Principles of Distributed Computing, PODC 2019. ACM (2019)

Tutorials

Better Late Than Never or: Verifying Asynchronous Components at Runtime

Duncan Paul Attard[1,2](✉) , Luca Aceto[2,3] , Antonis Achilleos[2] ,
Adrian Francalanza[1] , Anna Ingólfsdóttir[2] , and Karoliina Lehtinen[4]

[1] University of Malta, Msida, Malta
`{duncan.attard.01,afra1}@um.edu.mt`
[2] Reykjavík University, Reykjavík, Iceland
`{duncanpa17,luca,antonios,annai}@ru.is`
[3] Gran Sasso Science Institute, L'Aquila, Italy
`luca.aceto@gssi.it`
[4] CNRS, Aix-Marseille University and University of Toulon, LIS,
Marseille, France
`lehtinen@lis-lab.fr`

Abstract. This paper presents detectEr, a runtime verification tool for
monitoring asynchronous component systems. The tool synthesises executable monitors from properties expressed in terms of the safety fragment of the modal μ-calculus. In this paper, we show how a number
of useful properties can be flexibly runtime verified via the three forms
of instrumentation—inline, outline, and offline—offered by detectEr to
cater for specific system set-up constraints.

Keywords: Runtime verification · Instrumentation · Monitoring

1 Do You Want to Know a Secret

In the Cockaigne of software development, programs are verified using a smor-
gasbord of *pre-deployment* techniques, and executed only when their *correctness is ascertained*. Reality, however, tells a different story. Mainstream verification practices, including testing [47], only reveal the *presence* of errors [29].
Exhaustive approaches like model checking [41] are laborious to use, *e.g.* building
effective program models is non-trivial, and known to suffer from state explosion problems [27]. Other methods such as type systems [48] are intentionally
lightweight to prevent disrupting the software development lifecycle; this, in turn,
limits their precision since type-based analyses occasionally rule out well-behaved

Supported by the doctoral student grant (No: 207055-051) and the MoVeMnt project
(No: 217987-051) under the Icelandic Research Fund, the BehAPI project funded by the
EU H2020 RISE under the Marie Skłodowska-Curie action (No: 778233), the ENDEAV-
OUR Scholarship Scheme (Group B, national funds), and the MIUR project PRIN
2017FTXR7S IT MATTERS.

© IFIP International Federation for Information Processing 2021
Published by Springer Nature Switzerland AG 2021
K. Peters and T. A. C. Willemse (Eds.): FORTE 2021, LNCS 12719, pp. 207–225, 2021.
https://doi.org/10.1007/978-3-030-78089-0_14

programs. Present-day software poses even more challenges. Static verification often relies on having access to the program source code, which is not necessarily available when software is constructed from libraries or components subject to third-party restrictions. Moreover, modern applications are increasingly developed in decentralised fashion, where the constituent parts are not always known pre-deployment. This tends to increase both the complexity of software *and* the resources required to verify it, while at the same time, decreasing the time available to conduct its verification. Lately, with the availability of large data volumes, cutting-edge software components rely on machine learning to adapt their behaviour without the need to be explicitly programmed. Analysing these types of software artifacts statically is difficult, not least because their internal representation is notoriously hard to understand.

Although the proverbial correctness cake cannot be had and eaten, slices of it may still be savoured *after* the program has been deployed. In certain cases, post-deployment techniques such as Runtime Verification (RV) [20,37] can be used instead of, or in tandem with, static techniques to increase correctness assurances about a program or System under Scrutiny (SuS). RV uses *monitors*—computational entities consisting of logically-distinct *instrumentation* and *analysis* units—to observe the execution of the SuS. Analysers, *i.e.,* sequence recognisers [13], are typically synthesised automatically from formal descriptions of correctness properties expressed in a specification logic.

When devising a RV tool, substantial effort is focussed on the specification language that is used to describe correctness properties, and the synthesis procedure which generates the analysis that runtime checks these properties [6,20,45]. Arguably, less attention is given to the instrumentation aspect, particularly, how the SuS is equipped to run with monitors, and the manner in which the computation of the SuS is *extracted* and *reported* for analysis. There is no one-size-fits-all solution to these challenges. For instance, inline instrumentation—the *de facto* technique employed by state-of-the-art RV [20,34]—relies on access to the program source or unobfuscated binary, thus obliging the RV monitors to be expressed in the same language as that of the SuS. The hallmark of a *flexible* RV tool is, therefore, its ability to support various instrumentation techniques to cater for the different scenarios where RV is used. The RV tool should also provide a *common interface* for describing *what* properties should be verified, agnostic of the underlying instrumentation mechanism dealing with the technicalities of *how* the verification is performed. This contributes to lowering the learning curve of the tool and facilitate its adoption.

This paper presents the RV tool detectEr that addresses the *analysis* and *instrumentation* aspects of runtime monitoring. Our tool targets asynchronous component systems. It automatically synthesises *correct* analyser code from properties expressed in terms of the monitorable *safety* fragment of the modal μ-calculus. Since the correctness of the synthesised analysers is studied in prior work (see [1,5,17,39]), our account elaborates on the usability and instrumentation aspects of the tool. detectEr, developed on top of the Erlang [15,26] ecosystem, supports three instrumentation methods to cater for different SuS set-ups:

```
 1  start() → spawn(calc_server, loop, [0]).

 2  loop(Tot) →
 3    receive
 4      {Clt, {add, A₁, A₂}} →
 5        Clt ! {ok, A₁ + A₂},
 6        loop(Tot + 1);
 7
 8      {Clt, {mul, A₁, A₂}} →
 9        Clt ! {ok, A₁ * A₂},
10        loop(Tot + 1);
11
12      {Clt, stp} → % Stop service.
13        Clt ! {bye, Tot}
14    end.
```

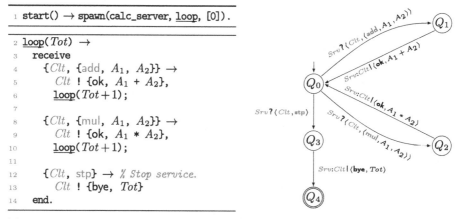

(a) `calc_server` implementation in Erlang   (b) Server-client interaction model

Fig. 1. Our calculator server and its abstraction in terms of symbolic actions

Inline: targets programs written in the Erlang language;
Outline: accommodates program binaries that are compiled for and run on the Erlang Virtual Machine (EVM), but whose source code is unavailable;
Offline: analyses recorded runs of programs that may execute outside the EVM.

We show how, from the same correctness specifications, detectEr is able to runtime monitor system components using these different instrumentation methods.

The paper is structured as follows. Section 2 introduces our running example that captures the typical interaction between concurrent processes, along with useful properties one may wish to runtime check on such systems. Sections 3 and 4 focus on the specification logic used by detectEr and how this is synthesised into executable analysis code. Section 5 summarises the role the instrumentation has with respect to the runtime analysis, and the mechanism detectEr employs to identify the SuS components in need of monitoring. Sections 6–8 overview the three instrumentation methods mentioned above, while Sect. 9 concludes.

2   A Day in the Life

We consider an idiomatic calculator server that handles client requests for arithmetic computation. Our server can be naturally expressed as an actor (process) [12] that blocks, and waits for client requests sent as *asynchronous* messages. These messages are addressed to the server using its unique process ID (PID), and deposited in its *mailbox* that buffers multiple client requests. The server unblocks upon consuming a message from the mailbox. In our client-server protocol, messages contain the *type* of operation to be executed on the server, its *arguments* (if applicable), and the client PID to whom the corresponding server reply is addressed.

Our calculator server is implemented as the Erlang module, `calc_server`, in Fig. 1a. The server logic is encapsulated in the function `loop(Tot)` that is *forked* to execute as an independent process by the launcher invoking `start()`, line 1. Processes in Erlang are forked via the built-in function `spawn()`, parametrised on line 1 by the *module name*, `calc_server`, the name of the *function* to spawn, `loop`, and the list of *arguments* accepted by `loop`, `[0]`. The server process reads messages from its mailbox (line 3), and *pattern-matches* against the three types of operations requested by clients, *Clt*: *(i)* addition (add) and multiplication (mul) requests carry the operands A_1 and A_2, lines 4 and 8, and, *(ii)* stop (stp) requests that carry no arguments, line 12. Pattern variables *Clt*, A_1 and A_2 in Fig. 1a are instantiated to concrete data in client request messages via pattern matching. Every request fulfilled by the server results in a corresponding reply that is sent to the PID of the client instantiated in variable *Clt*, lines 5, 9 and 13. Server replies carry the status *tag*, ok or bye, and the result of the requested operation. Parameter *Tot* of `loop()` is used by the server to track the number of client requests serviced, and is returned in reply to a stp request. The server loops on add and mul requests, incrementing *Tot* before recursing, lines 6 and 10; a stp request does not loop and *terminates* the server computation.

In the sequel, we focus on a system set-up consisting of one server and client to facilitate our exposition. The forked `loop(Tot)` function for some initial service count *Tot* induces a server runtime behaviour that can be abstractly described by the state transition model in Fig. 1b. Transitions between the states of Fig. 1b denote the computational steps that produce (visible) *program events* (*e.g.* event *Srv.Clt*!{bye , *Tot*} that carries the *concrete* payload values *Srv*, *Clt* and *Tot*). There are a number of correctness properties we would like such behaviour to observe. For instance, we do not control the value *Tot* that the server loop is launched with and, therefore, could require the invariant

$$\text{``The service request count returned on shutdown is } \underline{never} \text{ negative.''} \qquad (P_1)$$

Similarly, one would expect the safety properties

$$\text{``Replies are } \underline{always} \text{ sent to the client indicated in the request''} \qquad (P_2)$$

and *"A request for adding two numbers $\underline{always}$ returns their sum"* to hold, amongst others. The properties are data-dependent, which makes them hard to ascertain using static techniques such as type systems. Besides properties that reason on data, the implementation in Fig. 1a is expected to comply with control properties, such as,

$$\text{``Client requests are } \underline{never} \text{ serviced more than once'',} \qquad (P_3)$$

that describe the message exchanges between the server and client processes. All these properties are hard to ascertain without access to the source code.

3   I Want to Tell You

We overview the detectEr specification syntax, sHML [4,10,38], which is the *safety* logical fragment of the modal μ-calculus [43,44], and show how a selection of the properties in Sect. 2 can be formally specified in this logic.

The Logic. Specifications in sHML are defined over the states of transition models (such as the one of Fig. 1b), and are generated from the following grammar:

$$\varphi \in \text{sHML} ::= \text{tt} \quad (\text{truth}) \quad | \quad \text{ff} \quad (\text{falsehood}) \quad | \quad x \quad (\text{fix-point variable})$$
$$| \ \bigwedge_{i \in I} [\, p_i, c_i \,] \varphi_i \quad (\text{conj. necessities}) \quad | \quad \text{max} x. \varphi \quad (\text{max. fix-point})$$

sHML expresses recursive properties as maximal fix-point formulae $\text{max} x.\varphi$, that bind free occurrences of x in φ. A central construct to sHML is the universal modal operator, $[\, p, c \,] \varphi$. To handle reasoning over event data, sHML modalities are augmented with *symbolic actions* [10], consisting of event patterns $p \in \text{PAT}$, and *decidable* constraints, $c \in \text{BEXP}$. This is similar to how sets of actions are expressed in tools such as CADP [40] and mCRL 2 [22]. The pattern p contains data variables, $A, B, \ldots \in \text{VAR}$, that bind free data variables in c, along with any other free variables in constraints of the *continuation* φ. The pair (p, c) describes a *concrete set* of actions, $a \in \text{ACT}$ (*i.e.,* program events). An action a is in this set when: *(i)* p *matches* the shape of a, and maps the variables in p to the payload data in a as the substitution σ, and *(ii)* the *instantiated* constraint $c\sigma$ of p also holds. A state Q of the SuS (model) satisfies $[\, p, c \,] \varphi$ if the following holds: *whenever* Q transitions to state Q' with action a that is included in the set described by (p, c) with σ, then Q' *must* satisfy the instantiated continuation formula $\varphi\sigma$.

The logical variant [4,10] we use for detectEr combines necessities and conjunctions into one construct, $\bigwedge_{i \in I} [\, p_i, c_i \,] \varphi_i$, to denote $[\, p_1, c_1 \,] \varphi_1 \wedge \ldots \wedge [\, p_n, c_n \,] \varphi_n$, $I = \{1, \ldots, n\}$ being a finite index set. Conjunctions assume that every pair (p_i, c_i) describes a *disjoint* set of actions to facilitate the generation of deterministic monitors [35,36]. detectEr supports the five action patterns of Table 1 that capture the lifecycle of, and interactions between the processes of the SuS. A fork action is exhibited by a process when it creates a child process; its dual, init, is exhibited by the corresponding child upon initialisation. Process exit actions signal termination, while send and recv describe interaction. The labelled state transition model of Fig. 1b uses the actions send and recv from Table 1.

Example 1. Recall the SuS behaviour in Fig. 1b. Formula φ_0 with symbolic action (p, c) describes a property requiring that a state does *not* exhibit an output event that consists of $\langle Ack, Tot \rangle$, acknowledged with bye AND A NEGATIVE TOTAL, *Tot*.

$$\bigwedge \ [\ \overbrace{Srv.Clt! \langle Ack, Tot \rangle,}^{\text{pattern } p} \ \overbrace{Ack = \text{bye} \wedge Tot < 0}^{\text{constraint } c} \ ] \ \text{ff} \tag{φ_0}$$

Table 1. Trace event actions capturing the behaviour exhibited by the SuS

Action a	Action pattern p	Variables	Description
forkinit	$P_1 \rightarrow P_2,\ M{:}F(A)$ $P_1 \leftarrow P_2,\ M{:}F(A)$	P_1	PID P_1 of the parent process forking P_2
		P_2	PID P_2 of the child process forked by P_1
		$M{:}F(A)$	Function signature forked by P_1
exit	$P_1 \star\star Dat$	P_1	PID P_1 of the terminated process
		Dat	Exit data, *e.g.* termination reason, *etc.*
send	$P_1{:}P_2\ !\ Req$	P_1	PID P_1 of the process issuing the request
		P_2	PID P_2 of the recipient process
		Req	Request payload, *e.g.* integers, tuples, *etc.*
recv	$P_2\ ?\ Req$	P_2	PID P_2 of the recipient process
		Req	Request payload, *e.g.* integers, tuples, *etc.*

The universal modality states that, for *any* event satisfying the symbolic action (p, c) from a state Q, the state Q' it transitions to must then satisfy the continuation formula. No state can satisfy the continuation ff, and formula φ_0 can only be satisfied when Q *does not* exhibit the event described by (p, c). All the states in Fig. 1b *trivially* satisfy this property (as there are no outgoing state transitions on (p, c) of formula φ_0) with the exception of Q_3. If this state exhibits the *concrete* event $\mathrm{pid_1}{:}\mathrm{pid_2}\ !\ \langle\mathrm{bye}, -1\rangle$, it matches the pattern p, yielding the substitution $\sigma = \{Srv \mapsto \mathrm{pid_1},\ Clt \mapsto \mathrm{pid_2},\ Ack \mapsto \mathrm{bye},\ Tot \mapsto -1\}$. As $c\sigma$ also holds, then we can conclude that Q_3 violates formula φ_0. The formula φ_1 below extends φ_0 to one that is invariant for any state reachable from the current state; this formalises property $\mathrm{P_1}$ from Sect. 2.

$$
\mathrm{max}\underline{x}. \bigwedge \left(\overbrace{\underbrace{[\,Srv\,?\,Req,\ \top\,]\underline{x}}_{①},\ \overbrace{[\,Srv{:}Clt\,!\,\langle Ack, Tot\rangle,\ Ack = \mathrm{bye} \wedge Tot < 0\,]\,\mathrm{ff},}^{②}}^{} \atop \underbrace{[\,Srv{:}Clt\,!\,\langle Ack, Ans\rangle,\ Ack = \mathrm{ok} \vee (Ack = \mathrm{bye} \wedge Ans \geq 0)\,]\underline{x}}_{③} \right) \quad (\varphi_1)
$$

Whereas ② corresponds to formula φ_0, ① and ③ cover the other possible actions produced in Fig. 1b, recursing on the fix-point variable x. ■

Note that the formula variables Srv, Clt, Tot, *etc.* in Example 1 are different to the program variables of Fig. 1a bearing the same name. In our setting, program behaviour is observed as events, and formulae variables are used to pattern-match and reason about data in these events. We adopt the convention of naming formulae and program variables identically, merely to indicate the link between program and event data to readers.

The Tool. The syntax used by detectEr deviates minimally from sHML. Concretely, the comma symbol delimiting patterns and constraints is dropped in favour of the when keyword, whereas vacuous constraints, *i.e.,* when $\top$, may be omitted. The tool also supports a shorthand notation for patterns to specify atomic values directly; these are *implicitly matched* against action data, *e.g.* sub-formula ② from Example 1 can be abbreviated to $[\, Srv.Clt\,!\,\langle \mathsf{bye}, Tot\rangle$ when $Tot < 0\,]$. Moreover, redundant data variables can be replaced by the 'don't care' pattern, $(\_)$, that matches *arbitrary* data values. This sugaring enables us to rewrite φ_1 from Example 1 as:

$$\mathrm{max}\underline{x}.\bigwedge\left(\begin{array}{l}[\,\_?\_]\,\underline{x},\ [\,\_\_!\,\langle \mathsf{bye}, Tot\rangle \text{ when } Tot < 0\,]\,\mathsf{ff},\\ [\,\_\_!\,\langle Ack, Ans\rangle \text{ when } Ack = \mathsf{ok} \vee (Ack = \mathsf{bye} \wedge Tot \geq 0)\,]\,\underline{x}\end{array}\right)$$

Example 2. Property P$_2$ from Sect. 2 describes a fragment of the client-server interaction, asserting that server replies are always addressed to the clients issuing them. Unlike φ_1, this property induces *data dependency* across *nested formulae.*

$$\mathrm{max}\underline{x}.\bigwedge\left(\overbrace{[\,Srv_1\,?\,\langle Clt_1,\_\rangle\,]}^{①}\wedge\overbrace{\left(\underbrace{\begin{array}{l}[\,Srv_2\!:\!Clt_2\,!\_ \text{ when } Srv_1 = Srv_2 \wedge Clt_1 \neq Clt_2\,]\,\mathsf{ff},\\ [\,Srv_2\!:\!Clt_2\,!\_ \text{ when } Srv_1 = Srv_2 \wedge Clt_1 = Clt_2\,]\,\underline{x}\end{array}}_{③}\right)}^{②}\right) \quad (\varphi_2)$$

P$_2$ can be formalised as the formula φ_2, where the data dependency is expressed via the binders Srv_1 and Clt_1 in ①, which are then used in the constraint of sub-formulae ② and ③. The constraint $Srv_1 = Srv_2$ scopes our reasoning to a *single server* instance. Formula φ_2 is violated when $Clt_1 \neq Clt_2$ (since the continuation would need to satisfy ff), and recurs on x otherwise. Recall that the aforementioned comparisons between variable instantiations is possible since the substitution σ obtained from matching the symbolic action of modality ① *extends* to the context of sub-formulae ② and ③.   ∎

Example 3. Property P$_3$ specifies a control aspect of the client-server interaction, demanding that requests issued by clients are never serviced more than once. Formula φ_3 expresses this requirement via a guarded fix-point that recurs on x for sequences of send-recv actions; this captures normal server operation that corresponds to sub-formulae ① followed by ②, and then ④ followed by ②.

$$\bigwedge\overbrace{[\,\_?\_]}^{①}\mathrm{max}\underline{x}.\bigwedge\left(\underbrace{[\,Srv_1\!:\!Clt_1\,!\_]}_{②}\wedge\left(\overbrace{\begin{array}{l}[\,Srv_2\!:\!Clt_2\,!\_ \text{ when } Srv_1 = Srv_2 \wedge Clt_1 = Clt_2\,]\,\mathsf{ff},\\ \underbrace{[\,\_?\_]\,\underline{x}}_{④}\end{array}}^{③}\right)\right) \quad (\varphi_3)$$

Formula φ_3 is violated when a send action matched by ② is followed by a second send action that is matched by ③. The constraint $Clt_1 = Clt_2$ in sub-formula ③ ensures that duplicate send actions concern the same recipient.   ∎

Our earlier formula φ_2 of Example 2 does not account for the case where the server interacts with more than one client. It disregards the possibility of other interleaved events, that are inherent to concurrent settings where processes are unable to control when messages are received. For instance, while sub-formula ① matches an initial `recv` action, a second `recv` action (*e.g.* due to a second client C_2 that interacts with the server) matches neither ② nor ③. This does not reflect the requirement of our original property P_2. The problem can be addressed by augmenting formulae with clauses that 'eat up' non-relevant actions.

$$
\max\underline{x}.\wedge
\left(
[\,Srv_1\,?\,\langle Clt_1\,,\_\rangle\,]\,\overbrace{\max\underline{y}}^{①}.\wedge
\left(
\begin{array}{l}
[\,Srv_2\!:\!Clt_2\,!\_ \text{ when } Srv_1 = Srv_2 \wedge Clt_1 \neq Clt_2\,]\,\mathsf{ff},\\
[\,Srv_2\!:\!Clt_2\,!\_ \text{ when } Srv_1 = Srv_2 \wedge Clt_1 = Clt_2\,]\,\underline{x},\\
\underbrace{[\,\_\,?\,\_\,]\,\underline{y}}_{②}
\end{array}
\right)
\right)
$$

As the refinement of φ_2 above shows however, this bloats specifications, which is why we chose to scope our exposition to a single client-server set-up for the benefit of readers. Introducing the nested maximal fix-point ① and sub-formula ② filters `recv` actions by recursing on variable y; the rest of φ_2 is unaltered.

4   What Goes on

Often, post-deployment verification techniques such as RV, do *not* have access to the entire execution graph of a SuS, *e.g.* the transition model in Fig. 1b. Instead, these are limited to the *trace* of (program) events that is generated by the *current execution* of the SuS. For instance, an execution might generate the trace of events 'pid$_1$? $\langle$pid$_2$, stp$\rangle$. pid$_1$:pid$_2$! $\langle$bye, $-1\rangle$', that corresponds to the (finite) path traversal $Q_0 \rightarrow Srv\,?\,\langle Clt, \mathsf{stp}\rangle \rightarrow Q_3 \rightarrow Srv\!:\!Clt\,!\,\langle \mathsf{bye}, Tot\rangle \rightarrow Q_4$ in the transition model of Fig. 1b. In traces, events consist of concrete values instead of variable placeholders, *e.g.* pid$_1$ instead of Srv, *etc.* This limitation can be problematic when verifying specifications that reason about entire SuS transition models, *e.g.* properties expressed in the μ-calculus [11,43,44], CTL [19,41], and other branching-time logics. Recent studies show that finite traces suffice to adequately verify a practically-useful subset of these properties, as long as the verification is confined to *either* determining satisfaction *or* violation [1,5,7,38,39] (not both). This is more commonly referred to as *specification monitorability* [39]. sHML, used in Sect. 3 to encode properties P_1–P_3, has been shown to be a maximally-expressive subset of the μ-calculus for the runtime analysis of violations. This means that *(i)* any program that violates a property expressed as a sHML formula can be detected at runtime, *(ii)* any μ-calculus property whose violations can be detected at runtime can be expressed as a sHML formula.

From Specification to Analysis. detectEr synthesises *automata-like* analysers in Erlang from sHML; these inspect trace events *incrementally* and reach *irrevocable* verdicts. An analyser flags a *rejection* verdict when it processes a trace exhibiting the program behaviour that *violates* a property of interest—crucially, it never flags verdicts associated with the *satisfaction* of the property [1,7,39]. Intuitively, this is because the trace observed at runtime can never provide *enough information* to rule out the existence of violating behaviour in *other* execution branches of the program. The synthesised analyser code embeds this reasoning: when a trace event is *not* included in the set of actions denoted by the symbolic action of a necessity modality, an *inconclusive verdict* is flagged.

Our synthesis translates a sHML specification to Erlang code encoded as a higher-order function, tasked with the analysis of trace events. This function accepts an event as input, and returns a new function of the same kind that performs the residual analysis following the event just processed. The synthesis, $[\![-]\!]$, that maps sHML constructs to Erlang syntax is as follows:

$$[\![\mathsf{ff}]\!] \triangleq (\texttt{fun (\_)} \rightarrow \texttt{io:format("Rejection")} \texttt{ end) ()}$$

$$[\![\mathsf{max}\, x.\, \varphi]\!] \triangleq (\texttt{fun x()} \rightarrow [\![\varphi]\!] \texttt{ end) ()} \qquad\qquad [\![x]\!] \triangleq \texttt{x()}$$

$$[\![\bigwedge_{i\in I} [\, p_i, c_i \,]\, \varphi_i]\!] \triangleq \begin{cases} \texttt{fun}(p_1) \texttt{ when } c_1 \rightarrow [\![\varphi_1]\!] \\ \qquad \vdots \\ \quad (p_n) \texttt{ when } c_n \rightarrow [\![\varphi_n]\!] \\ \quad (\_) \rightarrow (\texttt{fun (\_)} \rightarrow \texttt{io:format("Stopped")} \texttt{ end) ()} \\ \texttt{end} \end{cases}$$

In $[\![-]\!]$, ff is translated into an anonymous function that flags rejection, modelling formula violations; tt is not synthesised into analysis code since it can never be violated. Maximal fix-point formulae are translated to *named* functions that can be referenced by $[\![x]\!]$. The conjunction of necessities, $\bigwedge_{i\in I} [\, p_i, c_i \,]\, \varphi_i$, maps naturally to a sequence of function clauses, where the pattern p_i matches the shape of the trace event, and the constraint c_i—expressed as an Erlang guard [26]—operates on variables bound in p_i *and* those instantiated in the parent function scope. Inconclusive verdicts are modelled via the catch-all clause (_) that matches any events *other than* those described by the clauses, $\texttt{fun}(p_i) \texttt{ when } c_i$. The order of clauses in Erlang *does* matter, and affects our synthesis in two ways: *(i)* it conveniently allows us to handle the inconclusive verdict case using a catch-all clause at the end of function definitions; *(ii)* the mutually-exclusive symbolic actions in a necessity conjunction allows us to synthesise them in the order specified, without affecting the commutativity of conjunctions. Note that the synthesis applies the generated functions, *i.e.*, (), to unfold them once.

Figure 2 depicts the behaviour of the analyser that is synthesised from formula φ_2 of Example 2. It consists of two states, Q_0, Q_1, the rejection verdict state ✗, and the inconclusive verdict state ?. The transition from Q_0 to Q_1 in Fig. 2 corresponds to the modality $[\, Srv_1\, ?\, \langle Clt_1\, , \_ \rangle\,]$, ① in formula φ_2, while the transitions between Q_1 and ✗, Q_1 and Q_0 express sub-formulae ② and ③. The auxiliary transitions from states leading to ? correspond to the catch-all (_) clause inserted by the synthesis for conjuncted necessities. Figure 2 illustratively labels these transitions by the complement of the set of actions from

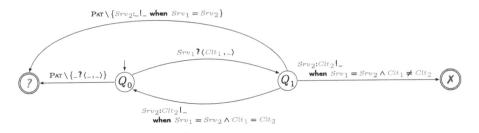

Fig. 2. Abstract model of the analyser synthesised from formula φ_2

a given state. For example, the symbolic action set $\text{PAT} \setminus \{\_? \langle \_, \_\rangle\}$ from Q_0 to $?$ matches *anything but* recv events; recv events are, in turn, matched by $Srv_1 ? \langle Clt_1, \_\rangle$ labelling the transition between Q_0 and Q_1. Verdict irrevocability, a prevalent RV requirement [6], is modelled by the detectEr synthesis in terms of final states ($\times$ and $?$ in Fig. 2). The analysis stops when a final state is reached.

Specification, in Practice. detectEr processes sHML formulae specified in plain text files. The syntax follows the one given in Sect. 3, albeit with two adaptations: *(i)* the keyword **and** is used in lieu of $\bigwedge$, and, *(ii)* we adopt the Erlang operators for writing boolean constraint expressions, *e.g.* =:= instead of =, **andalso** instead of $\wedge$, **orelse** instead of $\vee$, *etc.* Analysers resulting from the synthesis, $[\![-]\!]$, are compiled to binaries to be packaged with the SuS executables.

5   The Magical Mystery Tour

Instrumentation is central to RV. It refers to the extraction of the computation of interest in the form of a sequence of trace events from an executing program, *and* its reporting to the runtime analysis discussed in Sect. 3. Formulae can be rendered unverifiable at runtime when the program events they assume cannot be extracted and reported by the instrumentation. The instrumentation also plays a role in dropping *extraneous* events that can infiltrate the trace being observed and potentially, interfere with the analysis.

What to Monitor. We provide the meta keywords with and monitor to target the SuS component of interest for a particular specification. The with keyword picks out the signature of the function that is forked whereas the monitor keyword defines the property to be analysed. For example, to runtime verify the behaviour of the calculator server of Fig. 1a against formula φ_2, we write:

```
with
    calc_server : loop(_)
monitor
```
$$\max\underline{x}. \bigwedge \left(\begin{array}{l} [\, Srv_1 ? \langle Clt_1, \_\rangle \,] \bigwedge \\ \left(\begin{array}{l} [\, Srv_2\!:\!Clt_2 \,!\_ \text{ when } Srv_1 = Srv_2 \wedge Clt_1 \neq Clt_2 \,] \, \text{ff}, \\ [\, Srv_2\!:\!Clt_2 \,!\_ \text{ when } Srv_1 = Srv_2 \wedge Clt_1 = Clt_2 \,] \, \underline{x} \end{array} \right) \end{array} \right) \qquad (\varphi_2')$$

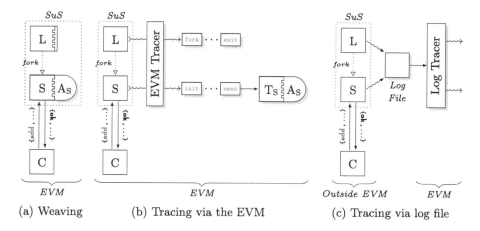

(a) Weaving   (b) Tracing via the EVM   (c) Tracing via log file

Fig. 3. Inline, outline and offline instrumentation methods offered by detectEr

From an instrumentation standpoint, with establishes the set of trace events corresponding to the SuS component it targets, thus enabling the specification to *abstract* from the events generated by other components. This helps to keep the size of specifications compact whenever possible. In using with, formula φ_2' need not account for superfluous events (*e.g.* those of another server component) that tend to make the specification exercise tedious and error-prone.

How to Monitor. detectEr offers three instrumentation methods, inline (Sect. 6), outline (Sect. 7), and offline (Sect. 8), to cater for different situations where the RV is conducted. These methods are depicted by the three set-ups of Fig. 3 that are instantiated with our calculator server, labelled by S, and its parent launcher process, labelled by L. Inlining *statically* instruments system components with the analyser, A_S, which then executes as *part of* the SuS. Outline instrumentation *decouples* the SuS components from the extraction and analysis of trace events by way of the tracing infrastructure provided by the EVM. The offline set-up extends the latter notion to a SuS that (possibly) executes *outside* the EVM, mirroring the same architecture of Fig. 3b to enjoy the same outline arrangement (elided from Fig. 3c). The with keyword directs detectEr to instrument the analyser code over the relevant SuS components regardless of the instrumentation method used, to ensure that the same set of trace events is reported to the runtime analysis. This makes the analysers generated by our synthesis of Sect. 4 agnostic of the underlying instrumentation.

6   Come Together

Inlining [34] is the most efficient instrumentation method detectEr offers. While it assumes access to the source code of the SuS, it carries advantages such as

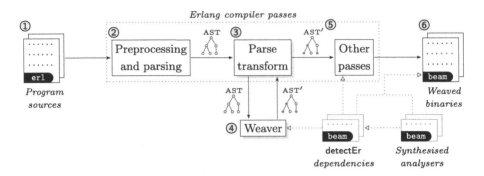

Fig. 4. Instrumentation pipeline for inlined program monitoring via weaving

low runtime overhead [31,32] and immediate detections [20]. detectEr instruments invocations to synthesised analysers via *code injection* by manipulating the program abstract syntax tree (AST). This procedure is depicted in Fig. 4. In step ①, the Erlang program source code is preprocessed and parsed into the corresponding AST, step ②. The Erlang compilation pipeline includes a *parse transformation* phase [26], step ③, that offers an optional hook to allow the AST to be processed externally, prior to code generation. Our custom-built weaver leverages this mechanism to transform the program AST in step ④ and produce the modified AST in step ⑤; this is subsequently compiled by the Erlang compiler into the program binary, step ⑥. The compilation phase depends on the detectEr core modules *and* analyser binaries, as does the SuS, once it executes.

Instrumentation, in One Go. Step ④ in Fig. 4 performs two transformations on the program AST. The first transformation initialises an analyser. It weaves code instructions that *store* the function encoding of the synthesised analyser (refer to Sect. 4) in the process dictionary (PD) of the instrumented process (PDs are process-local, mutable key-value stores with which Erlang actors are initialised). The weaver identifies spawn() calls that carry the function signature to be executed as a process. It then *replaces* the spawn() call with a counterpart which accepts an *anonymous wrapper* function that *(i)* stores the analyser function in the PD, and, *(ii)* applies the function specified inside the original spawn() call. Figure 5a recalls the function start() that forks our calculator server loop, line 1. The corresponding weaved version of its AST—given as Erlang code for illustration in Fig. 5a—performs the initialisation of *(i)* and *(ii)*. Line 3 contains (omitted) boilerplate logic that determines whether a particular spawn() call should be instrumented. The meta keyword with from Sect. 5 is used to this end: it results in the synthesis of auxiliary code that enables the weaver to effect this judgement. For example, the specification 'with calc_server:loop(_)...' of formula φ'_2 informs the weaver to initialise the analyser only for the function name loop forked by the invocation of spawn() on line 1 in Fig. 5a. In line 8, the encoding of the analyser function, AnlFun0, is stored in the PD. The signature used in the original spawn() call on line 1 is applied on line 10, where Mod0,

```
1  start() → spawn(calc_server, loop, [0]).
```

```
2  start() →
3    AnlFun0 = ... % Load analysis logic.
4    P1 = self(),
5    MFA = {Mod0, Fun0, Args0},
6    P0 = spawn(
7      fun() →
8        anl:embed(AnlFun0),
9        anl:dsp({init, self(), P1, MFA}),
10       apply(Mod0, Fun0, Args0)
11     end)
12   anl:dsp({fork, self(), P0, MFA}),
13   P0.
```

```
1  loop(Tot) →
2  receive
3    M2 = {Clt, {add, A₁, A₂}} →
4      anl:dsp(recv, self(), M2),
5      (P1 = Clt) ! M1 = {ok, A₁ + A₂},
6      anl:dsp(send, self(), P1, M1),
7      loop(Tot + 1);
8
9    M4 = {Clt, {mul, A₁, A₂}} →
10     ...
11
12   M6 = {Clt, stp} → % Stop service.
13     ...
14 end.
```

(a) Server initialised with analyser function   (b) Weaved analysis code in server loop

Fig. 5. Transformations to the AST of the `calc_server` program (shown as code)

Fun0, and Args0 are respectively instantiated to values `calc_server`, `loop`, and [0] by the boilerplate logic on line 3 (omitted).

The second transformation decorates the program AST with calls at points of interest: these correspond to the actions catalogued in Table 1. Each call constructs an intermediate trace event description that is *dispatched* to the analyser. Lines 9 and 12 in Fig. 5a construct events `init` and `fork`, and dispatch them to the analyser using the function `anl:dsp()` exposed by the core `detectEr` modules. The events `recv` and `send` are analogously handled on lines 4 and 6 in Fig. 5b.

Our weaver performs the two transformations outlined above regardless of whether monitoring is required by the SuS. This induces a *modular design* where the SuS is weaved *once*, while the analyser binaries may be *independently* regenerated, *e.g.* to refine or add sHML specifications. Updates in these binaries can afterwards be put into effect by restarting the weaved SuS. To determine whether to analyse a trace event, the dispatcher implementation `anl:dsp()` internally checks against the PD whether an analyser function has been initialised for the instrumented process. When the analyser function is initialised, `anl:dsp()` *applies* the function to the event, and *saves* the resulting unfolded analyser function back to the PD; otherwise, `anl:dsp()` discards the event. An irrevocable verdict is reached by the analyser function once its application to an event returns the internal value that encodes ✗ or ?.

Weaving makes it difficult to extract `exit` trace events, since abnormal termination due to crashes cannot be easily anticipated. This limits the ability to runtime check correctness properties concerning process termination. An instrumentation approach via external observation easily sidesteps this restriction.

7   Tell Me What You See

Outlining *externalises* the acquisition and analysis of SuS trace events. It relies on the tracing infrastructure provided by the EVM [26], and supports any software component that is developed for the EVM ecosystem, *e.g.* Erlang, Elixir [42] and Clojerl [33]. Figure 3b in Sect. 5 shows the outlined set-up for our calculator server example. Outlining uses *tracers*, actor processes tasked with the handling of trace events exhibited by the SuS. Tracers *register* with the EVM tracing infrastructure to be notified of process events in connection to the actions of Table 1. Our outlining algorithm instruments tracers *on-demand*, depending on what processes need to be analysed. This approach departs considerably from inline instrumentation in Sect. 6, and rather than weaving the SuS statically, outlining defers the decision of what to instrument until runtime.

While outline instrumentation tends to induce higher runtime overhead, it offers a number of benefits over inlining. It takes a *non-invasive* approach that leverages the EVM to trace components without modification, making it easy to enable and disable the runtime analysis without the need of restarts or redeployments. By decoupling the SuS and tracer components, outlining induces a degree of *partial failure*—a faulty analyser does not compromise the running system, nor does a crashed system component affect the external tracer. As a result of this arrangement, `exit` trace events can be detected, giving us the full expressiveness with respect to the system actions of Table 1. The implementation of an adequate outline monitoring set-up comes with its own set of challenges. For example, the instrumentation should be engineered to *scale* in line with the SuS, while the runtime analysis of trace events is underway. It has to contend with the race conditions (*e.g.* trace event reordering) that arise from the asynchronous execution of the SuS and tracer components. Scalability requires the instrumentation to *explicitly* manage garbage collection, where redundant tracer processes are discarded to minimise resource consumption. Inline instrumentation is spared these complications since the analysis logic is weaved directly in SuS processes. Although our outline instrumentation algorithm handles these aspects, we refrain from providing further detail in this presentation. Interested readers are encouraged to consult [8] for more details.

Instrumentation, as We Go. The EVM tracing infrastructure enables processes to register their interest in receiving trace event messages from other processes. Erlang provides the built-in function `trace()`, that processes may invoke to enable and disable process tracing dynamically at runtime. Our tracers from Fig. 3b leverage this functionality to fork other tracers, and scale the RV set-up as the SuS executes. We configure the EVM tracing to *automatically assign* the tracer of an already-traced SuS process to the children it forks [26]. Using this as a default setting allows us to analyse *groups* of processes as one component. The with keyword guides the targeting of which processes tracers need to track and analyse. By contrast to inlining—where the set of trace events of a component is *implicitly* determined as a byproduct of weaving—outlining must *actively isolate* processes from a group to assign dedicated tracers. Recall the specification 'with `calc_server:loop(_)`...' of formula φ_2'. This instructs our outline instrumentation to set up an independent tracer process for the calculator server loop forked by `spawn()` on line 1 in Fig. 5a.

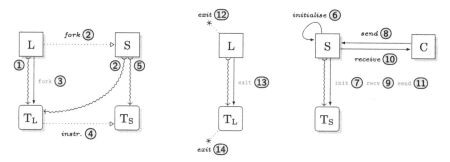

(a) S forked by L and traced by T_S (b) C interacting with S; events received by T_S

Fig. 6. Outline instrumentation for the calculator server (analysers omitted)

Tracers are programmed to react to `fork` and `exit` events in the trace. Figure 6 illustrates how the process creation sequence of the SuS is exploited to instrument a dedicated tracer for our calculator server. A tracer instruments other tracers whenever it encounters `fork` events. The initial RV configuration is shown in Fig. 6a, where the root tracer, T_L, is assigned to the launcher process, L, in step ①. L forks the server function `loop()` to execute as the process S which is automatically assigned the same tracer T_L, as steps ② both indicate. Subsequently, T_L instruments a *new* tracer, T_S, when it processes the `fork` trace event due to L in step ③. The data carried by `fork` contains the PID of the forked process (see Table 1) that designates the SuS process to be instrumented, S, in this case. At this point, T_S takes over the tracing of S from the root tracer T_L by invoking `trace()` to handle S independently of T_L, steps ④ and ⑤. T_S resumes its analysis of S, receiving the `init` event in step ⑦; this is followed by `recv` in step ⑨ as a result of the service request issued by the client, C, in step ⑧. In a similar way, the service reply sent by the server to C in step ⑩ results in `send` being exhibited by S and received by T_S in step ⑪. Process L eventually exits after the fork completes, step ⑫. The ensuing `exit` event in step ⑬ is interpreted by the root tracer T_L as the cue to self-terminate in step ⑭. This garbage collection measure maintains the lowest possible runtime overhead.

8   I'm Only Sleeping

We extend the notion of outline instrumentation to the offline case where the SuS may potentially run outside the EVM. To support offline instrumentation, detectEr implements a middleware that emulates the EVM tracing infrastructure, while preserving the configuration mentioned in Sect. 7, *i.e.*, where forked system processes automatically inherit the tracer assigned to their parent. This enables detectEr to employ the *same* outline instrumentation algorithm for offline monitoring. Offline set-ups are generally the slowest in terms of verdict detection, by comparison to the inline and outline forms of instrumentation. This stems from the dependence outline instrumentation has on the timely availability of pre-recorded runtime traces that are subject to external software entities such as files, databases, and the SuS itself. However, the outline set-up and SuS can reside on different hardware since they are mutually detached. Such an arrangement makes overhead issues secondary.

Figure 3c from Sect. 5 overviews our offline arrangement. It mirrors the set-up in Fig. 3b: the only difference lies in how the offline tracing infrastructure obtains events. Our Log Tracer component in Fig. 3c exposes a `trace()` function, providing the same EVM feature subset relevant to outlining. The implementation relies on log files as the medium through which the SuS can communicate trace events to the offline set-up. It can process log files with complete system executions, or *actively* monitor files for changes to dynamically dispatch events to tracers while the SuS executes and writes events to file. Offline tracing supports the event actions in Table 1; these carry the event data and are assumed to follow a pre-defined format. For instance, the offline event description `fork(pid₁, pid₂, {calc_server, loop, [0]})` is mapped to the action $pid_1 \rightarrow pid_2$, `calc_serverloop([0])` by the Log Tracer of Fig. 3c. Our file-based approach to collecting SuS events is motivated by the fact that file logging is widely-adopted in practice, and is offered by popular frameworks such as Lager [21] for Erlang, Log4J 2 [16] for Java [46], and the Python [49] logging facility. Besides logging, events may also be extracted from the SuS via other tracing frameworks, *e.g.* DTrace [23], LTTng [28], and OpenJ9 Trace [30].

9   Here, There and Everywhere

This paper presents detectEr, a RV tool that analyses program correctness *post-deployment* against properties expressed in a logic that has been traditionally used for *static* verification [14,22,40]. Sects. 5–8 describe how detectEr can flexibly runtime check the same specifications via three instrumentation methods. The tool can be found at https://duncanatt.github.io/detecter.

Future Work. We intend to asses the merits of our three instrumentation methods in terms of the multi-faceted overhead metrics proposed by [9]. We also plan to extend detectEr to handle sHML specifications where conjunctions and universal modalities can be treated as separate logical constructs [3,4,17, 18,24,25,39]. This facilitates the composition of properties via conjunctions, *e.g.* formulae φ_1–φ_3 from Sect. 3 can be combined as $\varphi_1 \wedge \varphi_2 \wedge \varphi_3$ to synthesise one global monitor. Although detectEr focusses on properties that are known to be runtime monitorable, new results argue that monitoring can be systematically extended to the entire class of regular properties, albeit, with possibly weakened detection guarantees [2,7]. We aim to incorporate these results within this tool.

References

1. Aceto, L., Achilleos, A., Francalanza, A., Ingólfsdóttir, A.: Monitoring for silent actions. In: FSTTCS. LIPIcs, vol. 93, pp. 7:1–7:14 (2017)
2. Aceto, L., Achilleos, A., Francalanza, A., Ingólfsdóttir, A.: A framework for parameterized monitorability. In: Baier, C., Dal Lago, U. (eds.) FoSSaCS 2018. LNCS, vol. 10803, pp. 203–220. Springer, Cham (2018). https://doi.org/10.1007/978-3-319-89366-2_11
3. Aceto, L., Achilleos, A., Francalanza, A., Ingólfsdóttir, A., Kjartansson, S.Ö.: On the complexity of determinizing monitors. In: Carayol, A., Nicaud, C. (eds.) CIAA 2017. LNCS, vol. 10329, pp. 1–13. Springer, Cham (2017). https://doi.org/10.1007/978-3-319-60134-2_1
4. Aceto, L., Achilleos, A., Francalanza, A., Ingólfsdóttir, A., Kjartansson, S.Ö.: Determinizing monitors for HML with recursion. JLAMP **111**, 100515 (2020)
5. Aceto, L., Achilleos, A., Francalanza, A., Ingólfsdóttir, A., Lehtinen, K.: Adventures in monitorability: from branching to linear time and back again. Proc. ACM Program. Lang. **3**(POPL), 52:1–52:29 (2019)
6. Aceto, L., Achilleos, A., Francalanza, A., Ingólfsdóttir, A., Lehtinen, K.: An operational guide to monitorability with applications to regular properties. Softw. Syst. Model. **20**(2), 335–361 (2021). https://doi.org/10.1007/s10270-020-00860-z
7. Aceto, L., Achilleos, A., Francalanza, A., Ingólfsdóttir, A., Lehtinen, K.: The best a monitor can do. In: CSL. LIPIcs, vol. 183, pp. 7:1–7:23 (2021)
8. Aceto, L., Attard, D.P., Francalanza, A., Ingólfsdóttir, A.: A choreographed outline instrumentation algorithm for asynchronous components. CoRR abs/2104.09433 (2021)
9. Aceto, L., Attard, D.P., Francalanza, A., Ingólfsdóttir, A.: On benchmarking for concurrent runtime verification. FASE 2021. LNCS, vol. 12649, pp. 3–23. Springer, Cham (2021). https://doi.org/10.1007/978-3-030-71500-7_1
10. Aceto, L., Cassar, I., Francalanza, A., Ingólfsdóttir, A.: On runtime enforcement via suppressions. In: CONCUR. LIPIcs, vol. 118, pp. 34:1–34:17 (2018)
11. Aceto, L., Ingólfsdóttir, A.: Testing Hennessy-Milner logic with recursion. In: Thomas, W. (ed.) FoSSaCS 1999. LNCS, vol. 1578, pp. 41–55. Springer, Heidelberg (1999). https://doi.org/10.1007/3-540-49019-1_4
12. Agha, G., Mason, I.A., Smith, S.F., Talcott, C.L.: A foundation for actor computation. JFP **7**(1), 1–72 (1997)
13. Alpern, B., Schneider, F.B.: Recognizing safety and liveness. Distrib. Comput. **2**(3), 117–126 (1987). https://doi.org/10.1007/BF01782772
14. Andersen, J.R., et al.: CAAL: concurrency workbench, Aalborg edition. In: Leucker, M., Rueda, C., Valencia, F.D. (eds.) ICTAC 2015. LNCS, vol. 9399, pp. 573–582. Springer, Cham (2015). https://doi.org/10.1007/978-3-319-25150-9_33
15. Armstrong, J.: Programming Erlang: Software for a Concurrent World. Pragmatic Bookshelf (2007)
16. ASF: Log4J 2 (2021). https://logging.apache.org/log4j/2.x
17. Attard, D.P., Francalanza, A.: A monitoring tool for a branching-time logic. In: Falcone, Y., Sánchez, C. (eds.) RV 2016. LNCS, vol. 10012, pp. 473–481. Springer, Cham (2016). https://doi.org/10.1007/978-3-319-46982-9_31
18. Attard, D.P., Francalanza, A.: Trace partitioning and local monitoring for asynchronous components. In: Cimatti, A., Sirjani, M. (eds.) SEFM 2017. LNCS, vol. 10469, pp. 219–235. Springer, Cham (2017). https://doi.org/10.1007/978-3-319-66197-1_14

19. Baier, C., Katoen, J.P.: Principles of Model Checking. MIT Press, Cambridge (2008)
20. Bartocci, E., Falcone, Y., Francalanza, A., Reger, G.: Introduction to runtime verification. In: Bartocci, E., Falcone, Y. (eds.) Lectures on Runtime Verification. LNCS, vol. 10457, pp. 1–33. Springer, Cham (2018). https://doi.org/10.1007/978-3-319-75632-5_1
21. Basho: Lager (2021). https://github.com/basho/lager
22. Bunte, O., et al.: The mCRL2 toolset for analysing concurrent systems. In: Vojnar, T., Zhang, L. (eds.) TACAS 2019. LNCS, vol. 11428, pp. 21–39. Springer, Cham (2019). https://doi.org/10.1007/978-3-030-17465-1_2
23. Cantrill, B.: Hidden in plain sight. ACM Queue 4(1), 26–36 (2006)
24. Cassar, I., Francalanza, A., Attard, D.P., Aceto, L., Ingólfsdóttir, A.: A suite of monitoring tools for Erlang. In: RV-CuBES, vol. 3, pp. 41–47. Kalpa Publications in Computing (2017)
25. Cassar, I., Francalanza, A., Said, S.: Improving runtime overheads for detectEr. In: FESCA. EPTCS, vol. 178, pp. 1–8 (2015)
26. Cesarini, F., Thompson, S.: Erlang Programming: A Concurrent Approach to Software Development. O'Reilly Media, Sebastopol (2009)
27. Clarke, E.M., Klieber, W., Nováček, M., Zuliani, P.: Model checking and the state explosion problem. In: Meyer, B., Nordio, M. (eds.) LASER 2011. LNCS, vol. 7682, pp. 1–30. Springer, Heidelberg (2012). https://doi.org/10.1007/978-3-642-35746-6_1
28. Desnoyers, M., Dagenais, M.R.: The LTTng tracer: a low impact performance and behavior monitor for GNU/Linux. Technical report, École Polytechnique de Montréal (2006)
29. Dijkstra, E.W.: Chapter I: notes on structured programming, p. 1–82. Academic Press Ltd. (1972)
30. Eclipse/IBM: Openj9 (2021). https://www.eclipse.org/openj9
31. Erlingsson, Ú.: The inlined reference monitor approach to security policy enforcement. Ph.D. thesis, Cornell University (2004)
32. Erlingsson, Ú., Schneider, F.B.: SASI enforcement of security policies: a retrospective. In: NSPW, pp. 87–95 (1999)
33. Facorro, J.: Clojerl language (2021). http://clojerl.org
34. Falcone, Y., Krstić, S., Reger, G., Traytel, D.: A taxonomy for classifying runtime verification tools. In: Colombo, C., Leucker, M. (eds.) RV 2018. LNCS, vol. 11237, pp. 241–262. Springer, Cham (2018). https://doi.org/10.1007/978-3-030-03769-7_14
35. Francalanza, A.: Consistently-detecting monitors. In: CONCUR. LIPIcs, vol. 85, pp. 8:1–8:19 (2017)
36. Francalanza, A.: A theory of monitors. Inf. Comput. 104704 (2021). https://doi.org/10.1016/j.ic.2021.104704
37. Francalanza, A., et al.: A foundation for runtime monitoring. In: Lahiri, S., Reger, G. (eds.) RV 2017. LNCS, vol. 10548, pp. 8–29. Springer, Cham (2017). https://doi.org/10.1007/978-3-319-67531-2_2
38. Francalanza, A., Aceto, L., Ingolfsdottir, A.: On verifying Hennessy-Milner logic with recursion at runtime. In: Bartocci, E., Majumdar, R. (eds.) RV 2015. LNCS, vol. 9333, pp. 71–86. Springer, Cham (2015). https://doi.org/10.1007/978-3-319-23820-3_5
39. Francalanza, A., Aceto, L., Ingólfsdóttir, A.: Monitorability for the Hennessy-Milner logic with recursion. FMSD 51(1), 87–116 (2017). https://doi.org/10.1007/s10703-017-0273-z

40. Garavel, H., Lang, F., Mateescu, R., Serwe, W.: CADP 2011: a toolbox for the construction and analysis of distributed processes. Int. J. Softw. Tools Technol. Transf. **15**(2), 89–107 (2013). https://doi.org/10.1007/s10009-012-0244-z

41. Clarke Jr., E.M., Grumberg, O., Peled, D.A.: Model Checking. MIT Press, Cambridge (1999)

42. Jurić, S.: Elixir in Action. Manning (2019)

43. Kozen, D.: Results on the propositional μ-calculus. In: Nielsen, M., Schmidt, E.M. (eds.) ICALP 1982. LNCS, vol. 140, pp. 348–359. Springer, Heidelberg (1982). https://doi.org/10.1007/BFb0012782

44. Larsen, K.G.: Proof systems for satisfiability in Hennessy-Milner logic with recursion. TCS **72**(2&3), 265–288 (1990)

45. Leucker, M., Schallhart, C.: A brief account of runtime verification. JLAP **78**(5), 293–303 (2009)

46. Loy, M., Niemeyer, P., Leuck, D.: Learning Java: An Introduction to Real-World Programming with Java. O'Reilly Media, Sebastopol (2020)

47. Myers, G.J., Sandler, C., Badgett, T.: The Art of Software Testing. Wiley, Hoboken (2011)

48. Pierce, B.C.: Types and Programming Languages. MIT Press, Cambridge (2002)

49. Python: Logging Facility for Python (2021). https://docs.python.org/3/library/logging.html

Tutorial: Designing Distributed Software in mCRL2

Jan Friso Groote and Jeroen J. A. Keiren$^{(\boxtimes)}$

Department of Mathematics and Computer Science,
Eindhoven University of Technology, Eindhoven, The Netherlands
{J.F.Groote,J.J.A.Keiren}@tue.nl

Abstract. Distributed software is very tricky to implement correctly as some errors only occur in peculiar situations. For such errors testing is not effective. Mathematically proving correctness is hard and time consuming, and therefore, it is rarely done. Fortunately, there is a technique in between, namely model checking, that, if applied with skill, is both efficient and able to find rare errors.

In this tutorial we show how to create behavioural models of parallel software, how to specify requirements using modal formulas, and how to verify these. For that we use the mCRL2 language and toolset (www.mcrl2.org/). We discuss the design of an evolution of well-known mutual exclusion protocols, and how model checking not only provides insight in their behaviour and correctness, but also guides their design.

Keywords: Model checking · Parallel software · Distributed software · mCRL2 toolset · Counterexamples

1   Introduction

Whoever designed parallel or distributed software and protocols must have found out how hard it is to get such software correct.[1] Distributed software defies testing, as some errors only occur very rarely, easily less than once in a million of runs. Yet, if such errors occur the software can go awry, with effects that range from confused internal administration, via crashing of the software, to losing control over safety-critical hardware.

The theoretical solution is to prove the correctness, for instance using assertional methods that have been under development since the advent of the first electronic computers [1]. These days these methods are supported by proof checkers such as Coq [3] and Isabelle [27], or integrated automatic provers for algorithms such as Dafny [24]. These techniques are unprecedented in locating software faults and are unbeatable if it comes to delivering correct software. However,

---

[1] In this paper, for the sake of brevity, we generally refer to parallel or distributed software just using the term distributed software. The techniques discussed in this paper apply equally in both situations.

© IFIP International Federation for Information Processing 2021
Published by Springer Nature Switzerland AG 2021
K. Peters and T. A. C. Willemse (Eds.): FORTE 2021, LNCS 12719, pp. 226–243, 2021.
https://doi.org/10.1007/978-3-030-78089-0_15

they have two important disadvantages. Proving the correctness of software can be very hard, as the proof may require tricky combinatorial arguments, and detailed bookkeeping. More importantly, it is very time consuming to provide a proof, even for a core algorithm, or a small distributed protocol. The net result of this is that proving correctness of actual software is hardly ever used in practical software development.

Fortunately, there is a method in between, namely model checking of models of the software. The idea is to use an abstract modelling language to model the essence of the distributed algorithm or protocol. Potential modelling languages with a powerful supporting model checking toolset are mCRL2 [18], LNT [14] and FDR3 [15] as they support behaviour with parallelism as well as all commonly used data types. Standard programming languages such as Java and C++ are less suitable for this purpose, as they are too versatile, and do not allow for concise mathematical formulation of protocols and algorithms. Domain Specific Languages to define automata based controllers such as ASD [23] and Dezyne [4] are suitable alternatives, with the advantage that they allow for code generation, but these languages generally provide limited verification possibilities.

Only formulating models of distributed algorithms already substantially improves the quality of a subsequent implementation. The reason is that models are more concise than implementations, and models tend to be studied and discussed more thoroughly than programs. Unfortunately, models still tend to contain errors and therefore, more needs to be done to increase the quality.

Improving the quality of software models further can be done by providing alternative independent views on the software and then comparing all views very precisely [8]. The probability to make the same mistake in all views is the product of the probabilities of making this mistake in each of the views. With a number of views the error probability drops dramatically, and error probabilities of 10^{-10} are attainable. In engineering such an approach is common where reliability is obtained due to redundancy. Even checking light-weight properties can already make a substantial difference [28].

We only discuss one such alternative view, namely formulating compact properties on the model and proving them using model checking. Other views are making alternative models, independently making an implementation, specifying tests, and carrying out field tests. The more alternative views, the higher the quality of the result, provided they are very precisely compared to each other. Formulating correctness and proving this with a proof checker is also a valid alternative view.

We use the modal mu-calculus with data as it is unsurpassed in expressivity [7,18]. Fairness can be expressed using alternating fixed-points, and, by using data, complex behaviour of the model can be tracked and analysed with modal formulas. Alternative property languages, such as CTL/LTL can be translated linearly to the modal mu-calculus with data [10].

Throughout this tutorial, we use mutual exclusion protocols for shared memory to illustrate the use of mCRL2. Such protocols are commonly studied using model checking tools. Even performance evaluation of mutual exclusion protocols has been studied using interactive Markov chains in CADP [25]. Mutual exclusion algorithms in the presence of time were verified using UPPAAL [9]. Note that recently mutual exclusion protocols modelled in process algebra became the topic of a more in-depth discussion. Some authors argue that mutual exclusion cannot be accurately captured using process algebra without extending the language due to fairness and justness problems [13]. Others argue that justness and fairness can be obtained on a case-by-case basis using the mu-calculus [6].

In this tutorial we first describe mCRL2 and the modal mu-calculus very compactly. Subsequently, we focus on the functional correctness of mutual exclusion protocols for shared memory and traverse through the development of such protocols, repeatedly identifying and repairing problems. We show how counterexamples help in identifying and understanding problems [31]. The modelling and analysis techniques described generalize to distributed algorithms in a straightforward way, using processes to model communication channels instead of shared variables. We thus hope that this tutorial will help in understanding how mCRL2 can be used to develop correct distributed algorithms and effectively obtain insight in their behaviour, which goes far beyond showing that they terminate with the right response.

2   mCRL2 Primer

In this section we give a concise description of mCRL2 and the modal mu-calculus. More information is available in [18]. The language mCRL2 is based on process algebra [2,26]. The modal mu-calculus is based on Hennessy-Milner logic [22] and fixed point equations [7].

Process algebraic modelling centers around the notion of an action, typically denoted as a, b, c. . ., representing some atomic activity of a modelled entity, such as a program. Sending a message, writing a variable or printing some text are typical examples. If actions must happen at exactly the same time we denote them as multi-actions. By writing a|b it is indicated that actions a and b happen at the same instant in time. Actions and multi-actions have the same properties, and therefore we generally only speak about actions in the sequel.

Using the sequential composition operator (.) actions can be put in sequence and the choice operator (+) expresses that the behaviour of either the left or the right operand can be done. A typical example is a.b + c.d saying it is possible to do either an a followed by a b or a c followed by a d.

Processes are specified by recursive equations. The equation **proc** P=a.P indicates that the process P can infinitely often do an a action. Using **init** P it is expressed that process P is the behaviour defined by the specification.

Actions and processes can carry data, and all common data types are available. The process equation

proc Adder(n:*Nat*)=*sum* m:*Nat*.add(m).Adder(n+m)

is an example. The sum indicates the choice over all natural numbers. This process can perform one of the actions add(m) for every number m and continues with the behaviour Adder(n+m). Behaviour can be executed conditionally on data using the if-then-else operator b->p<>q where b is a boolean expression and p and q are processes.

Processes are put in parallel using the parallel operator (||). Two parallel processes can communicate by synchronising their actions. This is denoted using the communication operator *comm*({a_s|a_r->a},p||q), expressing that if action a_s and a_r can happen simultaneously in processes p and q, these actions can happen together as a. We use the convention to write _s for send, and _r , after an action if they will be used for a communication. If actions a_s and a_r carry data they can only synchronise to a if the data in both actions are equal. Then a will have this data as parameter as well. To enforce that actions a_s and a_r must communicate, the allow operator is used. The process *allow* (a,p) expresses that only action a is allowed to happen in process p and all other actions are blocked.

The modal mu-calculus is an extension of propositional logic. Hence, we can use connectives such as &&, || and ! representing *and, or* and *not*, respectively. Writing <a>phi expresses that an action a can be done after which phi holds, and [a]phi expresses that if an action a is done, then phi must hold afterwards. Instead of an action a we can use *true* to represent any action, and !a to represent any action but a. We can use a Kleene star to indicate arbitrary sequences of actions. So, <!a*>phi indicates that it is possible to do a sequence of actions in which a does not occur such that afterwards phi holds. The formula [*true* *]phi expresses that phi is valid after each sequence of actions. All actions can carry data, and quantification over data using *exists* and *forall* is possible.

Using the minimal fixed point operator *mu* X.phi and the maximal fixed point operator *nu* X.phi recursive properties can be specified. By *nu* X.<a>X we express that an infinite sequence of actions a must be possible. The formula *mu* X.[!a]X&&<*true*>*true* says that the action a must be done on every path within a finite number of steps. Using nested fixed points fairness properties can be expressed.

The fixed point variables can also use data. To express that the total value offered to the adder will never exceed some maximum M the following formula can be used:

$$nu\ X(n{:}Nat{=}0).forall\ m{:}Nat.[\mathrm{add}(m)]X(n{+}m)\ \&\&$$
$$[!exists\ m{:}Nat.\mathrm{add}(m)]X(n)\ \&\&$$
$$\mathbf{val}(n{<}M)$$

Here the variable n, initially equal to 0, sums up all values of m occuring in actions add(m). The box modality with the exists expresses that whenever an action different from add is done, then checking proceeds with an unaltered parameter n. The condition n<M guarantees that the sum n never exceeds M. The keyword **val** is needed to let the parser distinguish between modal formulas and data expressions.

3   Mutual Exclusion

In this tutorial we study the mutual exclusion problem as we expect most of our readers to be familiar with it. This allows us to focus on how the mCRL2 toolset helps us to model and understand solutions for such a problem. The techniques we describe are equally applicable in other problem domains.

Dijkstra describes the mutual exclusion problem as follows [11]:

> "[...] consider N computers, each engaged in a process which, for our aims, can be regarded as cyclic. In each of the cycles a so-called 'critical section' occurs and the computers have to be programmed in such a way that at any moment only one of these N cyclic processes is in its critical section."

The first solution to the mutual exclusion problem has been known since 1959. It was first described by Dijkstra [12], who attributed it do Dekker. In this paper Dijkstra also shows two simpler solutions and discusses their incorrectness. A first solution for N processes is due to Dijkstra [11] and only much later the well-known solution by Peterson appeared [30].

From Sect. 3.1 onward we model Dijkstra's algorithms in increasing complexity. Subsequently, we investigate Peterson's mutual exclusion algorithm.

Requirements. Before modelling solutions, we ask ourselves what the properties are that a mutual exclusion protocol should have. In order to understand the requirements it is necessary to understand that we model mutual exclusion using three phases. First, a *wish* to enter the critical section is indicated, second access is granted indicated by *enter*, after which the process indicates that it left the critical section using *leave*.

Mutual exclusion. At any moment only one of the processes is in its critical section.

Always eventually request. Every process can always eventually wish to enter its critical section.

Eventual access. Whenever a process indicates a wish to enter its critical section, it is guaranteed to eventually get access to its critical section. This property is also referred to as starvation freedom.

Bounded overtaking. There is an absolute bound B such that, whenever a process indicates it wants to enter its critical section, at most B processes can enter their critical section, before this process enters its critical section.

It is natural to formulate mutual exclusion as a property. But mutual exclusion is insufficient, as it can easily be guaranteed by never letting a process enter the critical section. For a properly functioning mutual exclusion protocol the second and third properties are equally important. The last one is interesting especially in systems where execution of programs is not necessarily fair.

Memory Model. We assume that the mutual exclusion protocols are implemented on a platform with shared memory where variables are written and read atomically in some interleaved fashion by the parallel programs.

3.1   A Naive Algorithm for Mutual Exclusion

For two processes a naive solution of the mutual exclusion problem is Algorithm 1 suggested by Dijkstra [12]. It uses two global Boolean variables $flag[i]$ in which process i indicates that it is in its critical section. The algorithm, for process i now proceeds as follows. It first blocks until the flag of process $1-i$ becomes *false* using busy waiting. Once $flag[1-i]$ is *false*, the other process is not in its critical section. It then sets its own flag to *true* and enters its critical section. Once the work in the critical section is complete, it sets its flag to *false*.

Data: Global variables $flag[0], flag[1]: \mathbb{B}$
Init: $flag[0] := false$; $flag[1] := false$
while $flag[1-i]$ **do** /* Busy waiting */ **end**
$flag[i] := true$;
/* Critical section */
$flag[i] := false$;

Algorithm 1: A naive mutual exclusion algorithm for process i.

Below we go through a few steps to model this algorithm in mCRL2.

Shared Variables. A shared variable can be modelled as process that carries the current value of the variable as a parameter. It can perform a read action, in which it sends the current value of its parameter. Also, it can perform a write action for each possible value that can be stored in the variable. The array *flag* is modelled by the following process.

```
proc Flag(i:Nat, b:Bool)=
       sum b':Bool. set_flag_r(i, b').Flag(i, b') +
       get_flag_s(i, b).Flag(i, b);
```

The name of the process is `Flag`. The parameter `i:Nat` describes the index in the array, and `b:Bool` gives the current value of the variable. Using `sum b':Bool.set_flag_r(i,b').Flag(i,b')` we model that the process can receive any new value `b'` from another process, and store it to parameter `b`. The action `get_flag_s(i, b).Flag(i, b)` allows to send the current value to any process that requests the value.

Modelling the Busy Waiting Loop. The effect of the busy waiting loop is that the process can only continue when the guard becomes *false*, i.e., when $flag[1-i]$ has value *false*. We could model the busy waiting loop explicitly by a recursive process. However, in mCRL2, an action that participates in a communication blocks until the actions it communicates with can also be performed. We can model this loop by using `get_flag_r(1-i, false)`, that is, reading *false* from the flag of the other process. Since a subtraction results in an integer instead of a natural number, we need to add an explicit type conversion here, and write `get_flag_r(Int2Nat(1-i), false)`. As `i` is either 0 or 1, this is guaranteed to be natural number.

Model. We now combine this into an mCRL2 model. First, we define the behaviour of process i.

```
proc Mutex(i:Nat) =
        get_flag_r(Int2Nat(1-i), false).
        set_flag_s(i, true).
        enter(i).
        leave(i).
        set_flag_s(i, false).
        Mutex();
```

Note that the process is the sequential composition of the busy waiting loop, setting the flag of process i to *true* using set_flag_s(i, true), entering the critical section using action enter(i), leaving it using leave(i), and setting the flag to *false* again. At the end of the algorithm we write Mutex() to model that the critical section can repeatedly be entered. Writing Mutex() without parameters is a shorthand that leaves the current value of the parameters unchanged. Here it is thus equivalent to writing Mutex(i).

The system as a whole consists of two instances of Mutex and two shared variables, synchronising on get_flag and set_flag.

```
init allow({enter, leave, get_flag, set_flag},
        comm({get_flag_r | get_flag_s -> get_flag,
            set_flag_r | set_flag_s -> set_flag},
        Mutex(0) || Mutex(1) || Flag(0,false) || Flag(1,false)));
```

Here, the operator *comm* specifies that, get_flag_r and get_flag_s can synchronise. The result is named get_flag. It does the same for set_flag_r and set_flag_s. Writing *allow*{enter, leave, get_flag, set_flag} specifies that we are only interested in the result of the communication, essentially enforcing synchronisation. We also allow the actions enter and leave that are local to the processes, and hence do not participate in any synchronisation.

Verification. Now that we have a model of this first mutual exclusion algorithm, we focus on its correctness. How can we formalize the mutual exclusion property using the mu-calculus? Observe that we explicitly modelled entering and leaving the critical section. Process i is therefore in its critical section if it performed an enter(i) action, but has not yet done the corresponding leave(i). Mutual exclusion is then violated if we see two enter actions without an intermediate leave. This is captured in the following mu-calculus formula.

```
[true*][exists i1:Nat.enter(i1)]
        [!(exists i2:Nat.leave(i2))*][exists i3:Nat.enter(i3)]false
```

This formula expresses that invariantly ([true*]), after a process enters its critical section ([exists i1:Nat.enter(i1)]), as long as no leave action happened ([!(exists i2:Nat.leave(i2))*]), another process is not allowed to enter its critical section ([exists i3:Nat.enter(i3)]false).

We entered the model and the property in mcrl2ide, which is mCRL2's IDE that supports most basic uses of the mCRL2 toolset. A screenshot is shown in Fig. 1. By clicking the 'Verify' button (green triangle) of the mutual exclusion property, the tools will verify whether the property holds. In this case, it finds that the property is violated, and the 'Verify' button changes into a red

Fig. 1. Screenshot of mCRL2ide with naive mutual exclusion algorithm.

'C'. By clicking the red 'C', the tool shows a counterexample. In this case, the counterexample is the one shown in Fig. 2.

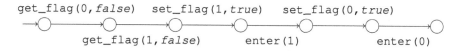

Fig. 2. Counterexample of the mutual exclusion property for the naive algorithm.

This counterexample is a trace where two processes execute an `enter` action without an intermediate `leave`. If we check the counterexample, it is immediately clear what is going on. Both processes check simultaneously that the other process is not in its critical section, concluding they can proceed to their critical section. Then, process 1 sets its flag and enters its critical section, immediately followed by process 0.

3.2 Fixing the Naive Algorithm

The problem with the naive algorithm is that each process first checks if the other process is in its critical section, and then sets its own flag. If this is done simultaneously, the processes do not observe that the other process is entering the critical section at the same time. We could potentially resolve this issue by first setting the flag, expressing the intent to enter the critical section, and then only proceed into the critical section if the flag of the other process is false. The improved algorithm is shown in Algorithm 2. It also stems from [12].

> **Data:** Global variables $flag[0], flag[1]: \mathbb{B}$
> $flag[i] := true;$
> **while** $flag[1-i]$ **do** /* Busy waiting */ **end**
> /* Critical section */
> $flag[i] := false;$

Algorithm 2: Improved naive mutual exclusion algorithm for process i.

Model. The change in the mCRL2 model is equally simple. We only exchange the first two lines of the Mutex process, which now becomes the following.

```
proc Mutex(i:Nat) =
    set_flag_s(i, true).
    get_flag_r(Int2Nat(1-i), false).
    enter(i).
    leave(i).
    set_flag_s(i, false).
    Mutex();
```

Verification. Changing the order of the program fixed the algorithm as it now satisfies the mutual exclusion property. So, we investigate the requirement that every process can always eventually wish to enter its critical section.

If process i sets its flag to *true* this means that it expresses the wish to enter its critical section. The property can then be expressed by saying that invariantly, for all processes i there is a path to a state in which process i can set its flag to *true*. This is expressed in the mu-calculus as follows.

```
[true*]forall i:Nat.val(i<=1) => <true*><wish(i)|set_flag(i, true)>true
```

Recall that [true*]phi is valid if phi holds in all reachable states. We express the property for all processes i using *forall* i:Nat with **val**(i<=1). The remaining formula <true*><set_flag(i, true)>true expresses that there is a path to a state in which set_flag(i, true) can happen.

When we verify this property, it turns out that it does not hold. The counterexample is shown in Fig. 3.

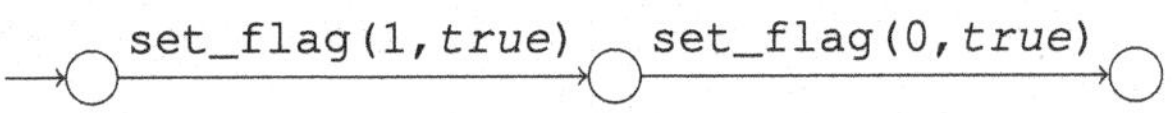

Fig. 3. Counterexample showing that a process cannot always eventually wish to enter.

This counterexample shows that if both processes wish to enter the critical section, they end up in a state in which at least one of the processes can never request access to its critical section again. A closer inspection reveals that both processes are waiting for the other process' flag to become *false* before being able to proceed into a critical section. Hence, the processes are stuck in a typical deadlock situation, and therefore neither process has a path to a state in which it can request to enter the critical section again.

3.3 Dekker's Algorithm

To resolve the deadlock in the previous mutual exclusion algorithm, Dekker's solution is to give priority to one of the two processes whenever both processes want to enter their critical section. We present it as Algorithm 3 [12].

> **Data:** Global variables $flag[0], flag[1] : \mathbb{B}$ and $turn : \mathbb{N}$
> $flag[i] := true;$
> **while** $flag[1{-}i]$ **do**
>   **if** $turn \neq i$ **then**
> $flag[i] := false;$
>   **while** $flag[1{-}i]$ **do** /* Busy waiting */ **end**
> $flag[i] := true;$
>   **end**
> **end**
> $turn := 1{-}i;$
> /* Critical section */
> $flag[i] := false;$

Algorithm 3: Dekker's algorithm for process i.

The idea behind the algorithm is as follows. First, compared to the previous attempts, the meaning of global Boolean variables $flag[i]$ changes, and now indicates whether process i wishes to access its critical section. Second, a new shared variable $turn$, which is 0 initially, indicates which process has priority when both processes want to enter their critical section.[2] The key idea now is that, while the other process $1{-}i$ wishes to enter its critical section, process i checks whether the other process has priority. If so, process i sets its flag to *false*, and then waits until process $1{-}i$ leaves its critical section and sets its flag to *false*. Then process i resets its flag to *true* and continues as before.

Model. To model this algorithm in mCRL2, we have to decide how to deal with the outer loop and the if-clause. We first model the outer while-loop.

In more general terms, we want to model a program $S_1;$ **while** b **do** S_2 **end**; S_3 in mCRL2. The most straightforward way to model this is to have two separate processes that are executed sequentially. The first process performs the behaviour of S_1 and hands execution over to the second process. The second process evaluates b. If b is *true* it executes the behaviour of S_2 and then executes itself, repeating the behaviour. Otherwise it executes the behaviour of S_3.

---

[2] In [12], variables LA and LB are used as flags, and a Boolean variable AP is used in the place of $turn$.

An if-then clause **if** b **then** S_1 **end**; S_2 can be modelled directly into the if-then-else construct b->p<>q of mCRL2. In this case p is the translation of $S_1; S_2$ and q is the translation of S_2. Note that both for the loop and the if-then clause, if the condition contains shared variables, their values must first be read.[3]

The outer loop of the algorithm is modelled as follows.

```
proc Dekker_outer_loop(i:Nat) =
        sum flag_other:Bool.get_flag_r(other(i), flag_other).
        flag_other -> (sum turn: Nat.get_turn_r(turn).
                        (turn != i) -> (set_flag_s(i, false).
                                        get_turn_r(i).
                                        set_flag_s(i, true).
                                        Dekker_outer_loop(i)
                                        )
                                    <> Dekker_outer_loop(i)
                        )
                <> (set_turn_s(other(i)).
                    enter(i).
                    leave(i).
                    set_flag_s(i, false).
                    Dekker(i);
                    );
```

Note that the shared variable `flag` of the other process is read, and its value is stored in `flag_other`. If the guard of the outer loop is true (`flag_other ->`), the loop is entered. In the body of the loop the turn variable is read, and it is decided whether the if-clause must be entered. In the body of the if the flag for this process is set to false, allowing the other process to enter its critical section. The process waits until the other process leaves its critical section. Note that here we use the construct we previously introduced for the busy waiting loop. Subsequently, the flag of this process is set to true, and the while loop is repeated.

If the guard of the outer loop is false, we jump to the else part starting with the lower <> symbol, from where the rest of the process is similar to our previous algorithms.

For the complete model, we also need a process modelling the global variable *turn*. This is done in a similar way as for the global array *flag*, where the variable is set and read using actions `set_turn` and `get_turn`, which are the results of synchronising `set_turn_r` and `set_turn_s`, and `get_turn_s` and `get_turn_r`. The parallel composition must be extended with the process modelling *turn*, as well as with an increased number of synchronising actions, and is given below. Some aspects of this process expression are explained in the next part on verification.

```
init allow({wish|set_flag, enter, leave,
            get_flag, set_flag, get_turn, set_turn},
        comm({get_flag_r | get_flag_s -> get_flag,
            set_flag_r | set_flag_s -> set_flag,
            get_turn_r | get_turn_s -> get_turn,
            set_turn_r | set_turn_s -> set_turn},
        Dekker(0) || Dekker(1) || Flag(0,false) || Flag(1,false) || Turn(0)));
```

---

[3] Note that, alternatively, the multi-actions in mCRL2 could be used to combine fetching the value and evaluating the condition, see, e.g., [5].

Verification. As the algorithm keeps the same logic guarding the critical section as before, mutual exclusion is still satisfied. This is easily verified using the modal formula given earlier.

However, to verify that we can always eventually request access to the critical section we need to be more careful. So far, we assumed that when a process sets its flag, this corresponds to expressing the wish to enter the critical section. However, as in Dekker's algorithm there are multiple places where the flag is set to true, we do not have this nice one-to-one correspondence. We therefore amend the model with an action wish(i) that makes the wish explicit the first time the process sets its flag. The main process therefore becomes the following.

```
proc Dekker(i:Nat) =
       wish(i)|set_flag_s(i, true).
       Dekker_outer_loop(i);
```

We here use a multi-action to model that wish and set_flag happen simultaneously. The set of allowed actions needs to be extended with wish|set_flag. We also need to modify the property to check for a such a multi-action instead of just the set_flag, hence the formula for always eventual request becomes the following.

```
[true*] forall i:Nat.val(i<=1) => <true*><wish(i)|set_flag(i, true)>true
```

This formula holds for Dekker's algorithm.

We now look at the property of eventual access. This says that, whenever a process wishes to enter its critical section, it inevitably ends up in the critical section. This can be formulated using the following mu-calculus formula.

```
[true*] forall i:Nat.val(i<=1)
       => [exists b:Bool.wish(i)|set_flag(i,b)]mu X.([!enter(i)]X && <true>true)
```

The formula says that invariantly, for every valid process i, when i wishes to enter its critical section (`[exists b:Bool.wish(i)|set_flag(i,b)]`), an enter(i) action inevitably happens within a finite number of steps (`mu X.([! enter(i)]X && <true>true)`). The conjunction `<true>true` ensures that the last part of the formula does not hold trivially in a deadlock state. Note that we know that wish(i) always appears simultaneously with a set_flag(i,b) because of the initialisation of the process. Using a bit more information of the model, we could observe that we never have wish(i)|set_flag(i,*false*), so we could simplify the corresponding modality to [wish(i)|set_flag(i,*true*)]. Here we choose not to in order to keep the requirement as general as possible. Verifying the property yields false, and we get the counterexample shown in Fig. 4.

Fig. 4. Counterexample of the eventual access property for Dekker's algorithm.

The counterexample is interesting. It describes the scenario where process 0 requests access to its critical section, by setting the flag. It then checks the guard of the outer loop, which is false, and sets $turn := 1$ just before the critical section. Next, process 1 indicates it wants to access the critical section. Since $flag[0]$ is *true*, process 1 enters the outer loop, and since $turn = 1$, it will not enter into the if-statement, so it will keep cycling here until $flag[0]$ becomes *false*. What we see here is that, because process 1 is continuously cycling through the outer loop, process 0 never gets a chance to actually enter into its critical section. This is a typical fairness issue.

We could try to alter the formula in such a way that unfair paths such as in the counterexample satisfy the property, and are thus, essentially, ignored. In this case, we can do so by saying that each sequence not containing an `enter(` i`)` action ends in an infinite sequence of `get_flag` and `get_turn` actions. This results in the following formula.

```
[true*]forall i:Nat.val(i<=1) =>
  [exists b:Bool . wish(i)|set_flag(i,b)]
    nu X.mu Y.
      ([!enter(i) && !(exists i1:Nat.get_flag(i1, true) || get_turn(i1))]Y &&
       [exists i1:Nat.get_flag(i1, true) || get_turn(i1)]X)
```

Unfortunately, if we verify this property, we find it also does not hold. We get a different counterexample, which is shown in Fig. 5.

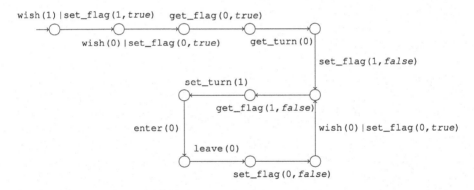

Fig. 5. Counterexample of the eventual access property under fairness for Dekker's algorithm.

What we see is that after process 1 wishes to enter its critical section, process 0 can come and enter the critical section infinitely many times, preventing process 1 from entering the critical section. A closer inspection reveals that this is because, to allow process 0 to enter, process 1 sets its flag to false, and then waits until process 0's flag becomes false. However, again, since we do not have any fairness guarantees, after setting its flag to false, process 0 can immediately request access to its critical section again, before process 1 observes that the flag became false.

We could, of course, try to change the property once more to exclude also this unfair execution, or investigate whether the counterexample is a 'just' execution, and thus indicates a real issue with Dekker's algorithm. However, instead we change our focus to Peterson's mutual exclusion protocol, as it is simpler, and therefore easier to analyse.

3.4   Peterson's Mutual Exclusion Algorithm

Some of the issues in Dekker's algorithm, particularly regarding eventual access, are alleviated by Peterson's mutual exclusion protocol [30]. We previously presented a model of this algorithm in [16]. We describe this in Algorithm 4.

> **Data:** Global variables $flag[0], flag[1]$: $\mathbb{B}$ and $turn$: $\mathbb{N}$
> $flag[i] := true$;
> $turn := 1-i$;
> **while** $flag[1-i] \wedge turn = 1-i$ **do** /* Busy waiting */ **end**
> /* Critical section */
> $flag[i] := false$;

Algorithm 4: Peterson's algorithm for process i.

In Algorithm 4, the *turn* variable is used differently from Dekker's algorithm. When a process requests access to the critical section by setting its flag, it will behave politely, and let the other process go first. It waits until either the other process does not ask for access to the critical section, i.e. $flag[1-i]$ is *false*, or the other process arrived later, in which case $turn = i$.

Model. Peterson's algorithm can be modelled in mCRL2 using the same principles we have used before. The structure of the initialization is completely analogous to that of the previous models. A single process executing Peterson's algorithm can be modelled as follows.

```
proc Process(i:Nat) =
        wish(i)|set_flag_s(i, true).
        set_turn_s(other(i)).
        (get_flag_r(other(i), false) + get_turn_r(i)).
        enter(i).
        leave(i).
        set_flag_s(i, false).
        Process(i);
```

Note that we use the fact that the negation of the guard of the loop is $\neg flag[1-i] \vee turn=i$, hence we can still use communicating actions to block until the guard becomes *false*.

Verification. This model satisfies all properties we investigated so far, including eventual access. This confirms the intuition we presented when introducing the algorithm. Let us now switch our attention to bounded overtaking, which we have not investigated yet.

Bounded overtaking says that if one process indicates its wish to enter, other processes can enter the critical section at most B times before this process is allowed to enter. It can be expressed as follows.

```
[true*] forall i:Nat.[exists b:Bool.wish(i)|set_flag(i,b)]
               (nu Y(n:Nat = 0).val(n<=B) &&
                              [!(exists i1:Nat.enter(i1))]Y(n) &&
                              [enter(other(i))]Y(n+1) )
```

In this formula, for all processes i, whenever process i wishes to enter its critical section, we start to count the number of times the other process enters its critical section using the parameter n. All actions other than enter maintain the current value. Meanwhile, the property asserts that n<=B, i.e., the bound is satisfied.

Intuitively, we may expect that whenever a process wishes to enter its critical section, the other process may enter once first. However, if we check bounded overtaking with B=1, we get the counterexample shown in Fig. 6.

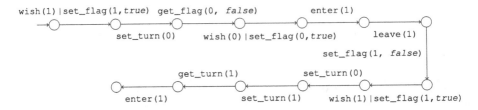

Fig. 6. Counterexample: Peterson does not satisfy bounded overtaking for $B=1$.

Let us take a close look at the counterexample. First, process 1 wishes to enter its critical section; it sets its flag, sets the turn to 0 and then checks the flag of process 0, which is currently *false*. At this point, process 1 is allowed to enter its critical section. However, before process 1 enters, process 0 also wishes to enter its critical section, and sets its flag. Subsequently, process 1 actually enters the critical section, sets the turn to process 0, and only then process 0 sets the turn to 1, ultimately allowing process 1 to enter a second time. Hence, because process 0 is stalled after setting its flag, but before setting the turn to process 1, process 1 can overtake process 0 and enter a second time.

This leads to the question of whether overtaking for higher values of B is also possible. By reverifying the formula for B = 2, we find that the formula is valid. Bounded overtaking for Peterson's mutual exclusion protocol is limited to at most 1 times.

In [16] we investigated a version of Peterson's algorithm where, initially, one of the flags is set to *true* instead of *false*. This alternative initialisation was, at some point, described on Wikipedia [29]. It turns out that, also for that version, all four properties discussed above hold. However, as process 0 will set the turn to 1 when it wants to enter the critical section, it will need the cooperation of process 1 to be allowed to enter for the first time.

This raises the question whether our properties are sufficient to cover the desired properties of mutual exclusion protocols. In particular one might want to verify the property that a process can always eventually request entry, without the other process having to perform any action. This is done by the following formula, which distinguishes Peterson's algorithm with and without correct initialisation.

```
[true*]forall i:Nat.val(i<=1) =>
        <!(wish(other(i))|set_flag(other(i), true)*>
        <wish(i)|set_flag(i, true)>true
```

4 Epilogue

We went through several versions of mutual exclusion algorithms and showed that their correctness can be formulated and investigated using modal formulas. Although it requires skill and experience to write down process algebraic specifications, and in particular modal formulas with data, they provide a powerful pair of tools to investigate and design protocols and distributed algorithms. We used it to study and design many systems varying from games [19,20] to core protocols for embedded systems [21].

When the systems that are modelled become more complex, the state space grows, and verification of modal formulas becomes more time consuming, up to a point where the state space cannot be handled by contemporary tools. It turns out that the style of modelling has a substantial influence on how complex systems can become. In [17] 7 different specification guidelines are presented to keep the state space small.

References

1. Apt, K.R., Olderog, E.: Fifty years of Hoare's logic. Formal Aspects Comput. **31**(6), 751–807 (2019). https://doi.org/10.1007/s00165-019-00501-3
2. Bergstra, J.A., Klop, J.W.: The algebra of recursively defined processes and the algebra of regular processes. In: Paredaens, J. (ed.) ICALP 1984. LNCS, vol. 172, pp. 82–94. Springer, Heidelberg (1984). https://doi.org/10.1007/3-540-13345-3_7
3. Bertot, Y., Castéran, P.: Interactive Theorem Proving and Program Development - Coq'Art: The Calculus of Inductive Constructions. Texts in Theoretical Computer Science. An EATCS Series, Springer, Heidelberg (2004). https://doi.org/10.1007/978-3-662-07964-5
4. van Beusekom, R., et al.: Formalising the Dezyne modelling language in mCRL2. In: Petrucci, L., Seceleanu, C., Cavalcanti, A. (eds.) FMICS/AVoCS -2017. LNCS, vol. 10471, pp. 217–233. Springer, Cham (2017). https://doi.org/10.1007/978-3-319-67113-0_14
5. Bouwman, M., Luttik, B., Schols, W., Willemse, T.A.C.: A process algebra with global variables. In: Dardha, O., Rot, J. (eds.) Proceedings Combined 27th International Workshop on Expressiveness in Concurrency and 17th Workshop on Structural Operational Semantics, EXPRESS/SOS 2020, and 17th Workshop on Structural Operational Semantics. EPTCS, vol. 322, pp. 33–50 (2020). https://doi.org/10.4204/EPTCS.322.5
6. Bouwman, M., Luttik, B., Willemse, T.A.C.: Off-the-shelf automated analysis of liveness properties for just paths. Acta Informatica **57**(3–5), 551–590 (2020). https://doi.org/10.1007/s00236-020-00371-w

7. Bradfield, J.C., Stirling, C.: Modal mu-calculi. In: Blackburn, P., van Benthem, J.F.A.K., Wolter, F. (eds.) Handbook of Modal Logic, Studies in Logic and Practical Reasoning, vol. 3, pp. 721–756. North-Holland (2007). https://doi.org/10.1016/s1570-2464(07)80015-2

8. van den Brand, M., Groote, J.F.: Software engineering: redundancy is key. Sci. Comput. Program. **97**, 75–81 (2015). https://doi.org/10.1016/j.scico.2013.11.020

9. Cicirelli, F., Nigro, L., Sciammarella, P.F.: Model checking mutual exclusion algorithms using UPPAAL. In: Silhavy, R., Senkerik, R., Oplatkova, Z.K., Silhavy, P., Prokopova, Z. (eds.) Software Engineering Perspectives and Application in Intelligent Systems. AISC, vol. 465, pp. 203–215. Springer, Cham (2016). https://doi.org/10.1007/978-3-319-33622-0_19

10. Cranen, S., Groote, J.F., Reniers, M.A.: A linear translation from CTL* to the first-order modal μ-calculus. Theor. Comput. Sci. **412**(28), 3129–3139 (2011). https://doi.org/10.1016/j.tcs.2011.02.034

11. Dijkstra, E.W.: Solution of a problem in concurrent programming control. Commun. ACM **8**(9), 569 (1965). https://doi.org/10.1145/365559.365617

12. Dijkstra, E.W.: Over de sequentialiteit van procesbeschrijvingen (Undated, 1962 or 1963)

13. Dyseryn, V., van Glabbeek, R.J., Höfner, P.: Analysing mutual exclusion using process algebra with signals. In: Peters, K., Tini, S. (eds.) Proceedings Combined 24th International Workshop on Expressiveness in Concurrency and 14th Workshop on Structural Operational Semantics, EXPRESS/SOS 2017, Berlin, Germany, 4th September 2017. EPTCS, vol. 255, pp. 18–34 (2017). https://doi.org/10.4204/EPTCS.255.2

14. Garavel, H., Lang, F., Mateescu, R., Serwe, W.: CADP 2011: a toolbox for the construction and analysis of distributed processes. Int. J. Softw. Tools Technol. Transf. **15**(2), 89–107 (2013). https://doi.org/10.1007/s10009-012-0244-z

15. Gibson-Robinson, T., Armstrong, P.J., Boulgakov, A., Roscoe, A.W.: FDR3: a parallel refinement checker for CSP. Int. J. Softw. Tools Technol. Transf. **18**(2), 149–167 (2016). https://doi.org/10.1007/s10009-015-0377-y

16. Groote, J.F., Keiren, J.J.A., Luttik, B., de Vink, E.P., Willemse, T.A.C.: Modelling and analysing software in mCRL2. In: Arbab, F., Jongmans, S.-S. (eds.) FACS 2019. LNCS, vol. 12018, pp. 25–48. Springer, Cham (2020). https://doi.org/10.1007/978-3-030-40914-2_2

17. Groote, J.F., Kouters, T.W.D.M., Osaiweran, A.: Specification guidelines to avoid the state space explosion problem. Softw. Test. Verification Reliab. **25**(1), 4–33 (2015). https://doi.org/10.1002/stvr.1536

18. Groote, J.F., Mousavi, M.R.: Modeling and Analysis of Communicating Systems. MIT Press (2014). https://mitpress.mit.edu/books/modeling-and-analysis-communicating-systems

19. Groote, J.F., de Vink, E.P.: Problem solving using process algebra considered insightful. In: Katoen, J.-P., Langerak, R., Rensink, A. (eds.) ModelEd, TestEd, TrustEd. LNCS, vol. 10500, pp. 48–63. Springer, Cham (2017). https://doi.org/10.1007/978-3-319-68270-9_3

20. Groote, J.F., Wiedijk, F., Zantema, H.: A probabilistic analysis of the game of the goose. SIAM Rev. **58**(1), 143–155 (2016). https://doi.org/10.1137/140983781

21. Groote, J.F., Willemse, T.A.C.: A symmetric protocol to establish service level agreements. Log. Methods Comput. Sci. **16**(3) (2020). https://lmcs.episciences.org/6812

22. Hennessy, M., Milner, R.: Algebraic laws for nondeterminism and concurrency. J. ACM **32**(1), 137–161 (1985). https://doi.org/10.1145/2455.2460

23. Hopcroft, P.J., Broadfoot, G.H.: Combining the box structure development method and CSP for software development. Electron. Notes Theor. Comput. Sci. **128**(6), 127–144 (2005). https://doi.org/10.1016/j.entcs.2005.04.008

24. Leino, K.R.M., Wüstholz, V.: The Dafny integrated development environment. In: Dubois, C., Giannakopoulou, D., Méry, D. (eds.) Proceedings 1st Workshop on Formal Integrated Development Environment, F-IDE 2014, Grenoble, France, 6 April 2014. EPTCS, vol. 149, pp. 3–15 (2014). https://doi.org/10.4204/EPTCS. 149.2

25. Mateescu, R., Serwe, W.: Model checking and performance evaluation with CADP illustrated on shared-memory mutual exclusion protocols. Sci. Comput. Program. **78**(7), 843–861 (2013). https://doi.org/10.1016/j.scico.2012.01.003

26. Milner, R.: Communication and concurrency. PHI Series in Computer Science. Prentice Hall, Upper Saddle River (1989)

27. Nipkow, T., Wenzel, M., Paulson, L.C. (eds.): Isabelle/HOL. LNCS, vol. 2283. Springer, Heidelberg (2002). https://doi.org/10.1007/3-540-45949-9

28. Osaiweran, A., Schuts, M., Hooman, J.: Experiences with incorporating formal techniques into industrial practice. Empir. Softw. Eng. **19**(4), 1169–1194 (2014). https://doi.org/10.1007/s10664-013-9251-2

29. Peterson's algorithm, May 17. https://en.wikipedia.org/wiki/Peterson

30. Peterson, G.L.: Myths about the mutual exclusion problem. Inf. Process. Lett. **12**(3), 115–116 (1981). https://doi.org/10.1016/0020-0190(81)90106-X

31. Wesselink, W., Willemse, T.A.C.: Evidence extraction from parameterised boolean equation systems. In: Benzmüller, C., Otten, J. (eds.) Proceedings of the 3rd International Workshop on Automated Reasoning in Quantified Non-Classical Logics (ARQNL 2018) affiliated with the International Joint Conference on Automated Reasoning (IJCAR 2018), Oxford, UK, July 18, 2018. CEUR Workshop Proceedings, vol. 2095, pp. 86–100. CEUR-WS.org (2018). http://ceur-ws.org/Vol-2095/paper6.pdf

Author Index

Printed in the United States
by Baker & Taylor Publisher Services